Materials Forming, Machining and Tribology

Series editor

J. Paulo Davim, Aveiro, Portugal

More information about this series at http://www.springer.com/series/11181

Kapil Gupta
Editor

Advanced Manufacturing Technologies

Modern Machining, Advanced Joining, Sustainable Manufacturing

Editor
Kapil Gupta
Department of Mechanical and Industrial
 Engineering Technology
University of Johannesburg, Doornfontein
 Campus
Johannesburg
South Africa

ISSN 2195-0911 ISSN 2195-092X (electronic)
Materials Forming, Machining and Tribology
ISBN 978-3-319-56098-4 ISBN 978-3-319-56099-1 (eBook)
DOI 10.1007/978-3-319-56099-1

Library of Congress Control Number: 2017937294

© Springer International Publishing AG 2017
This work is subject to copyright. All rights are reserved by the Publisher, whether the whole or part of the material is concerned, specifically the rights of translation, reprinting, reuse of illustrations, recitation, broadcasting, reproduction on microfilms or in any other physical way, and transmission or information storage and retrieval, electronic adaptation, computer software, or by similar or dissimilar methodology now known or hereafter developed.
The use of general descriptive names, registered names, trademarks, service marks, etc. in this publication does not imply, even in the absence of a specific statement, that such names are exempt from the relevant protective laws and regulations and therefore free for general use.
The publisher, the authors and the editors are safe to assume that the advice and information in this book are believed to be true and accurate at the date of publication. Neither the publisher nor the authors or the editors give a warranty, express or implied, with respect to the material contained herein or for any errors or omissions that may have been made. The publisher remains neutral with regard to jurisdictional claims in published maps and institutional affiliations.

Printed on acid-free paper

This Springer imprint is published by Springer Nature
The registered company is Springer International Publishing AG
The registered company address is: Gewerbestrasse 11, 6330 Cham, Switzerland

Preface

The more stringent requirements for enhanced quality of engineered products, especially those of miniature size and that have typical features and made from difficult-to-machine materials; accelerated global competitiveness; and strict environmental regulations, have been responsible for the development and subsequent wide use of *advanced manufacturing technologies*.

In general, advanced manufacturing technologies encompass modern manufacturing techniques and advancements in the existing or conventional manufacturing. They offer excellent part quality, cost-effectiveness, and high productivity with less environmental footprint. Some of these technologies such as modern machining methods, advanced repair and joining processes, and sustainable manufacturing techniques are primarily focused in this book.

Part I is about modern machining and consists of four chapters (Chaps. 1–4). Chapter 1 describes various advanced processes and a hybrid process developed for fabrication of micro cutting tools. Chapter 2 provides an overview of advanced machining of glass materials. Chapter 3 is dedicated to thermal assisted machining of titanium alloys. Chapter 4 sheds light on advanced machining of composite materials by abrasive water jet technique.

Part II includes four chapters (Chaps. 5–8) on advanced repair and joining techniques. Chapter 5 provides an insight into advanced joining and welding techniques. Chapter 6 describes laser-based advanced repair of dies, molds, and gears. Chapter 7 details the important aspects of friction stir welding technique. Chapter 8 highlights the novel aspects of ultrasonic spot welding for transport applications.

Part III is dedicated to sustainable manufacturing and comprises four chapters. Chapter 9 discusses various important aspects of green manufacturing. Chapter 10 details the experimental work on environment-friendly machining of super alloys. Chapter 11 describes dry and near dry type sustainable electrical discharge machining processes. Chapter 12 discusses the remanufacturing of engineered products using laser metal deposition process.

The present book is intended to facilitate the researchers, engineers, technologists, and specialists who are working in the field of advanced manufacturing, production and sustainable engineering by offering the theoretical background, novel aspects, research advances, and applications of advanced manufacturing technologies. It will also enable and encourage the researchers to explore the field and develop the technology further with an objective to find the solutions for the major industrial problems, and to work for the societal benefits.

I sincerely acknowledge Springer for this opportunity and for their professional support. Finally, I would like to thank all the chapter authors for their availability and valuable contribution.

Johannesburg, South Africa Kapil Gupta
May 2017

Contents

Part I Modern Machining

1 **Fabrication of Micro-cutting Tools for Mechanical Micro-machining** 3
M. Ganesh, Ajay Sidpara and Sankha Deb

2 **Machining of Glass Materials: An Overview** 23
Asma Perveen and Carlo Molardi

3 **Thermal-Assisted Machining of Titanium Alloys** 49
O.A. Shams, A. Pramanik and T.T. Chandratilleke

4 **Abrasive Water Jet Machining of Composite Materials** 77
Sumit Bhowmik, Jagadish and Amitava Ray

Part II Advanced Repair and Joining

5 **Advanced Joining and Welding Techniques: An Overview** 101
Kush Mehta

6 **Laser-Based Repair of Damaged Dies, Molds, and Gears** 137
Sagar H. Nikam and Neelesh Kumar Jain

7 **Friction Stir Welding—An Overview** 161
Arun Kumar Shettigar and M. Manjaiah

8 **Ultrasonic Spot Welding—Low Energy Manufacturing for Lightweight Fuel Efficient Transport Applications** 185
Farid Haddadi

Part III Sustainable Manufacturing

9 **Perspectives on Green Manufacturing** 213
Varinder Kumar Mittal

10 **Experimental Investigation and Optimization on MQL-Assisted Turning of Inconel-718 Super Alloy**............ 237
Munish K. Gupta, P.K. Sood, Gurraj Singh and Vishal S. Sharma

11 **Dry and Near-Dry Electric Discharge Machining Processes** 249
Krishnakant Dhakar and Akshay Dvivedi

12 **Laser Metal Deposition Process for Product Remanufacturing** 267
Rasheedat M. Mahamood, Esther T. Akinlabi and Moses G. Owolabi

Index 293

Part I
Modern Machining

Chapter 1
Fabrication of Micro-cutting Tools for Mechanical Micro-machining

M. Ganesh, Ajay Sidpara and Sankha Deb

Abstract Micro-cutting processes are very effective manufacturing methods for complex micro-parts used in MEMS, micro-dies, micro-structured surfaces on nonconductive materials, etc. Main challenge in employing conventional machining methods for fabrication of micro-parts and features is the unavailability of the smaller tools. Popular method like grinding is failed in miniaturizing the cutting tools because of rigidity problems. This became a driving force to research the alternative processes. Different processes like electro-discharge machining, wire electro-discharge machining, laser beam machining, focused ion beam machining, etc., are evolved to accomplish the need of new fabrication methods for micro-cutting tools. Each process has showed its capabilities and limitations through various machining experiments. In this chapter, an overview of the micro-tool fabrication processes along with their characteristics is presented. New micro-end mill tool geometry and a tool fabrication method are also presented.

Keywords Micro-cutting · Cutting tool · EDM · Wire-EDM · Focused ion beam · Laser beam machining

1.1 Introduction

The history of human kind often recognized as the evolution of tools used by the species for survival. The ability to shape the bones of animals to make newer weapons gave an upper hand for the species in the food chain. Need of harder weapons drove them to the invention of metals. Making ploughs for farming demanded some tools for controlled removal of material from wood or metal in different shapes. These elements all together demanded immediate technological advancements in cutting tools. The search of new cutting tools and manufacturing methods started there.

M. Ganesh · A. Sidpara (✉) · S. Deb
Mechanical Engineering Department, Indian Institute of Technology, Kharagpur, India
e-mail: ajaymsidpara@mech.iitkgp.ernet.in; ajaysidpara4u@gmail.com

© Springer International Publishing AG 2017
K. Gupta (ed.), *Advanced Manufacturing Technologies*, Materials Forming,
Machining and Tribology, DOI 10.1007/978-3-319-56099-1_1

Conventional metal cutting processes are those in which material removed as chips by direct contact between the workpiece and the defined cutting edges of the cutting tool. The necessary condition to be satisfied to realize this is the superior hardness of the cutting tool material. The force required to move the tool over the surface of the workpiece can be reduced by properly designing the cutting tool angles. The cutting force, relative speed between the tool and workpiece, tool angles, the presence of special fluids for lubrication and cooling, etc., are the main factors affecting the dimensional accuracy and surface finish of the machined workpiece. Despite the invention of more and more techniques for shaping metals, conventional machining technologies held its place because of high material removal rate (MRR) and less metallurgical effects on the metals.

The increasing demand for miniaturized components became a strong motive for the advancements in micro-manufacturing technologies. Popular nonconventional machining processes like electro-discharge machining (EDM), laser machining, electrochemical machining (ECM), and lithographic processes, gradually found an inevitable space in the micro-machining regime. Conventional mechanical machining methods were struggling to extend its capabilities for machining of components in micro-dimensions. This was because of unavailability of a high speed spindle which is necessary to cut the near grain size dimensions and process limitations in the fabrication of micro-tools. Though high speed air spindles solved the issue of cutting speed, difficulty in manufacturing of micro-cutting tools remained as a hindrance to further advancements. Improvements in the alternate micro-machining methods and positioning mechanisms helped to precisely fabricate conventional micro-cutting tools with sharp cutting edges.

Figure 1.1 shows different types for micro-cutting processes and their applications. Micro-cutting processes can be employed for the fabrication of parts for MEMS devices [7], biomedical devices [8], dies for micro-casting applications [9], and micro-EDM electrodes [1], etc. Conventional micro-machining processes are not just the scaling down of the processes they named after. The main factors that differentiate between micro- and macro-machining processes are the uncut chip thickness, dimensions of the fabricated features, geometry and dimension of cutting tools and the mechanics of material removal. In micro-machining, the uncut chip thickness is comparable to the cutting edge dimension which is responsible for some unusual behavior during machining, broadly known as size effects. Size effects influence the specific cutting energy, surface finish, and burr formation, etc. Uncut chip thickness can vary from 200 to 10 μm according to the accuracy of machine tool and positioning systems used for machining [10, 11]. The components in which the fabricated features fit in the range of 1–1000 μm and make use of the conventional machine tools for processing can be included in micro-cutting. The machining capability is measured in terms of dimensional accuracy and surface finish which can be in submicron range for micro-cutting [11]. Most of the tools used in micro-cutting have dimensions in the range of 25–1000 μm. However, some exceptionally smaller tools with dimensions below 25 μm are also fabricated using various conventional processes but their industrial usage is limited due to rigidity issues.

1 Fabrication of Micro-cutting Tools for Mechanical Micro-machining

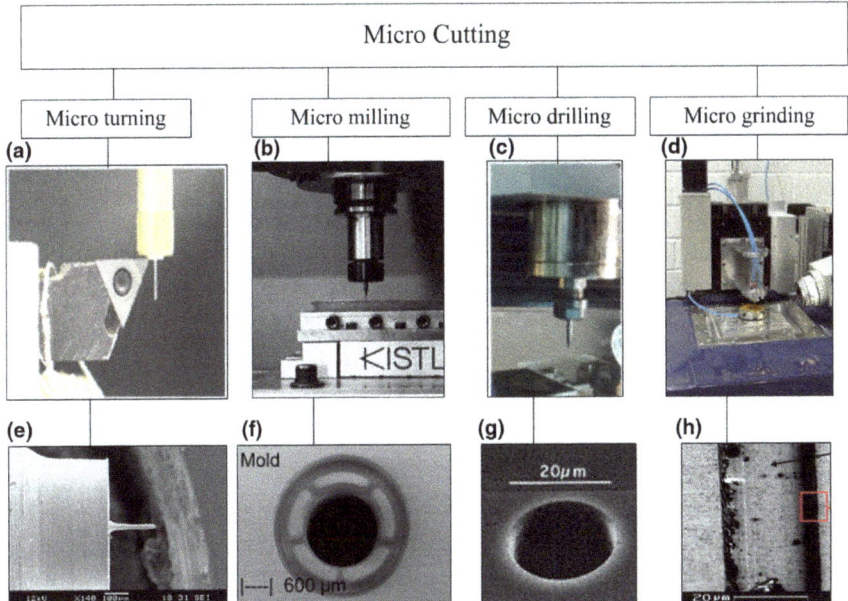

Fig. 1.1 Types of micro-cutting process **a** micro-turning [1] **b** micro-milling [2] **c** micro-drilling **d** micro-grinding [3] **e** 22 μm diameter micro-rod fabricated by turning [4] **f** micro-mold fabricated by milling [5] **g** hole drilled by micro-drilling process [6] **h** micro-slit fabricated by grinding [3] (with kind permissions from Elsevier)

Compared to the other micro-machining methods, micro-cutting strategies exhibit higher degree of dimensional accuracy in the fabrication of 3D surfaces along with higher MRR and surface finish. The relative accuracy can be in the order of 10^{-3}–10^{-5} and roughness values can go below 100 nm which can go down to submicron accuracy and nanometric surface finish with diamond cutting tools [11]. There is no restriction of material properties such as electric or thermal conductivity except hardness.

Despite being a promising technique for fabrication of micro-components, the capabilities of conventional micro-machining process is shortened because of the unavailability of small and rigid cutting tools. The biggest challenge is to keep the rigidity and sharpness of the cutting tool for a longer period of time. For example, conventional end mills with two flute geometries have a problem in rigidity because of the lesser solid part and the high tool runout to tool diameter ratio which results in engagement of single cutting edge during machining. It exhibits a variation in cutting force and cutting edge breaks eventually. So miniaturization of the exact tool geometries to smaller dimensions will not serve the purpose. Some special tool geometries which are optimized in rigidity and MRR have to be realized [12].

In this chapter, characteristics of different cutting tool fabrication process used for micro-machining (in Sect.1.2), and the performance analyses of micro-tools with essential characteristics (in Sect.1.3) are discussed.

1.2 Processes Used for Fabrication of Micro-cutting Tools

Fabrication and effective use of micro-tools is the primary step in employing conventional machining process in micro-regime. The processes for fabrication of micro-cutting tools have diverse range of capabilities and limitations. Figure 1.2 shows various processes used for fabrication of micro-cutting tool. The fabricated micro-cutting tool must have sharp cutting edges which will help to minimize the cutting force and uncut chip thickness [13]. Flank of the cutting tool has to be designed to avoid contact of side walls to the workpiece and the fabrication process should be fast and reliable to make the tool less expensive.

1.2.1 Grinding

Grinding is the predominant process used to make cutting tools for conventional macro-machining. Considering the case of micro-cutting tools, using grinding process is little tricky to accomplish the size reduction and dimensional accuracy. The process exerts large amount of cutting force on the workpiece. This cannot be appreciated if we are dealing with tools of micrometer dimensions which lack in structural rigidity. Furthermore, when cutting tools are fabricated using grinding process, a considerable quantity of scrap is generated due to breakage of the tools [11]. These limitations can be overcome by using vibration assisted grinding process. The advantage of using ultrasonic vibration assisted grinding is of reducing grinding force thus reduces the chance of tool breakage during tool machining. Ultrasonic vibration assisted grinding helped to reduce the diameter of cutting tools to 10–20% and to increase the aspect ratio to 50% higher compared to the ordinary tool grinding process [14].

A micro-end mill tool down to a size of 10 μm is fabricated using three-axis tool grinding machine with a vibration isolated granite base [13]. Figure 1.3 shows the tool grinding machine with parts. Figure 1.4 shows fabricated end mill cutters using this process.

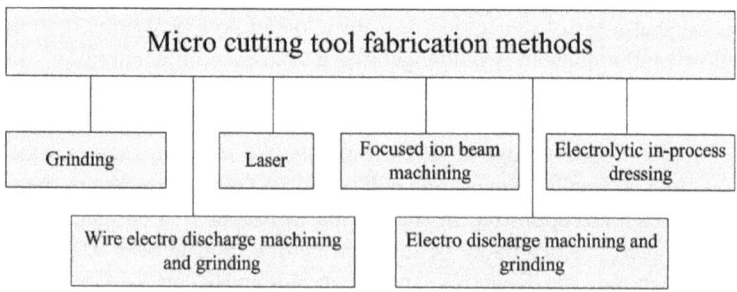

Fig. 1.2 Different processes used for fabrication of micro-cutting tool

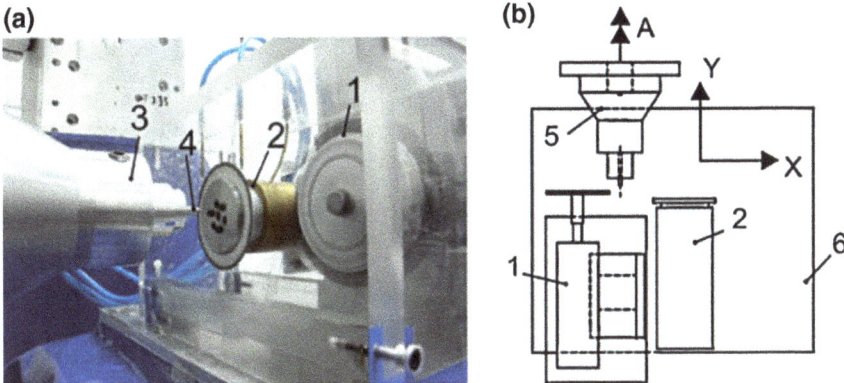

Fig. 1.3 **a** Photograph of tool grinding machine (*1*—pre grinding spindle, *2*—fine grinding spindle, *3*—clamping device) **b** Schematic diagram of grinding machine (*4*—tool shank, *5*—rotational axis, *6*—XY table) [13] with kind permission from Elsevier

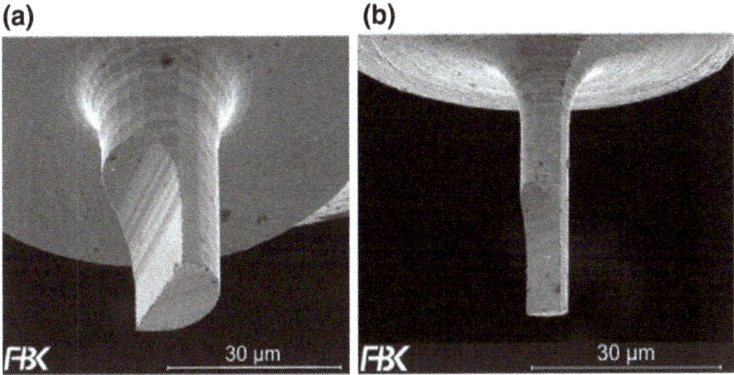

Fig. 1.4 **a** End mill of 20 μm diameter **b** end mill of 10 μm diameter fabricated by grinding [13] with kind permission from Elsevier

Milling operations are performed using milling cutters of 20 and 10 μm diameter. Despite being successful in fabrication of tools of very small diameter, this encountered a problem in making high aspect ratio tools. Tools which are small and long, could not withstand the grinding force. To achieve better machining performance, the assistance of electrolytic in-process dressing (ELID) is introduced for making micro-tools [15, 16]. Cylindrical tools with tip diameter of 1 μm are machined using this process. However, this process faces difficulty in achieving uniform tool diameter and accurate tool dimensions because of corner wear of grinding wheel [15]. Figure 1.5 shows the ELID grinding system and tool fabricated using this process.

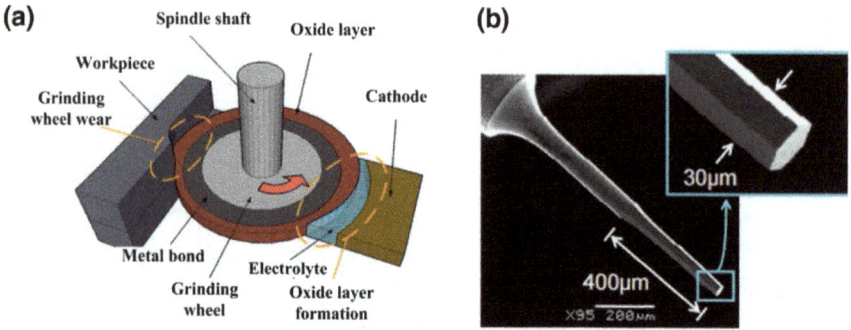

Fig. 1.5 a Schematic diagram of ELID grinding system [17] **b** hexagonal micro-tool fabricated by ELID grinding [16] with kind permission from Elsevier

1.2.2 Electro-Discharge Machining (EDM) and Its Variants

EDM and its variants have the great advantage of force free machining, so that micro-features with high aspect ratio can be made using simple fixtures. This characteristic of EDM variants is very much useful when machining cutting tools with lower rigidity. The variants of EDM differ in tool electrode shape, tool feed mechanism, etc.

1.2.2.1 Micro-Electro-Discharge Grinding (μ-EDG)

Electro-discharge machining and electro-discharge grinding (EDG) are the two processes popularly used for micro-component fabrication for different applications including micro-fluidic devices, etc. [8]. EDG is an important variant of EDM which is basically the hybridisation of EDM and mechanical grinding. When this process is employed without grinding wheel, material is removed by melting and vaporization only. Poly-crystalline diamond tools, which are very difficult to fabricate by other machining processes, are easily fabricated using EDM variants. Micro-tools for grinding of BK7 glass with different geometries (D-type, triangular, and square) are made by using micro-EDG [18]. Figure 1.6 shows the set up to machine PCD cutting tools using EDG process.

Figure 1.7 shows different cutting tool geometries fabricated by EDG. Tungsten carbide tools with a diameter down to 31 μm are also fabricated using methods of micro-EDM [19]. The tool geometry is very primitive in nature and it is advised to keep the axial engagement and feed to minimum to avoid burr formation during machining with micro-tools made by EDM.

1 Fabrication of Micro-cutting Tools for Mechanical Micro-machining

Fig. 1.6 Photograph of the set up used for fabrication of PCD micro-tools using EDG [18]

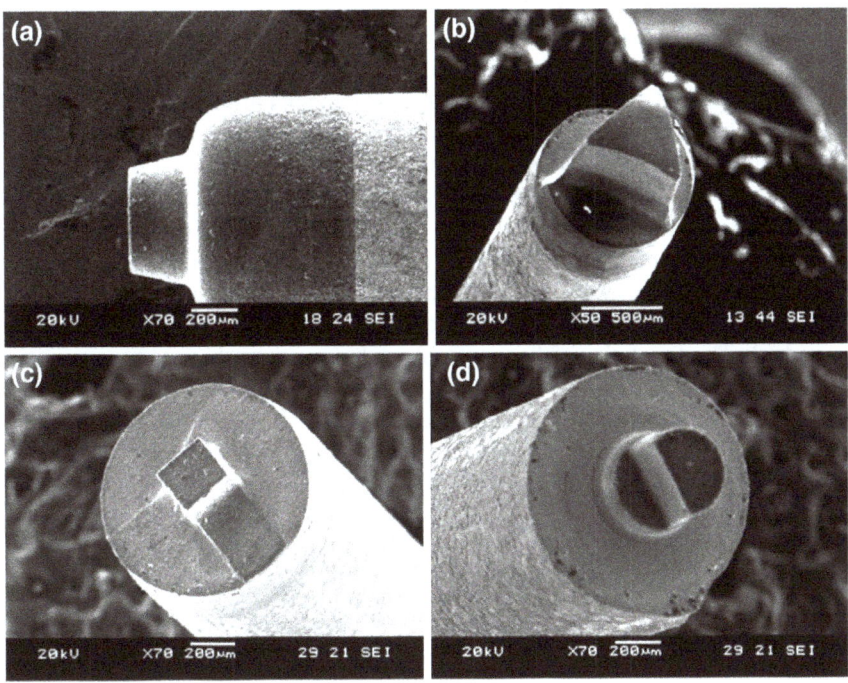

Fig. 1.7 a Circular b triangular c square d D-type PCD tools fabricated by EDG [18]

1.2.2.2 Wire Electro-Discharge Grinding (WEDG)

Compared to tool electrodes used in EDG, thin wires which are used as tool electrode in WEDG are capable of producing smaller features with higher dimensional accuracy as the wear of wires is significantly lower to the electrodes used in classical EDG. Gun barrel type cemented tungsten carbide drilling tool of 3 μm diameter is made using the principle of wire EDG [20]. A 100 μm wire with wire guides to avoid vibration is used for tool fabrication. The contact free process establishes high aspect ratio and good degree of dimensional accuracy. The tool is employed to cut a micro-slot with 3 μm depth and 4 μm width. Figure 1.8 shows the tool fabricated using wire EDG on cemented carbide with 90 nm grain size. PCD tools with form accuracy and edge sharpness in the order of 1 μm are fabricated by this process [19, 21]. Wire electro-discharge grinding induced graphitization of diamond grains at low energy condition is mainly responsible for such a high accuracy. Figure 1.9 explains the wire EDG set up used for fabrication of PCD micro-cutting tools.

Figure 1.10 presents the mechanism of graphitization, dissolution, and material removal from diamond tool surface by wire EDG. Usually, the process of machining PCD tools using EDM variants is done with high-energy electrical discharges. In this, the material removal occurs because of the dissolution of cobalt binder and falling off of the diamond grains. However, use of low discharge energy will help to remove the cobalt binder in a controllable manner which results in graphitization and thermo chemical reactions on diamond instead of falling off [21]. The fabricated end mills are utilized for making micro-grooves on tungsten carbide dies which gave nearly polish surface finish around 2 nm Ra.

A large variety of PCD tools are fabricated using wire EDG [22]. The machining parameters for wire EDM in PCD micro-tool fabrication is optimized to get good repeatability in tool production [23]. PCD micro-tools are very efficient in machining of non-ferrous materials because of its high hardness and low friction coefficient. The problem arises when the workpiece material is ferrous and the machining temperature is very high. The dissolution of diamond tool results in poor machining characteristics. $Ti(C_7N_3)$-based cermet micro-end-milling tool is successfully machined using wire EDG [24]. The wear mechanism of fabricated tool is studied and superior wear resistance of $Ti(C_7N_3)$-based cermet micro-end-milling tool is reported.

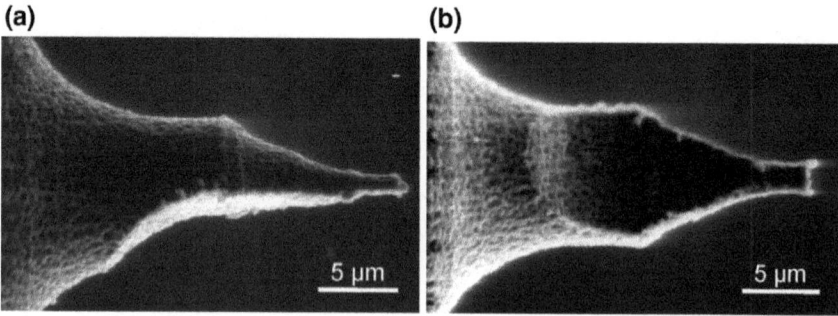

Fig. 1.8 **a** Front view **b** bottom view of the CTC tool fabricated by wire EDG [19] with kind permission from Elsevier

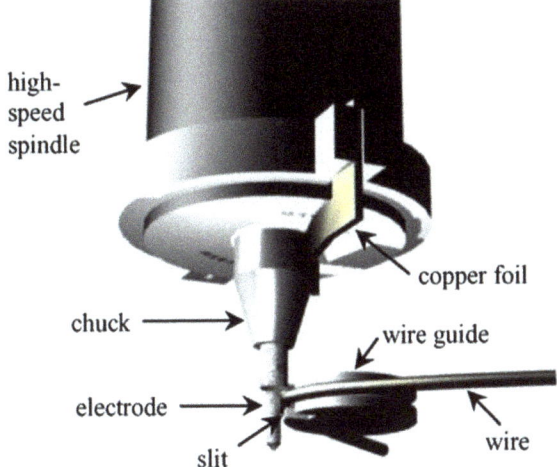

Fig. 1.9 Illustration of wire EDG set up for the fabrication of PCD micro-tools [20] with kind permission from Elsevier

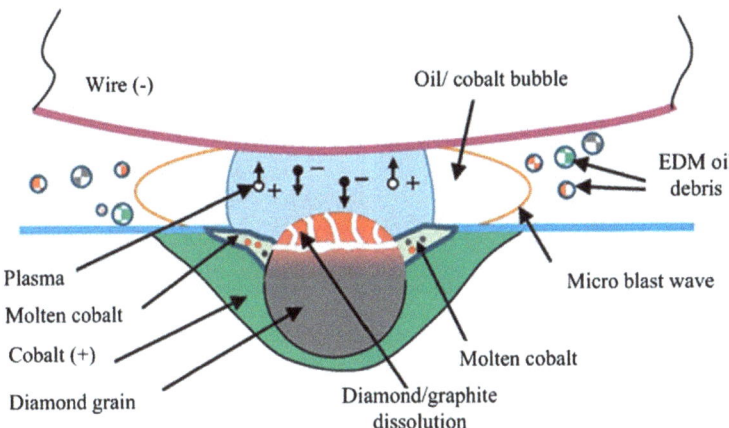

Fig. 1.10 Graphitization, dissolution and gradual removal of diamond grain from the workpiece surface by wire EDG [21] with kind permission from Elsevier

1.2.3 Laser Beam Machining

Single crystalline diamond (SCD) tools are popularly used for machining of ceramic molds. Usually, the tools are fabricated using laser machining [25]. The SCD chip is machined to cylindrical shape using IR YVO_4 laser and bonded to cemented carbide shank. This is further shaped with the laser to a milling tool with defined cutting edges and flutes in a work table with three axes control. Fabricated end mill tool has 10 cutting edges with −40° rake angle and 0.5 mm edge radius. This tool is used for ductile mode machining of tungsten carbide dies. Figure 1.11 explains the principle of micro-tool

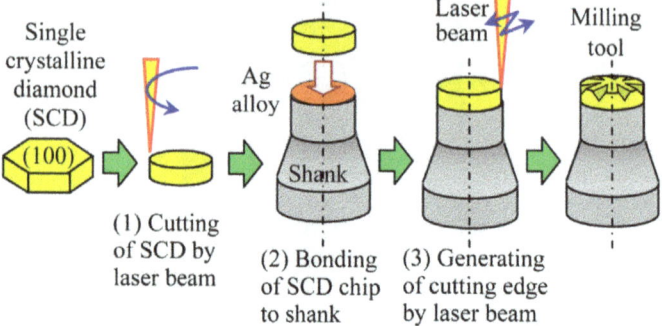

Fig. 1.11 Procedure to fabricate SCD micro-tools using laser machining [25] with kind permission from Elsevier

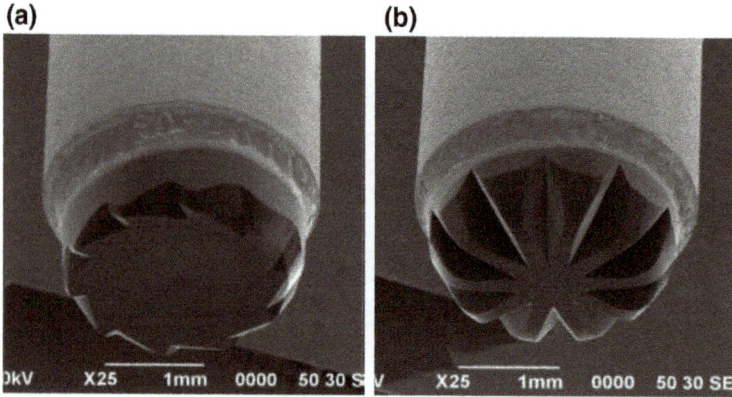

Fig. 1.12 **a** Micro-end mill tool with sharp edges **b** tool with 0.5 mm edge radius fabricated using laser machining [25] with kind permission from Elsevier

fabrication by laser machining. The tool is capable of producing a machined surface of surface roughness 12 nm Ra. Figure 1.12 shows the micro-end mill tool machined using laser beam. These micro-tools are used to machine ceramic molds for micro-glass lenses.

1.2.4 Focused Ion Beam Machining (FIB)

Among the processes used in micro-tool fabrication, focused ion beam (FIB) machining is capable to machine cutting tools with the smallest possible features. Micro-cutting tools with high profile accuracy can be made using FIB by controlling the sputtering time on the specific areas of tool sputtering [26]. Figure 1.13a explains the procedure of FIB machining of sharp edged single point micro-tools.

FIB milling is usually used to machine cutting tools with dimensions in the range of 15 μm to 100 μm. Cutting tools with rectangular, triangular, and other complex tool geometries are machined using FIB [27]. Figure 1.14 shows the diamond micro-threading tool made with FIB sputtering. The rake and relief angle of the cutting edge is precisely controlled and finished using focused ion beam. One of the main advantages of machining with FIB sputtering is to observe the tool during its

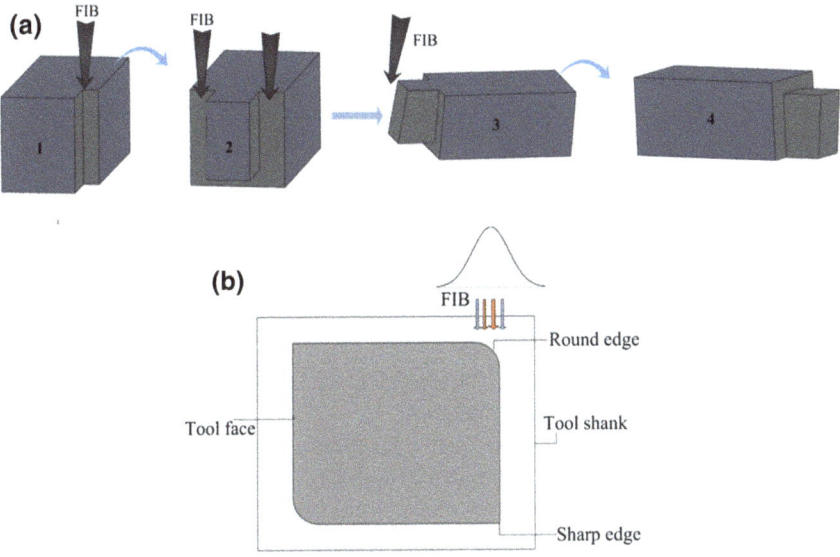

Fig. 1.13 a Procedure for fabrication of micro-cutting tool using FIB (*1*—machining of rake surface, *2*, *3*—machining of clearance surfaces, *4*—final tool) **b** Illustration of edge formation during FIB machining

Fig. 1.14 Micro-threading tool fabricated using FIB [27] with kind permission from Elsevier

fabrication. The cutting edge dimensions can be brought down to nanometric range by precisely controlling the ion sputtering [26].

Micro-end mills with 25 µm diameter are successfully fabricated and used for precision machining [28]. Sub-micrometer cutting edge radius is accomplished using this method. FIB machining is extensively adopted for cutting tool fabrication in tungsten carbide, CBN, and high speed steel [29, 30]. The process is capable of realizing more complex geometries, better dimensional resolution, and repeatability for the tool. Despite of all these advantages, FIB has its own limitations. Due to the Gaussian energy distribution of the beam, more material removal occurs at the edge close to the ion source results in edge or corner rounding as explained in Fig. 1.13b. Along with that, the sputtering process which removes material as atom by atom from the tool surface is very slow in nature [31].

Cutting tool geometry has an important role in controlling the cutting force and tool rigidity. Miniaturizing the conventional tool geometries exhibits some problems as explained in the previous sections. The triangular type and D-type end mills shows good rigidity compared to the two flute mills. However, the quality of machined surface is poor due to large negative rake angle due to small uncut chip thickness [32]. PCD tools made with different processes are analyzed by varying the cutting speed and feed [33]. The optimum range of depth of cut and feed range of PCD cutting tool are found, beyond which ploughing effects will be predominant and the surface quality reduces. After conducting several performance tests with different cutting tool geometries, the important considerations for fabrication of micro-end mill tools for grooving applications have summarized as below [34]

- To reduce the burr formation, the interference between tool peripheral and groove wall surfaces has to be reduced.
- To increase the surface finish and tool life, rubbing between tool end and groove bottom surfaces has to be minimized and the space between the tool and workpiece has to be increased for efficient chip removal.

1.3 Fabrication of End Mill Tool by Compound Machining of EDM Milling and Die-Sinking EDM

Although the machining processes discussed in the previous section are used for fabrication of micro-tools but they also possess certain demerits in terms of processing time, dimensional accuracy, capital investment, difficulty in producing complex and high aspect ratio tools, etc. As explained in Sect. 1.2.2, EDM is a promising technique for cutting tool fabrication because of its 'force free' nature during machining. The EDM variants popularly used so far for the fabrication of micro-cutting tools are EDG, wire EDG, wire EDM, block EDM, etc. EDM milling and die-sinking EDM which are comparatively faster variants of EDM are not effectively used for micro-cutting tool fabrication so far. Apart from that,

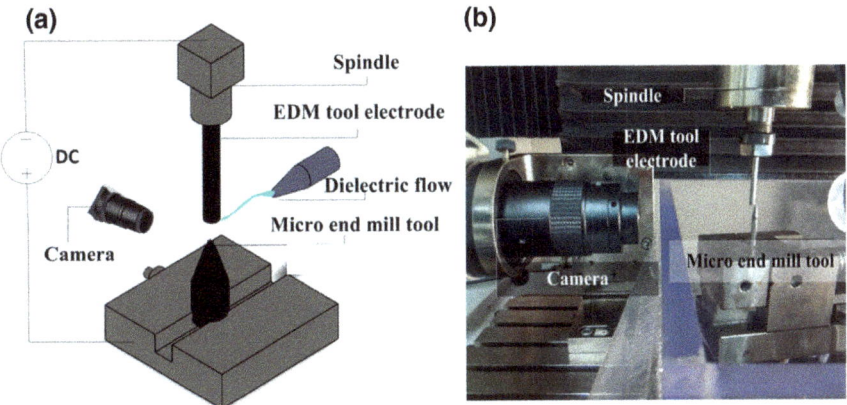

Fig. 1.15 **a** Schematic diagram of set up for fabricating micro-cutting tools by EDM **b** Photograph of machining set up for micro-tool fabrication

commercially available tools exhibit frequent tool breakages due to rigidity issues. To address this problem, more rigid tool geometries have to be realized. Therefore, an attempt has been made by the authors to combine different EDM variants for fabrication of micro-end mill tool.

End mill tool with a diameter approximately 200 μm is machined on the top of a 1 mm tungsten carbide (WC) end mill tool by combined operation of EDM milling and die-sinking EDM. As shown in Fig. 1.15, the general arrangement for machining includes the EDM tool electrode, workpiece, dielectric supply, power supply, and two cameras. The cameras will help to align workpiece with tool and help to observe and measure the micro-tool dimensions during machining.

The process includes following steps: (a) flattening the workpiece surface, (b and c) fabrication of end mill tool by EDM milling, (d) die-sinking EDM to remove material from the side surface, (e) testing of micro-tool on poly methyl methacrylate (PMMA) material by machining micro-channels. Figure 1.16 explains the process steps.

A WC rod is used to flatten the top surface of 1 mm end mill cutter. On this finished surface, the same WC electrode is used to fabricate a square of almost 200 μm diameter by EDM milling. Utilizing the cameras placed in different directions, the dimension of this machined micro-feature is measured during machining. Figure 1.17 shows the SEM image and photograph of machined micro-feature. This has four sharp edges which are capable of removing material or in other words, can be used as a micro-cutting tool.

This is referred as the straight edge micro-end mill tool. However, this straight edge tool has high negative rake angle during machining which deteriorate the surface quality. To solve this problem, die-sinking EDM is employed to make a curve edged tool which reduces the effective rake angle during machining.

The EDM tool electrode is changed to a 300 μm tungsten carbide electrode. A small area of the electrode is precisely moved down at the flanks of the straight

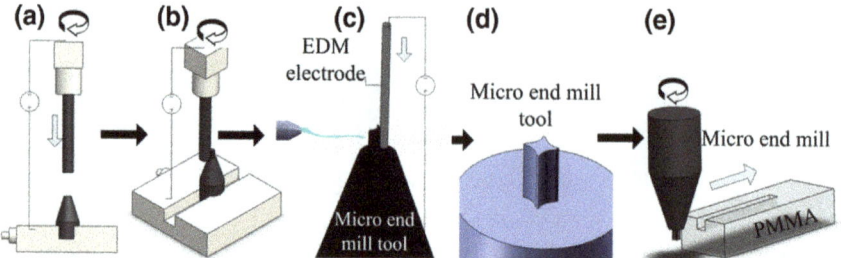

Fig. 1.16 **a** Flattening of WC workpiece **b** fabrication of square edge micro-end mill tool by EDM milling **c** fabrication of curve edge micro-end mill tool by die-sinking EDM **d** curved edge micro-end mill tool **e** machining of micro-channels on PMMA material by fabricated micro-tool

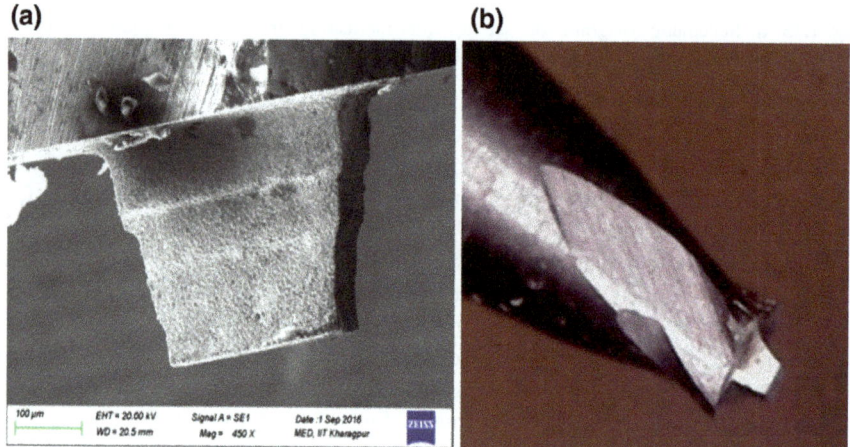

Fig. 1.17 **a** SEM image of micro-tool and **b** photograph of straight edge micro-tool fabricated by EDM

edged tool to make a flute on the side surface. The removed area produced a cutting edge with reduced rake angle. Figure 1.18 shows the curve edged tool fabricated by EDM milling followed by die-sinking EDM. The same EDM arrangement is followed with two cameras for setting up the electrode and tool inspection.

To test the micro-tool fabricated by this compound machining process, machining experiments are done on PMMA material to cut micro-channels. The micro-tool is carefully fixed to the spindle and rotated at 3000 rpm. A thin brass sheet is employed to set the top surface reference point. Figure 1.19 explains the arrangement for machining experiments along with the 3D image of machined micro-channel captured using optical profilometer.

After the machining experiments, the channel is examined using 3D noncontact optical profilometer. The side walls are straight. The machined micro-channels had burrs in the surface which cannot be removed or polished by ordinary finishing

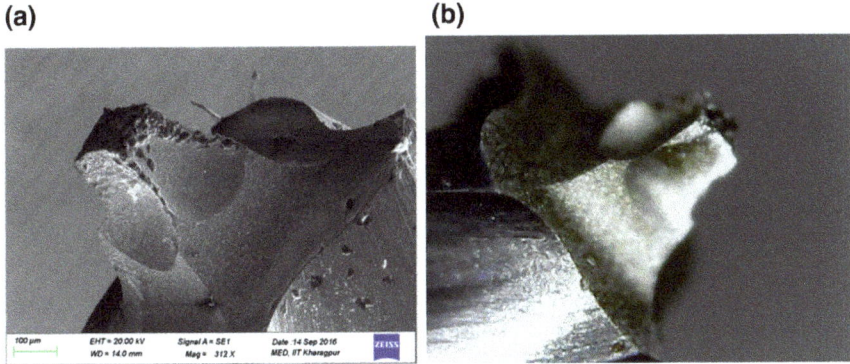

Fig. 1.18 **a** SEM image of curve edge micro-tool and **b** photograph of curve edge micro-tool fabricated by EDM

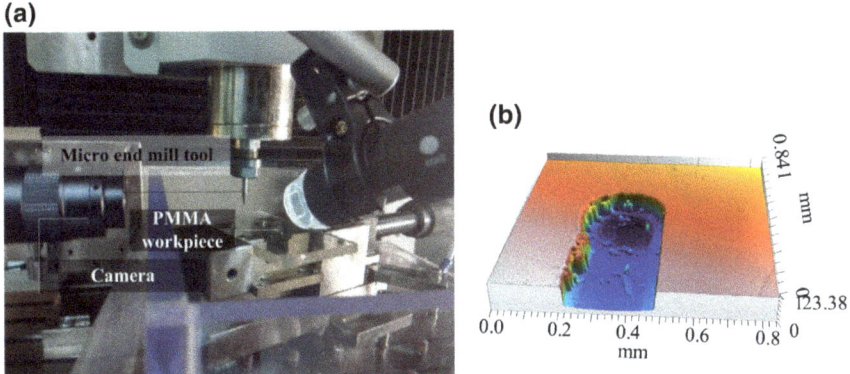

Fig. 1.19 **a** Set up for machining micro-channels on PMMA workpiece **b** 3D profile of machined micro-channel

methods. Burrs formed during the machining are clearly visible in Figs. 1.19b and 1.21a. To remove these burrs, magneto rheological finishing (MRF) process is used. Figure 1.20 explains different arrangements and strategies for MR finishing. This process produced highly finished micro-channels on PMMA material. In Fig. 1.21b, the burrs are almost removed after the finishing process.

Finally, to satisfy the criteria for good cutting tool, rubbing of top middle area of the cutting tool with the bottom surface of channel has to be reduced. To accomplish this, die-sinking EDM can be again used to remove material from the center of the cutting tool. This may produce micro-channels with less burrs and more finished and uniform bottom surface. Figure 1.22 shows the proposed design of micro-end mill tools with reduced negative rake angle and minimized rubbing of bottom surface. Availability of micro end mill tools with higher rigidity will help to

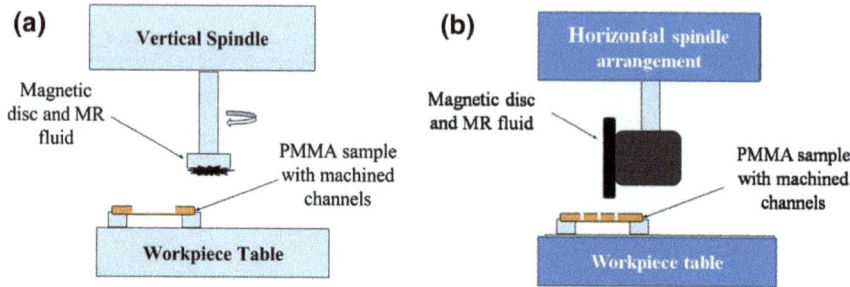

Fig. 1.20 Finishing and Deburring of micro-channel by MR Finishing **a** vertical spindle **b** horizontal spindle

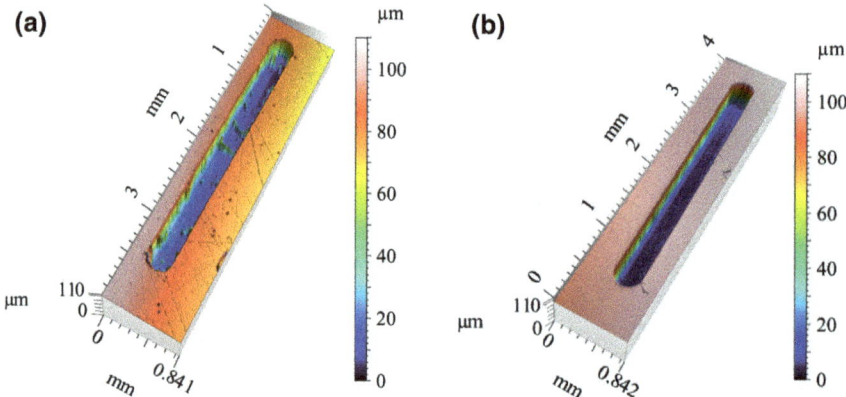

Fig. 1.21 3D profile of the channel machined using micro-end mill tool **a** before finishing **b** after finishing

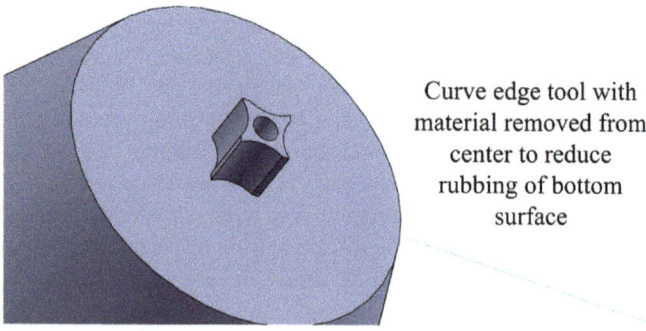

Fig. 1.22 Proposed tool geometry of micro-end mill tool

machine micro channels with higher dimensional accuracy for Lab-on-chip (LoC) applications including inertial micro fluidic devices.

1.4 Summary

Fabrication of micro-cutting tools is an important step in conventional mechanical micro-machining. To accomplish this, different advanced machining techniques have to be carefully studied and employed. Understanding the capabilities and limitations of each process will help to select an appropriate method of tool fabrication. It also helps to advance the research in mechanical micro-machining. A variety of machining techniques including grinding, EDM, WEDM, focused ion beam machining, electrolytic in-process dressing, laser beam machining are discussed and their capabilities have been presented. Grinding process faces difficulty in machining smaller tools because of rigidity problems which can be handled by introducing vibration assistance. WEDM/WEDG is successfully applied for fabricating PCD micro-tools. Laser beam machining is employed for fabrication of SCD micro-end mill tools with complex tool geometry. FIB is a promising machining technique for very small cutting tools but Gaussian energy distribution of the beam results in edge rounding. A new method of compound machining of EDM milling and die-sinking EDM for cutting tool fabrication is presented. A new micro-end mill tool is successfully fabricated using compound machining method and tool characteristics are studied with performance tests. This encourages exploring this technology further for fabrication of high quality micro-tools.

Acknowledgements The authors acknowledge the funding support from the Institute under ISIRD and SGDRI grants, Science and Engineering Research Board (SERB) under young scientist scheme and Board of Research in Nuclear Sciences (BRNS) under young scientist research award.

References

1. Rahman M, Lim HS, Neo KS, Kumar AS, Wong YS, Li XP (2007) Tool-based nanofinishing and micromachining. J Mater Process Technol 185(1):2–16
2. Rahman M, Kumar AS, Prakash JR (2001) Micro milling of pure copper. J Mater Process Technol 116(1):39–43
3. Aurich JC, Engmann J, Schueler GM, Haberland R (2009) Micro grinding tool for manufacture of complex structures in brittle materials. CIRP Ann Manuf Technol 58(1):311–314
4. Asad AB, Masaki T, Rahman M, Lim HS, Wong YS (2007) Tool-based micro-machining. J Mater Process Technol 192:204–211
5. Weule H, Hüntrup V, Tritschler H (2001) Micro-cutting of steel to meet new requirements in miniaturization. CIRP Ann Manuf Technol 50(1):61–64

6. Egashira K, Mizutani K (2002) Micro-drilling of monocrystalline silicon using a cutting tool. Precis Eng 26(3):263–268
7. Chae J, Park SS, Freiheit T (2006) Investigation of micro-cutting operations. Int J Mach Tools Manuf 46(3):313–332
8. Ali MY (2009) Fabrication of microfluidic channel using micro end milling and micro electrical discharge milling. Int J Mech Mater Eng (IJMME) 4(1):93–97
9. Zhan Z, He N, Li L, Shrestha R, Liu J, Wang S (2015) Precision milling of tungsten carbide with micro PCD milling tool. Int J Adv Manuf Technol 77(9–12):2095–2103
10. Masuzawa T, Tönshoff HK (1997) Three-dimensional micromachining by machine tools. CIRP Ann Manuf Technol 46(2):621–628
11. Huo D (2013) Micro-cutting: fundamentals and applications. Wiley, New York
12. Fang FZ, Liu K, Kurfess TR, Lim GC (2006) Tool-based micro machining and applications in MEMS. In: MEMS/NEMS 2006; 678–740. Springer, US
13. Aurich JC, Reichenbach IG, Schüler GM (2012) Manufacture and application of ultra-small micro end mills. CIRP Ann Manuf Technol 61(1):83–86
14. Onikura H, Ohnishi O, Take Y, Kobayashi A (2000) Fabrication of micro carbide tools by ultrasonic vibration grinding. CIRP Ann Manuf Technol 49(1):257–260
15. Lee HW, Choi HZ, Lee SW, Choi JY, Jeong HD (2002) A study on the micro tool fabrication using electrolytic in-process dressing. J Korean Soc Precis Eng 19(12):171–178
16. Ohmori H, Katahira K, Naruse T, Uehara Y, Nakao A, Mizutani M (2007) Microscopic grinding effects on fabrication of ultra-fine micro tools. CIRP Ann Manuf Technol 56(1):569–572
17. Biswas I, Kumar AS, Rahman M (2010) Experimental study of wheel wear in electrolytic in-process dressing and grinding. Int J Adv Manuf Technol 50(9–12):931–940
18. Perveen A, San WY, Rahman M (2012) Fabrication of different geometry cutting tools and their effect on the vertical micro-grinding of BK7 glass. Int J Adv Manuf Technol 61(1–4):101–115
19. Egashira K, Hosono S, Takemoto S, Masao Y (2011) Fabrication and cutting performance of cemented tungsten carbide micro-cutting tools. Precis Eng 35(4):547–553
20. Chern GL, Wu YJ, Cheng JC, Yao JC (2007) Study on burr formation in micro-machining using micro-tools fabricated by micro-EDM. Precis Eng 31(2):122–129
21. Zhang Z, Peng H, Yan J (2013) Micro-cutting characteristics of EDM fabricated high-precision polycrystalline diamond tools. Int J Mach Tools Manuf 65:99–106
22. Zhan Z, He N, Li L, Shrestha R, Liu J, Wang S (2015) Precision milling of tungsten carbide with micro PCD milling tool. Int J Adv Manuf Technol 77(9–12):2095–2103
23. Fonda P, Katahira K, Kobayashi Y, Yamazaki K (2012) WEDM condition parameter optimization for PCD microtool geometry fabrication process and quality improvement. Int J Adv Manuf Technol 63(9–12):1011–1019
24. Xu K, Zou B, Wang Y, Guo P, Huang C, Wang J (2016) An experimental investigation of micro-machinability of aluminum alloy 2024 using Ti (C 7 N 3)-based cermet micro end-mill tools. J Mater Process Technol 30(235):13–27
25. Suzuki H, Okada M, Fujii K, Matsui S, Yamagata Y (2013) Development of micro milling tool made of single crystalline diamond for ceramic cutting. CIRP Ann Manuf Technol 62(1):59–62
26. Wu W, Li W, Fang F, Xu ZW (2015) Micro tools fabrication by focused ion beam technology. In: Handbook of manufacturing engineering and technology 2015. Springer, London, pp 1473–1511
27. Picard YN, Adams DP, Vasile MJ, Ritchey MB (2003) Focused ion beam-shaped microtools for ultra-precision machining of cylindrical components. Precis Eng 27(1):59–69
28. Adams DP, Vasile MJ, Benavides G, Campbell AN (2001) Micromilling of metal alloys with focused ion beam–fabricated tools. Precis Eng 25(2):107–113
29. Adams DP, Vasile MJ, Mayer TM, Hodges VC (2003) Focused ion beam milling of diamond: effects of H2O on yield, surface morphology and microstructure. J Vac Sci Technol B Microelectron Nanometer Struct 21(6):2334–2343

30. Adams DP, Vasile MJ, Krishnan AS (2000) Microgrooving and microthreading tools for fabricating curvilinear features. Precis Eng 24(4):347–356
31. Ding X, Lim GC, Cheng CK, Butler DL, Shaw KC, Liu K, Fong WS (2008) Fabrication of a micro-size diamond tool using a focused ion beam. J Micromech Microeng 18(7):075017
32. Fang FZ, Wu H, Liu XD, Liu YC, Ng ST (2003) Tool geometry study in micromachining. J Micromech Microeng 13(5):726
33. Nakamoto K, Katahira K, Ohmori H, Yamazaki K, Aoyama T (2012) A study on the quality of micro-machined surfaces on tungsten carbide generated by PCD micro end-milling. CIRP Ann Manuf Technol 61(1):567–570
34. Ohnishi O, Onikura H, Min SK, Aziz M, Tsuruoka S (2007) Characteristics of grooving by micro end mills with various tool shapes and approach to their optimal shape. Memoirs of the Faculty of Engineering, Kyushu University 67(4):143–151

Chapter 2
Machining of Glass Materials: An Overview

Asma Perveen and Carlo Molardi

Abstract Being hard and brittle, glass is considered one of the very difficult-to-cut materials. This chapter is primarily concerned about the glass types, properties, and applications as well as associated difficulties in their machining. Following that, this chapter explores various conventional and nonconventional technologies available till today for machining of glass material. Furthermore, it also sheds light on cutting mechanisms of glass materials for both single and multiple cutting edge tools in case of conventional machining. Aspects of tool wear and surface/subsurface damage relevant to glass machining are also discussed.

Keywords Abrasive water jet machining · Discharge · Electrochemical · Glass · Laser · Ultrasonic machining · Ultraprecision · Wear

2.1 Glass and Its Machining

2.1.1 Introduction

In the recent past, to fulfill the request for extreme applications, there have been tremendous interests in improved ceramics and glasses such as, silicon nitride and carbide, alumina, BK7, and soda lime glass materials with unique metallurgical properties. Even if these materials are relatively harder, less heat sensitive and resistant to chemicals, fatigue, and corrosion, such materials impose significant challenges to machining [1]. These materials, known as brittle and difficult to cut, are commonly used not only in the electronic industries but also in the public welfare industries like biomedical and optics [2]. For electronic industries, glass is popular functional material in microelectromechanical system (MEMS) device

---

A. Perveen (✉)
Mechanical Engineering Department, Nazarbayev University, Astana, Kazakhstan
e-mail: asma.perveen@nu.edu.kz

C. Molardi
Information Engineering Department, University of Parma, Parma, Italy

packaging, microelectronic packaging, and microfabricated device like, solid oxide fuel cell for portable electronic device, pump, and reactors [3]. In case of biomedical industry, glass has found its applications in microfluidic device, DNA array, micro-valve, micro-flow sensor, biomedical parts, and biological instrumentations [4]. Glass also acts as substrate for bonding with wafer in semiconductor applications. On top of these areas, glass is also contributing in optics industry for the fabrication of telescope lenses, microscopic slides, optical fiber alignment, and mini-vision system [5, 6]. In addition, it has extensive applications in automotive industry for windscreen, and cockpit windows.

There are various types of glasses available in the market at present. Commercially available glasses can be categorized based on their chemical composition as below:

- Soda lime glass
- Lead glass
- Borosilicate glass
- Aluminosilicate glass
- 96%-silica glass
- Fused silica glass.

Glass is widely used in micro technology system, due to its beneficial and functional properties. Examples can be retrieved from the field of microfluidic devices, used in biotechnical applications [7, 8]. In micro-total-analysis-system (μTas), glass offers several benefits, much more than silicon like optical transparency which is useful as an example for visual inspection, and real time optical detection. The dielectric property of glass can be effectively used to resist strong voltages which are necessary in electrokinetical separation and flow driving. Furthermore, glass presents other interesting properties such as stability to strong temperature gradient and chemical inertness. Such outstanding properties make glass the perfect material to build substrate for DNA arrays. Some of the excellent properties associated with glass materials are quoted as follows [7–10]:

- High hardness
- Homogeneity
- Optical transparency
- Isotropy
- Various refractive indices
- Low thermal and electrical conductivity
- High dielectric strength
- High chemical resistivity.

In the area of machining, advanced working of ceramics and glass materials has earned primary attention. Like most other materials, glass undergoes plastic deformation as seen from indentation test if the depth of cut can be maintained under certain critical value which will avoid brittle fracture [11]. Hence, the possibility to machine hard to cut brittle materials, using ductile strategy avoiding fractures, has become real. This fact leads to the new possibilities of machining brittle materials with optical surface finish by exploiting traditional machining processes which eliminates the necessities of the secondary finishing process. Machining processes like grinding, milling, turning, abrasive jet machining, ultrasonic machining, laser machining have already been quite successfully employed to machine glass. The excellent mechanical properties of these hard to machine materials, i.e., glasses doubtlessly increase their usage in several important applications while opening up the new windows of their applications due to their treatments by ductile machining.

2.1.2 Challenges in Machining of Glass

The family of optical glasses, which are hard and brittle, represents huge challenge to mechanical machining. For optical applications, glass shaping is usually obtained by grinding, followed by a series of polishing processes to remove the damage caused by grinding. However, it is quite challenging to generate favorable microstructure imparting necessary precision characteristics in glass for microfluidics devices. This difficulty to build glass structure is quite evident, among large number of micromachining techniques, both conventional and nonconventional.

Significant researches have been conducted on micro fabrication technology which mainly involves photolithography, as well as chemical etching methodology. Glass, showing isotropic nature can be wet etched using hydrofluoric acid with a nondirectional strategy. Dry etching by chemical action is also possible in typical SF6 plasma which is hindered due to slow etching rate [12]. Etching techniques involve hazardous problems, since the etching material contains lead or sodium which produces nonvolatile halogen compounds as reaction products. Reactive ion etching, which uses special plasma source, in order to produce high density plasma at low pressure, has been exploited for silicon channels generation. Nevertheless, such techniques are not enough developed to be used in complex glass structures.

Laser machining also offers potential risk of micro-cracking, debris formation, and other damages, because of the high fragility and the poor thermal behaviors of many types of glass [5]. Mechanical machining techniques, such as abrasive jet machining can be used for brittle materials. This technique, which consists of mechanical removal by the use of high-speed jet particles, allows obtaining eroded structure with complex shapes. However, jet machining is hindered due to the rough surface and limited to larger component [13, 14].

Machining of metals to achieve high quality surface finish is relatively easier considering the high ductility. But as the utilization of nonmetallic materials like glass is growing rapidly in manufacturing of precision products ranging from jet engine parts to the mold manufacturing for precision casting, there is an equivalent growth in the demand for ductile mode machining of these materials. This ductile regime cutting has been experimented, mostly using single point process where a single-crystal diamond is employed to remove the brittle material without any fracture. However, generation of asymmetrical features and complex shapes are beyond the capability of turning. With the increasing demand of miniaturization, the challenge of achieving complex shapes and structures on miniaturized devices is also rising. This challenge appeals for versatile machining processes, like milling, to produce such complex shapes. Nevertheless, a multi-cutting edge process creates even more difficult scenario to control the cutting conditions to achieve ductile machining. Brittle materials are machined with difficulties by mechanical cutting process such as milling. The reason can be searched in the damages due to the material removal induced by brittle fracture, leading to a nonuniform surface which requires further polishing. High-speed milling, widely used by industry of metallic mold, is hard to apply in the machining of glass because of the relevant strength and hardness associated with this material. Such technical difficulties highlighted in the above mentioned machining techniques, and the need to reduce the high cost associated with the machining procedures, push the industry to foster newer approaches for machining glass materials [6].

2.2 Cutting Mechanisms

Underpinning knowledge associated with brittle to ductile transition and optimum cutting conditions may offer significant improvement in the machining technology of brittle materials. This section will give an overview of various cutting mechanisms developed over time relevant to glass and prerequisite conditions for ductile mode material removal.

2.2.1 Ductile Mode Machining

Precision applications specific to the tight tolerances have opened up a huge opportunity to explore the new ways and techniques to machine glass materials. The property of ductility indicates the extent of permanent deformation without fracture. The term plastic deformation refers to the ability of the material to shape permanently under loading. All materials have some extent of ductility no matter how brittle they are. So fracture in all materials is preceded by the manifestation of more or less ductility. However, the extent of ductility is diverse for different materials. The scale of consideration is an important factor to assess the plastic deformation probability. Material like glass that is perfectly brittle at macro scale, may exhibit plastic deformation at microscale.

Extensive researches have been pursued over the past two decades, to evaluate the plastic deformation characteristics of brittle materials like glass through indentation, scratching, grinding, and machining. Dolev et al. [15] observed microplasticity phenomenon, where the glass exhibits ductile or plastic behavior, when indented with a concentrated load. Finnie et al. [16], with the help of

Fig. 2.1 Schematic representation of elastoplastic indentation [18, 19] (Hydrostatics core is depicted as *black* area. Plastic zone is depicted as *gray* color. Elastic matrix is represented by *black arrows*) with kind permission from Elsevier

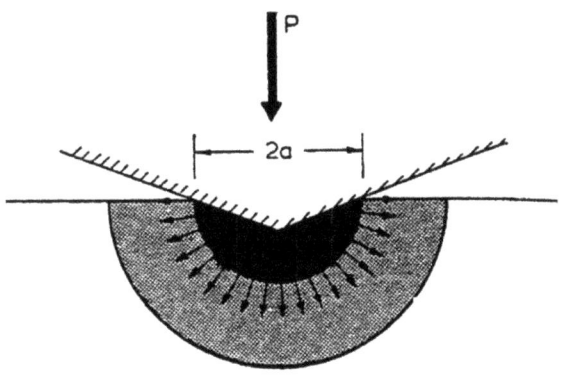

Auerbach's law, explained that the use of small indenter can exploit the transition from brittle to ductile mode, where the cracking load is having linear proportional relation with the indenter diameter. Lawn et al. [17] quantified the results obtained by indenting soda lime glass with Vickers's Pyramid indenter at different loads (see Fig. 2.1) and reported that cracking is a favorable mechanism above a critical load. Below this point, there are no cracks or fracture. This resulted in a conclusion that significantly lower loads can help to create clear impressions using indenter, however, the possibility of crack initiation increases with increase of load value. Under such controlled environment, possibility of machining glass without any crack in ductile mode has got new dimension. Ductile mode machining can happen in any material so long the deformation keeps in smaller range.

Giovanola and Finnie (1980) performed the first feasible work in order to study the plastic deformation of glass. According to this study, glass can be processed, similar to metal, using ductile phenomenon as long as the cut thickness is kept significantly small [20]. Later on, after a thorough investigation on ductile mode grinding of several ceramics, Bifano [21] stated that the critical undeformed chip thickness, which governs the brittle to ductile transition, is influenced by the intrinsic properties of material like elastic modulus, hardness, and fracture toughness. Blackely (1991) also derived an analytical model using processing parameters of diamond turning in order to evaluate the critical parameters for achieving ductile mode cutting. Figure 2.2 shows a schematic of turning tool with cutting parameters.

This research recommended a critical value of undeformed chip thickness below micron range. According to this model, for achieving ductile surface, it is not necessary that material removal is completely done by ductile mode, rather a combination of ductile brittle mechanism may work well so long the fractured surface remains far from final surface. This situation can be created by using relatively smaller feed rate which keeps the subsurface damage depth far from the critical limit as given in Fig. 2.2 Considering the critical feed rate, the critical undeformed chip thickness, will occur significantly well above the cut surface so that subsurface damage will not be able to reach below the cut surface. As a result, a damage-free surface will be generated even though the brittle fracture will still be

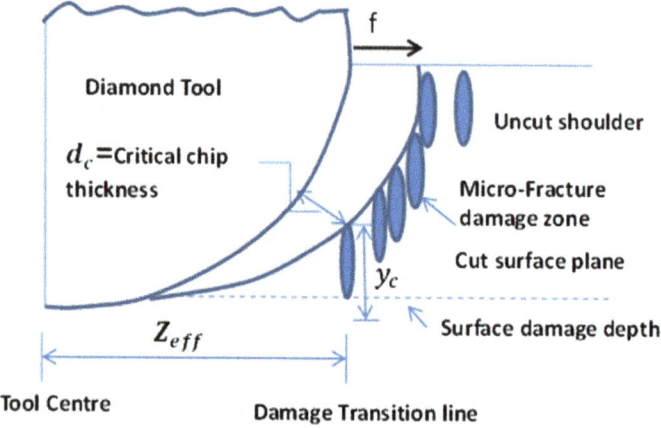

Fig. 2.2 Schematic cutting model for critical depth of cut [11] with kind permission from Elsevier

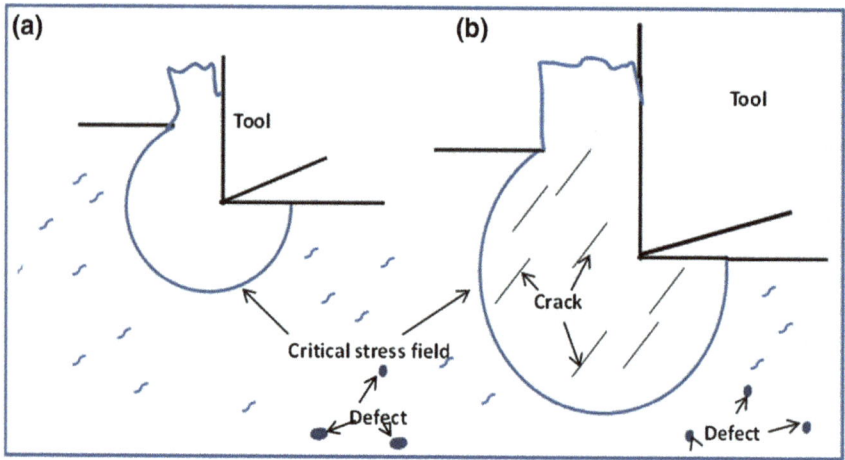

Fig. 2.3 Schematic of defect distribution and chip removal. Stress field generated by **a** small depth of cut **b** large depth of cut [22] with kind permission from Elsevier

there in the machining. Alternatively, higher feed rate causes the critical undeformed chip thickness to occur adjacent to the cut surface. Consequently, damage will propagate beneath the cut surface, contributing in brittle mode machining [11].

Brittle to ductile transition can be alternatively explained by the analysis of cleavage fracture using the existence of defects [22]. Density of defects limits the plastic deformation and the cleavage. Generally, the critical value of fracture is determined by stress field size due to the smaller size of defect density in brittle materials. Figure 2.3 shows the size effects phenomenon. Smaller uncut chip thickness, causing the generation of smaller stress field, avoids the cleavage fracture which in turn shifts the cutting mode from brittle to ductile.

Therefore, there exists two different approaches of material removal, namely brittle fracture and plastic deformation for brittle materials machining. The former one occurs when the cleavage plane coincides with the maximum shear stress plane. The later occurs on the slip plane coinciding with the maximum tensile stress plane. Considering the case where the critical shear stress is overtaken by the tensile stress applied in slip direction, before the occurrence of cleavage, the workpiece undergoes plastic deformation in the small stress field associated with the cut depth. Conversely, when the applied tensile stress, which is normal to the plane of the cleavage, manages to overcome the critical tensile stress before the plastic deformation onset, in such case, the scenario is dominated by the brittle fracture [23].

Cai et al. [24] narrated about another possibility that can help in ductile regime machining. It is the case when the value of uncut chip thickness is significant; in particular, its value is less than the size of cutting edge. Under this circumstance, the cutting force becomes lower than the thrust force in the machining process which squeezes the material beneath the cutting edge and a highly compressive hydrostatic force is prevailed in the regime of chip formation which arrests the propagation of fractures. Therefore, material removal mechanism is dominated by plastic deformation over brittle fracture. Ultraprecision machining using higher negative rake angle tools usually provide enough hydrostatic pressure for inducing plastic deformation. Moreover, careful look into a grinding wheel reveals the fact that grinding grains possessing higher negative rake angle cause the cutting force to reduce half of the thrust force. A deep analysis of ultraprecision technique for brittle materials suggested that higher negative rake angle can be achieved by selecting the value of cut depth less than tool radius. Consequently, plastic deformation or ductile mode machining takes place, which also prevents median cracks initiation because of huge hydrostatic pressure created by the tool edge [25].

2.2.2 Material Removal Mechanism in Glass

Machining of glass in ductile manner could be considered similar as the machining of metal. However, high brittleness arises from the irregularity of atoms structure inside the glass material. On the other hand, atoms inside the metal maintains static regular pattern as indicated by Miller indices contrary to the amorphous material [26]. The status of atomic bond indicates the kind of material removal mechanism involved [27, 28]. Metals, having metallic bonds, generally experience ductile manner machining. Brittle mode machining of glass can cause vertical cracks which results in substantial amount of subsurface damage during loading whereas during unloading it can cause lateral cracks to generate which consequently results in material removal. Nevertheless, choosing very low depth of cut can avoid the situation of brittle mode cutting even for brittle material like glass. Blackley [11] suggested the depth of cut to be below 10 nm in order to machine glass in ductile manner. Therefore, mechanics of material removal involves brittle fracture and plastic deformation in glass. Lateral and median cracks are consequence of brittle fracture created while indenting on brittle substrates. Cutting load and work piece properties govern the brittle mode fracture. On the other hand, plastic deformation is similar to chip formation involving scratching, ploughing, and formation of chips during the grinding process of metal. Theory of ductile mode machining suggests that all materials irrespective of its ductility will experience brittle ductile transition phenomenon below the critical depth of cut for that particular material. Below this critical depth of cut limit, the energy for crack propagation seems to be greater than the energy for plastic deformation which makes the plastic deformation mechanism as predominant [29]. When the grain makes impact on the material during grinding, in the contact region; heat accumulation, caused by the poor thermal dissipation, along with the large compressive stress causes plastic deformation [30]. Therefore, ductile regime machining is influenced altogether by the tool shape, feed rate, and the critical depth. However, clean ductile mode only appears along the tool apex as the actual depth of cut is found to be lower than the critical depth [31]. During turning, material response is governed by the stress field size and value, along with the cutting combination. While investigating the transition between brittle and ductile phenomenon, it has been understood that generated stress field can be arranged in four different zones as given below [32]:

1. Zone I: Chemical/temperature effect on top of mechanical action removes materials in this region (Machining unit below 10^{-6} mm).
2. Zone II: The material behaves as an ideal crystal without any dislocation. Initiation of dislocation appears just prior to brittle fracture which leads the crystal to next zone (Machining unit between 10^{-6} and 10^{-4} mm).
3. Zone III: This zone exhibits plastic deformation. Typically, cracks appear in this stage (Machining unit between 10^{-4} and 10^{-2} mm).
4. Zone IV: Crack/brittle mode is dominating at this region (Machining unit above 10^{-2} mm).

Therefore, it can be comprehended that, material removal occurs due to erosion/chemical action while using extremely small depth of cut, then it follows plastic deformation/microfracture, depending on the conditions.

2.3 Machining of Glass

2.3.1 Conventional Machining

2.3.1.1 Turning

Single point diamond turning (SPDT) is an ultraprecision cutting technique to generate nano surfaces with submicrometer level of form accuracy while product manufacturing [33]. The capability of machining optical substrates, for operation in infrared and visible region, has given widely acceptance to SPDT process as a fabrication technology for optics [34]. In the recent past, the availability of high-precision turning machines with nano-level tool positioning accuracy was considered to be a prerequisite for machining brittle materials in ductile mode. As mentioned earlier, there is this critical value of feed rate and depth of cut for every material which should not be exceeded; otherwise it will cause the ductile mode to switch into brittle mode and initiates several different crack systems (as shown in Fig. 2.2) [11]. A study on ultraprecision turning of optical glass ZKN7, conducted by Fang et al. [35] reported that undeformed chip thickness with the size of submicrometer level can produce mirror finish (R_a = 14.5 nm) with ductile mode cutting. The force generated during ductile mode cutting of ZKN7 glass is found to be around 1 N, however, the growth of shear stress continues until it becomes too enormous to cause speedy tool wear, due to the submicron size of undeformed chip thickness.

Previously, ultraprecision machine tools, providing high accuracy and stiffness, was used mainly to conduct ductile mode machining because of the difficult situation arises from the smaller cutting depth and lower feed rate. Research performed to investigate ductile machining of glass proved to be not so practical due to the high wear rate of tools. Therefore, more researches on vibration assisted turning of brittle materials were undertaken with a hope to elevate this critical value [36,37,38] and the results suggested the promising capability of cutting fragile materials by ultrasonic vibration in ductile manner. Although, specially controlled vibration assisted machining was able to offer mirror finish on soda lime glass, the mechanism behind this vibration assisted machining was not well established back then. Tool wear was still quite large compared with that in the conventional machining of soft metals. Therefore, Wang et al. [39] performed another study to examine the cutting process of glass by varying the impact of vibration in a diamond tool and suggested that the ratio of the cutting speed associated with the workpiece and the maximum vibration speed of the tool is a key factor to determine the critical depth of cut. In cutting, assisted by ultrasonic vibration, the reduction of cutting force is influenced by the combination of dynamic friction and aerodynamic lubrication. The increasing tendency of critical depth of cut is influenced by the cutting force reduction. Chalcogenide glass used for infrared imaging purpose has been attempted to machine recently using diamond turning and experimental result reported the change in the cutting mechanics to ductile mode at uncut chip thickness of 1 micron [40].

Investigation on diamond turning of glass to examine tool wear mechanism recommended that some wear mechanism are predominant to degrade the operation of the tool, in particular these mechanisms are identified to be cleavage and micro chipping. The beneficial application of ultrasonic vibration to the cutting tool suggested the significant improvement in cutting performance. Ultrasonic vibration not only lessened the tool wear rate compared to the one without vibration, but also suppressed brittle fracturing in glass cutting zone [41]. Bakkal et al. [42] also reported severe tool wear condition like chip welding and micro chipping of polycrystalline cubic boron nitride tool while turning of bulk metallic gas.

2.3.1.2 Grinding

Grinding is the most widely used process for machining hard–brittle materials like optical glass. The behaviors of fracture in optical glasses is now a days a subject of intense research. Processes like grinding or indentation and in general, all the processes that involve the use of an abrasive grain to interact on the material surface are far to be completely understood. Malkin and Hwang [43] have reviewed all these contributions and reported two major crack systems that exist in these indentation processes of brittle material. One of them is the formation of lateral cracks which is accountable for material to be removed to generate new surface. The other one instead is produced by the cracks located in radial and median position which work as the origin of strength degradation. However, the improvement of high-precision machine tools ensured the suppression of these cracks using appropriate machining parameters. As explained in [27], material will undergo plastic deformation only when the penetration depth can be kept below the critical, resulting in lower converted energy which is insufficient for cracks to form. This described mechanism is usually referred as ductile regime grinding [27].

Grinding mechanism and all the processes comparable to the material removal by grinding are pushed forward by the growing interest given by industrial sectors. Chen et al. [44] investigated the optimum settings for partial optical glasses and microcrystalline glasses to reach the transition from brittle to ductile using indentation tests and reported that exceptionally smooth surface is possible to generate using diamond wheel with grain size below 10 μm. Brinksmeier et al. [45] found that the modes of material removal for optical glass are not only affected by the cutting depth but also by the fraction of feed rate over cutting depth. Stephenson et al. [46] studied the grinding properties of Bk7 glass using both cross and parallel grinding strategies, and revealed that parameters like roughness of surface and depth of damage under the surface are directly triggered by the setting of grinding modes. Surface grinding on horizontal plane and induced cracks have been analyzed by the use of kinematic strategies. The results suggest dividing grinding mode into four different subfamilies: brittle, semi-brittle, semi-ductile, and eventually ductile mode as shown in Fig. 2.4. These four different grinding modes also influence output parameters like surface roughness as well as the depth of damages under the surface [47].

Zhong and Venkatesh [48] emphasized that the machining of brittle materials in ductile fashion will be constantly growing as rigorous research arena after a through summarization of grinding development. The main reasons behind this involve rapidly increasing industrial demands which requires underpinning knowledge of the ductile mode grinding mechanisms. Engineered wheel for grinding optical glass has also been proposed by Heinzhel et al. [49]. Perveen et al. [50] conducted the effect of various tools geometry such as circular, d-shaped, triangular, and square in the case of Bk7 glass. They found that better performance is obtained by d-shaped micro-grinder, showing a better surface finish, by limiting tool wear and cutting force [51]. In their work, focused on micro grinding of Bk7 glass, Perveen et al. [46] also investigated on PCD micro-grinder wear. According to their research, two

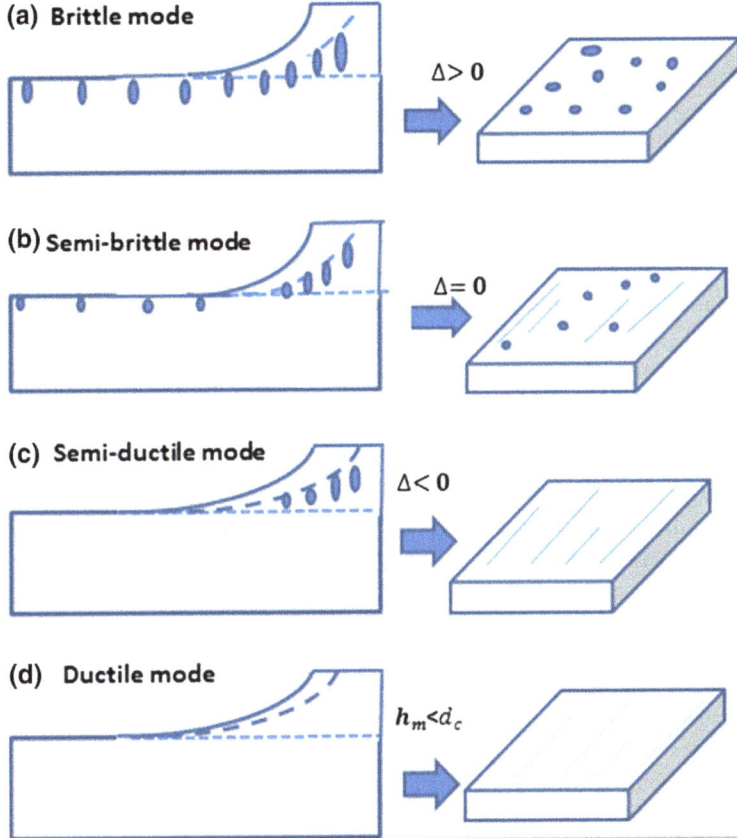

Fig. 2.4 Four different modes of grinding during horizontal surface grinding (maximum penetration depth h_m and critical depth d_c) [45] with kind permission from Elsevier

basic types of tool wear were noticed. First one is edge chipping which appears near to tool rim and contributes on the reduced diameter of tool. Second, abrasive wear occurring by the fast grain boundary breaks down, also initiates some intergranular cracks to propagate. This propagation makes its path either through a group of grains or through the grain boundary causes the grain spalling thus forming wear zone [46]. By applying taper polishing technique, surface cracks formed during grinding processes have been characterized as near-surface lateral and deeper trailing indent type fractures on silica glass. Both the average crack length and surface roughness shows linear relation with the maximum SSD depth. In addition, these relationships can be used to identify and measure the SSD depth without destructing the part [47].

2.3.1.3 Micro-Milling

Recently, increasing advancement in microelectronics has raised the necessity for micro-components using complicated geometries and three dimensional features. In order to fulfill this demand, investigation on the material removal using plastic deformation technique with the use of multi-cutting edge tool has started to get more attention, due to the limitation associated with single point tool. Milling is basically an interrupted cutting operation where teeth of milling tool remove materials in an intermittent manner with tool rotation. Milling process can be either up milling or down milling type. In up milling, direction of cutter rotation opposes feed motion so that maximum chip thickness is at the end of the cut as shown in Fig. 2.5. Contrarily, for down milling, cutter rotation is in the same direction with feed motion and chip thickness remains maximum at the beginning of the cut. Takeuchi et al. [48] used ultraprecision milling machine to conduct ultraprecision 3D micromachining of glass. They demonstrated the possibility of manufacturing 1 mm diameter glass mask with roughness value of 50 nm using pseudo ball-end mills. Matsumura et al. [49, 51, 52] achieved fracture-free machining of glass by exploiting different geometry tool, in particular milling tool with flat end and ball-end. The machining of micro-channels in glass is also possible in ductile manner by multi-edge cutting tool if the cutting edge is sharp enough. It is also suggested that ductile mode machining involves smooth signal of cutting force while brittle mode machining is characterized by sharp fluctuations in the cutting force because of the occurrence of repeated fracture above the critical value of chip thickness. As reported by Foy et al. [2], improvements in surface finish during milling of glass are possible by tilting the ball-end mill along the direction of feed. It is also demonstrated that ductile mode machining of glass is possible with carbide end mill if the edge roundness is at submicron level. It is suggested that cooling time during the cut increases by tilting the end mill which suppresses the tool wear considerably.

End milling process can either be side or slot milling type. For up milling orientation in case of side milling, uncut chip thickness varies from zero to maximum with tool rotation. However, when this chip thickness reaches critical limit imposed by the transition from ductile to brittle mode, fracture begins to appear.

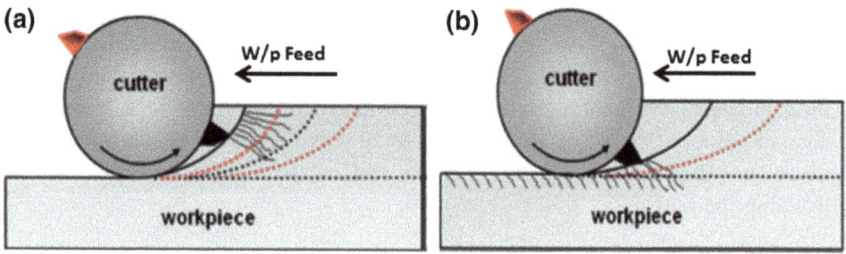

Fig. 2.5 **a** Low feed per edge ductile mode milling process; **b** High feed per edge brittle mode milling process [54]

Again, tool feed per edge will determine whether this brittle fracture point should be far (low feed per edge) or near to the finished surface (high feed per edge) as shown in Fig. 2.5a, b. If the specific brittle fracture limit is distant from the finished surface, the fractured part will be eliminated by the cutter rotation later on as depicted in Fig. 2.5a, contrarily brittle fracture that cannot be eliminated by cutter rotation reaches to the surface as shown in Fig. 2.5b [6, 53]. Cutting force acting tangential to the tool axis is proportional to the undeformed chip area. Therefore, domination of plastic deformation causes the cutting force to be increased linearly within ductile region, whereas fluctuation of cutting forces referring the initiation of fracture when the uncut chip thickness reaches the critical value [54].

For slot milling, undeformed chip thickness remains very small both at the starting point and the final point of cut, and the maximum value can be found at the cut center. If the fracture occurs due to larger contact angle between cutter and work piece, it can be eradicated by succeeding cutter flute (Fig. 2.6a) so long the depth of fracture remains lower than the depth of axial cut. However, the damage will sustain on the final surface if the axial cutting depth is overtaken by the depth of fracture. Damage will also retain for larger axial depth of cut (Fig. 2.6b).

Fracture will occur at smaller contact angle for higher feed compared to that of lower feed. Therefore, smaller feed rate and low axial cutting depth are considered for fracture-free slots. In addition, up milling occurs at one side of the cut and down milling occurs at the other side of the cut as shown in Fig. 2.7. Due to this typical cutting mechanism, the two ends of the slot may experience different surface quality [6].

While investigating the capability of the end milling of glass, Arif et al. [53] also identified two different types of tool wear. The first type of wear is abrasion wear which has been found on the flank face created under the condition of smaller uncut chip thickness. A second type of wear is basically constituted by the formation of grooves on tool cutting edge which in turn causes continuous deformation of cutting edge after longer machining time. Shifting of cutting mode from ductile to brittle can be heavily influenced by this tool wear.

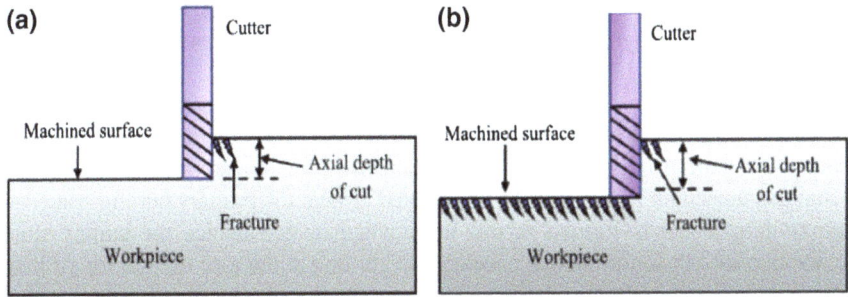

Fig. 2.6 **a** Slot milling using small axial depth of cut **b** Slot milling using larger axial cutting depth [6] with kind permission from Elsevier

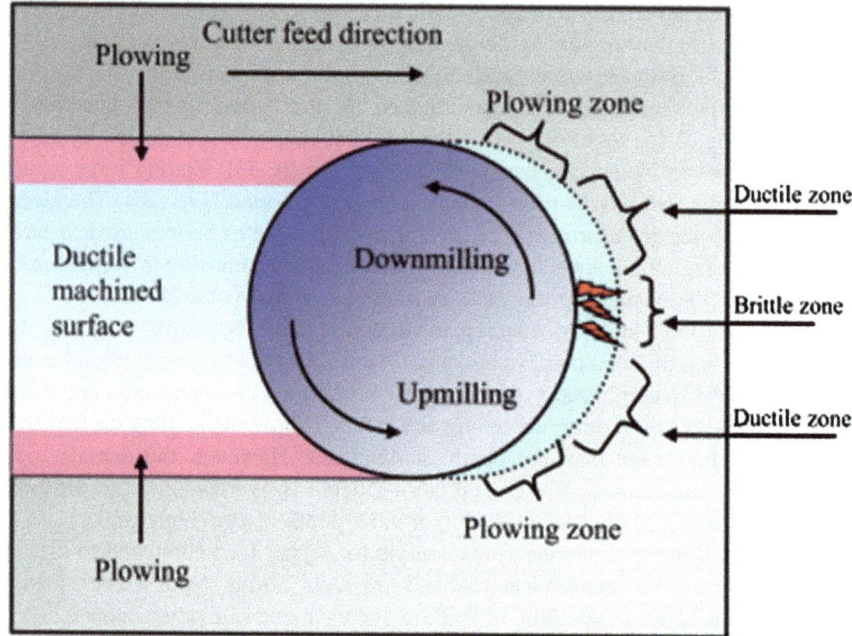

Fig. 2.7 Different regimes machining in slot machining of glass [6] with kind permission from Elsevier

2.3.2 Nonconventional Machining

2.3.2.1 Ultrasonic Machining (USM)

Ultrasonic machining is a popular machining technique for the materials with low ductility and high hardness (above 40 HRC), irrespective of their conductivity. In USM, mechanical vibrations of greater than 20 kHz with 5–50 μm amplitude are obtained by converting high frequency electrical energy using transducer, booster, and horn. To avoid the resistance of cutting action, controlled static load is maintained on the tool. A mixture of SiC/B_4C with water is hammered on the workpiece surface by the tool vibration and thus removes the material by microchipping [55].

Very poor thermal conductivity of glass and the friction generated between glass and tool during conventional cutting are great causes of temperature accumulation around the cutting edge of the tool. By applying an ultrasonic assisted cutting technique, a vacuum region may be constituted around the primary cutting zone and the pumping effect is formed in this area, which may enhance the cutting fluid penetration into the cutting zone accelerating cooling at the tool-tip. On top of that, ultrasonic assisted cutting technique also converts the engagement type of cutting tool-chip into an interaction manner of a small amount of vibration. Thus, appropriate application of ultrasonic assisted technique process can actually improve the cutting performance [56]. Applying workpiece vibration as illustrated in Fig. 2.8a, a new method of micro-ultrasonic machining (MUSM) has been developed which is considered one of the important achievement of USM so far. With the help of this method, microholes with diameter smaller than 5 µm has been obtained in fused quartz as seen in Fig. 2.8b. The high tool wear associated with MUSM is addressed using a sintered diamond tool [57]. A combination of MUSM and Micro-EDM represents an interesting hybrid process for manufacturing extremely precise holes with the advantage of higher aspect ratio.

Another improvement of USM, process can be achieved by addition of chemical assistance with the use of HF particularly in low concentration. With this technique, typical drawbacks like deteriorated roughness and small-scale removal can be overcome. The mechanism of this action is given by the interaction between HF solution and glass. The superficial silicon molecules experience reduced bond strength due to the acidic action, thus improving the effectiveness of USM method. Significant enhancement in surface quality (40%) and improvement in MRR (200%) were reported by Choi et al. [58] while micro- and macro-drilling of glass by ultrasonic machining technique (Fig. 2.9). However, due to the hole enlargement occurred in chemically assisted USM; below 5% HF solution is suggested.

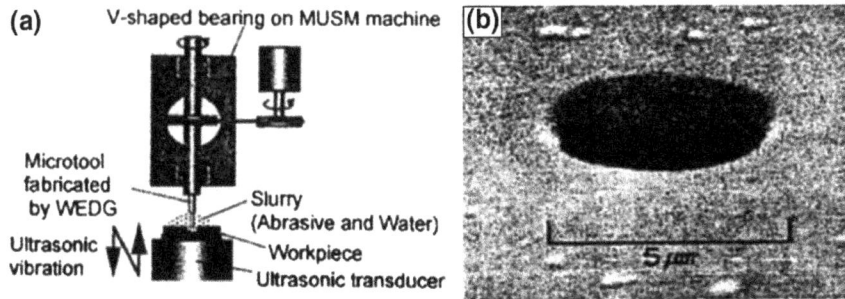

Fig. 2.8 **a** MUSM process **b** Micro hole of 5 µm diameter (quartz glass, depth: 10 µm, amplitude: 0.4 µm, tool diameter: 4 µm, machining load: 29–59 µN) [57] with kind permission from Elsevier

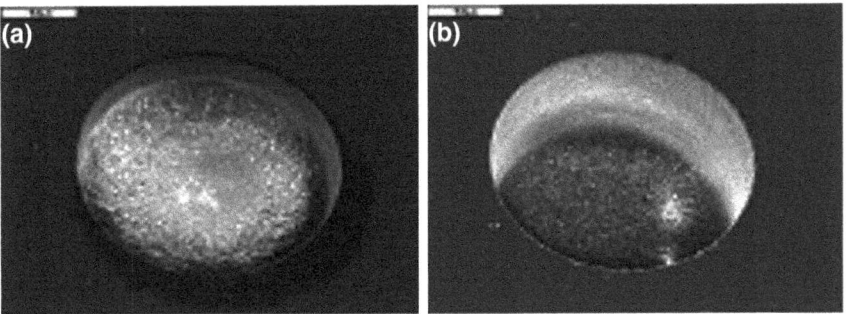

Fig. 2.9 Comparison of material removal by 1.5 mm diameter tool using **a** USM (550 µm depth) and **b** CUSM (950 µm depth) [58] with kind permission from Elsevier

2.3.2.2 Abrasive Jet Machining (AJM)

Abrasive jet machining (AJM) operates on materials without producing shock and heat. In this method, the surface to machine is bombarded with an intense jet composed of a mixture of gas/air and abrasive particles (Fig. 2.10). At the nozzle exit, the pressure of the carrier fluid induces a jet of particles with high kinetic energy, which impacts on the surface to remove materials using micro cutting action along with brittle fracture. AJM has the capability to perform primary operations like machining, drilling, and surface finishing efficiently [59].

Another type of AJM, namely abrasive water jet machining (AWJM), makes use of a water jet mixed with abrasive particles instead of gas. Cutting mechanism involves the solid particles to impact and erode on the surface for material removal either by cutting wear or by deformation wear. Erosion taking place at smaller impact angle is known as cutting wear whereas erosion at higher impact angle with repeated bombardment is known as deformation wear [60]. While investigating erosion behavior of borosilicate glass, Aich et al. [61] observed brittle fracture on the top surface of glass due to the high velocity of abrasives particles that impinge on the workpiece. Significant turbulence is created at the lower zone of cutting due to the damping of axial velocity of water which causes changes in radial energy distribution. Therefore, material removal is dominated by plastic deformation.

One of the major advantage offered by AWJM is the possibility of suppressing heating during the material removal, in particular the area of machined surface remains unaffected by heat. Azmir et al. [62] established that it is possible to produce better surface by playing with kinetic energy of AWJM process, in particular they found that an increase of energy is beneficial for machining glass composites reinforced with plastic fibers. According to their research, machining performances like

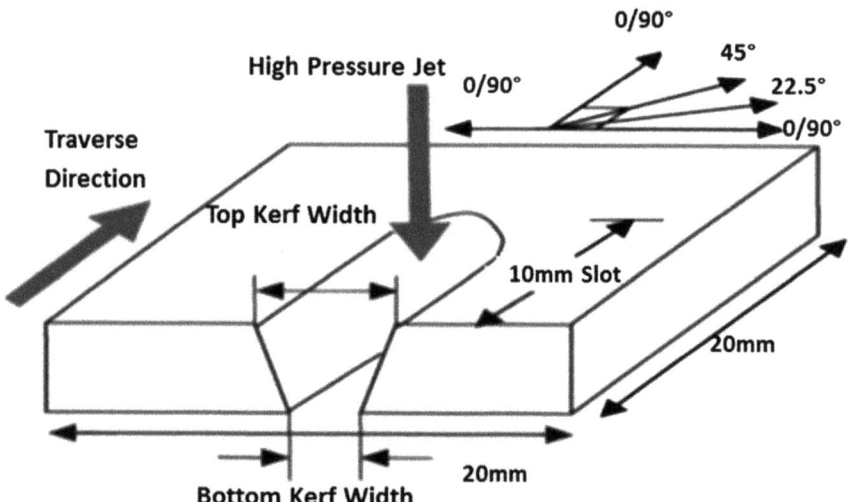

Fig. 2.10 Schematic of AWJ machining on glass/epoxy laminate [62] with kind permission from Elsevier

roughness and kerf taper ratio can be enhanced by increased hydraulic pressure and abrasive mass flow rate. However, increasing the standoff distance and traverse rate affects the performance in opposite manner. Khan et al. [63] observed several important facts while analyzing the cutting performance of three different types of abrasives such as garnet, SiO_2 and, Al_2O_3 particles on glass. According to their study, the taper of cut slot increases if the standoff distance and work feed rate are increased, however, it decreases with higher nozzle pressure. Garnet causes the taper of cut to be bigger followed by Al_2O_3 and SiC. In addition, SiC offers higher average width of cut followed by Al_2O_3 and garnet due to its higher hardness. Width of cut is also influenced positively by the increased stand off distance of the nozzle from the work due to divergence shape of the abrasive water jet.

2.3.2.3 Electrochemical Discharge Machining (ECDM)

The needs for nonconducting materials machining process have led to the development of another module of high-energy density machining process such as ECDM, which aims to remove material by using the combined effect of anodic dissolution and electrical sparks [64]. In ECDM (Fig. 2.11), difference of potential as low as 20 V is applied between tool electrode and counter electrode, while workpiece is dipped inside the electrolytic solution (NaOH, KOH) [65]. A chain of micro-explosions occurs on the workpiece surface layer due to electrical sparks following electrochemical reaction once the applied voltage reaches above the critical value; thus, removing material in small quantity with the occurrence of melting, or vaporization or both of these phases [66].

Along with this removal, chemical etching also adds on to the machining process. Figure 2.12a, b show images of holes entrance curved in Pyrex glass [68].

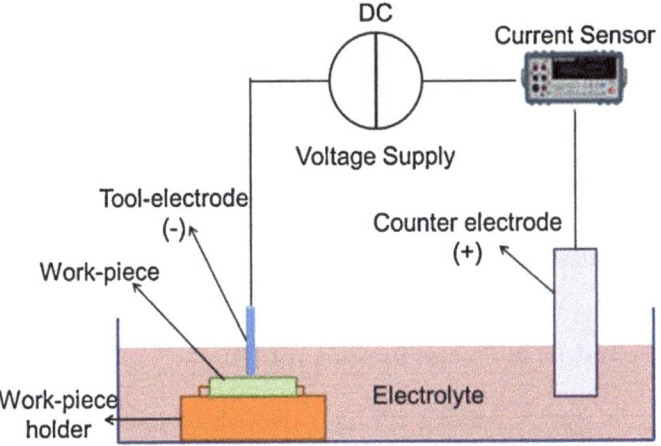

Fig. 2.11 Schematic diagram of ECDM setup [67] with kind permission from Elsevier

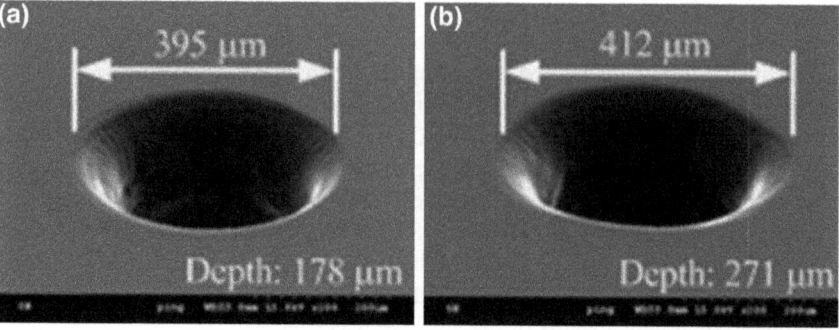

Fig. 2.12 Entrance of holes machined on Pyrex glass **a** 40 V and **b** 45 V (tool diameter 200 μm) [68] with kind permission from Elsevier

Fig. 2.13 a Contour machining using a textured tool (R_a = 1.5 μm, V = 28 V, DC) [70] with kind permission from Elsevier; **b** micro-pillar on soda lime; **c** micro-pyramid on Pyrex glass [71] with kind permission from Elsevier

Similarly, Fig. 2.13a shows a contour scribed in soda lime glass obtained with the help of textured tool [69]. An interesting feature shaped as pillar with height of 55 μm has also been manufactured by the same process (Fig. 2.13b). Furthermore, this process was successfully applied to machine pyramid on Pyrex glass in Fig. 2.13c [68].

While features less than 100 μm have been successfully machined on glass, machining high aspect ratio features becomes a challenge due to overcut and tool wear. Holes with the aspect ratio about 11 or more can be achieved by ECDM using micro-tools and low concentration electrolyte. Moreover, significant reduction in overcut (by 22%), hole taper (by 18%), and tool wear (by 39%) were observed during ECDM, thus increasing the aspect ratio of the microholes [67].

2.3.2.4 Electrolytic in Process Dressing (ELID)

Electrolytic in process dressing (ELID) is a viable substitute of conventional grinding which offers an efficient grinding of mechanical components and parts made of different materials including glass [72]. This technique offers extreme of finish at nano-level finishing on difficult to cut and brittle materials. In ELID, the surface of wheel is electrolyzed with a new oxide layer, following the grain worn out, which removes bonding material from the wheel surface, to protrude new grit. This dressing action becomes active due to the reduction of oxide layer and remains inactive when there is sufficient oxide layer on wheel surface. This way it can always maintain the fresh grain protrusion on the wheel surface. ELID process can overcome the limitation of conventional grinding by lessening the grinding force using in process dressing, however, due to electrolytic action, grinding wheel experiences rust deposition.

Ohmori [68] worked on ELID of silicon wafers, silicon nitride, and BK7 glass. His observations recommend the use of this process to obtain mirror finishing for difficult to cut materials [73]. Figure 2.14 depicts the mechanism of ELID process. Electrolytic action takes place in the separation gap created between wheel surface and cathode which is maintained from 100 to 500 μm based on the process conditions. The electrolyte, flushed into the electrode gap also serves the purpose of coolant. With the correct DC high voltage (preferably 60–120 V), electrolysis initiates and causes the formation of insulating layer of anodic oxide using removed metal bond from the wheel. Softness and brittleness of the layer result in the easy wear off of the anodic oxide during interaction which exposes sharp abrasives of the wheel. In addition, chips and blunt abrasives are removed along with oxide layer removal. Since the oxide layer acts as insulation, the reduction of layer thickness causes the resistance to drop off and dressing current to go high to form more oxide layer. Successive grinding again removes the oxide layer, and the cycle keeps going on as shown in Fig. 2.14. Therefore, oxide layer formation before the initiation of grinding serves as precondition for ELID grinding to take place efficiently. This operation is known as predressing only if it continues for 10–90 min [68, 73].

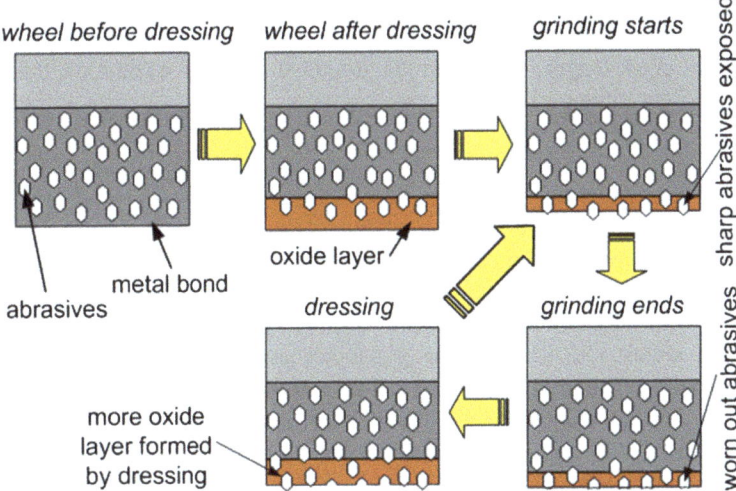

Fig. 2.14 Mechanism of electrolytic in process dressing grinding [68] with kind permission from Elsevier

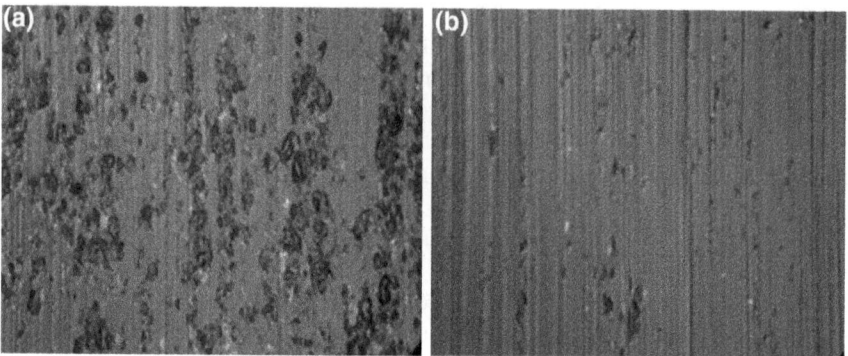

Fig. 2.15 SEM image of ground surface after **a** grinding; **b** ELID [69] with kind permission from Elsevier

The SEM images, as shown in Fig. 2.15, state that significant improvement in terms of surface roughness of BK7 glass is obtained by the use of ELID grinding over traditional grinding. The reason behind this worse result generated by traditional grinding process is the continuous reduction of the active grain per area. On the other hand, ELID offers an enhanced surface integrity and surface roughness by keeping the active grain same per unit area using continuous dressing. The ELID technique also improves the grindability by contributing on the reduction of the bonding strength of contact surface. Although, higher current duty ratio results in better roughness, black strip formed on the surface limits its increment up to certain value. The ELID grinding system developed has provided practical applications of the process such as macro lens fabricated on glass rod etc. [69].

While investigating surface and subsurface integrity of fused silica and fused quartz using ELID grinding, Zhao et al. [74] reported the existence of median and cone cracks that propagate underneath the ground surface with a depth ≤ 0.5 μm. The normal crack length underneath the ground silica surface can have the size of 0.3–0.8 μm depending on different crack system.

2.3.2.5 Laser Machining

MEMS components, integrated vision systems, as well as the family of photonic devices require high standards in terms of edge control, surface quality, and drilling precision, while maintaining the absence of micro-cracks in the glass material. An interesting solution to improve the glass processing could be the use of laser technology [75]. Different laser types can be considered for glass micromachining, depending on the kind of process and physical mechanism which leads the laser light to interact with the material. In this context, pulsed lasers, such as GaN diode lasers or excimer lasers, emitting in near-UV region, can be effective. With different approach, near-IR emitting lasers can also be considered. This list includes crystal based bulk lasers, such as Yb:YAG and Ti:sapphire lasers, as well as Yb-doped fiber lasers [76].

The interaction with transparent materials can be carried out with two different strategies, related to the electron–phonon coupling time τ_c, whose value is around 1 ps in most of the silica-based glasses. This value represents the necessary time to release excited electrons energy to the surrounded lattice, through the onset of a vibrational quantum, usually called phonon. By the use of pulsed lasers, with different pulse duration, the light-material interaction can be longer or shorter with respect to τ_c [77]. In case of longer pulses, in the order of magnitude of ns and more, electrons and lattice reach thermodynamic equilibrium. Therefore, the material removal, which is primarily caused by thermal effects, is obtained by melting and evaporation [78]. With the large spectrum of possibilities given by laser technologies in glass machining, several type of machining with different level of precision can be obtained, ranging from fast cutting and deep drilling to finer and accurate 3D-machining of micro-channels in integrated optical devices or shaping of micro-lenses for vision systems [79].

Nikumb et al. [5] investigated on nanosecond, femtosecond and laser induced plasma to obtain high quality surface on the micro-features on glass. Nanosecond and femtosecond laser can be exploited with the proper control of thermal process to produce crack free, clean surface. Fabrication of microscale shallow features with greater surface quality can be obtained with the use of laser induced plasma (Fig. 2.16a) [5]. Wlodarczyk et al. [80] investigated on the feasibility test of cutting and drilling of thin flex glass substrates of thicknesses 50 and 100 μm by using picosecond laser operating at wavelengths of 1030, 515, and 343 nm (Fig. 2.16b). Experimental results suggested that the highest effective cutting speed can be attained with the wavelength of 1030 nm; however, the quality of the cuts at this wavelength deteriorates. The best cutting outcome is achieved with the wavelength of 343 nm using the lowest cutting speed. The maximum drilling speeds were

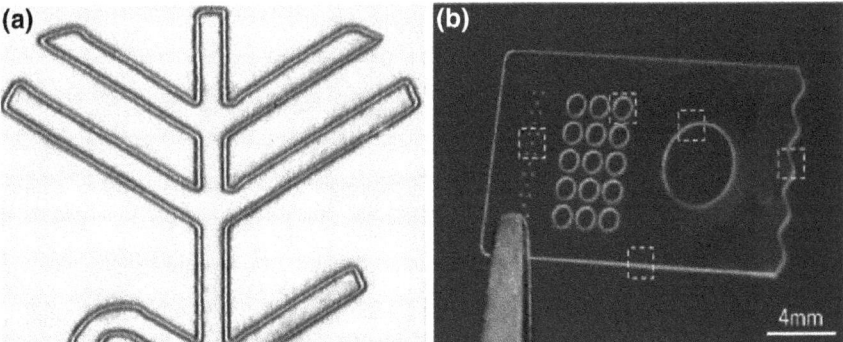

Fig. 2.16 **a** Laser induced plasma machining was used to fabricate micro-channels on glass surface with a width of 8 μm. Laser pulse repetition rate was 1 kHz and scanning speed was 400 μm/s [5]. **b** An example of the laser-cut structure in a 100 μm thick AF32® Eco thin Glass substrate [80] with kind permission from Elsevier

found to be approximately 2 microholes per second for 343 nm and 8 microholes per second for 515 nm, nevertheless presence of debris and heat affected zone were also observed [80].

In this section, conventional and nonconventional machining processess involving glass machining are discussed primarily. Associated process mechanisms and relevant research contributions are illustrated prudently. On top of that tool wear aspect and surface/subsurface integrity aspects are considered as well to provide a deep understanding.

2.4 Summary

This chapter attempts to provide an overview on various conventional and nonconventional techniques used to machine glass materials. Being fragile, manufacturing processes dedicating for glass must consider the probability of cracks generation and breakage. Although most of the conventional mechanical machining seems to be working for thick glass parts, thin glass machining can be conducted using noncontact processes such as nonconventional processes. Substantial research efforts have been dedicated in the past years, in order to reveal the underpinning knowledge related to glass cutting mechanism and development of various machining technologies. An important criterion to take consideration for conventional glass machining is brittle to ductile transition phenomenon. Although, nanofinish in glass surface can be achieved using this phenomenon with proper control of process parameters, tool wear suffers greatly due to abrasive nature of glass and downscaling of chip thickness. Thus research direction of conventional glass machining should direct toward tool wear in order to achieve more economical production process. Another important fact to consider in ductile mode machining process is the influence of temperature. The future work may also pay attention on development of numerical methodology in order to assess the total temperature in the cutting zone and its effect especially on machined surface of glass.

References

1. Agarwal S, Rao PV (2008) Experimental investigation of surface/subsurface damage formation and material removal mechanisms in SiC grinding. Int J Mach Tools Manuf 48(6):698–710
2. Foy K, Wei Z, Matsumura T, Huang Y (2009) Effect of tilt angle on cutting regime transition in glass micromilling. Int J Mach Tools Manuf 49(3–4):315–324
3. Tölke R, Bieberle-Hütter A, Evans A, Rupp JLM, Gauckler LJ (2012) Processing of Foturan® glass ceramic substrates for micro-solid oxide fuel cells. J Eur Ceram Soc 32(12):3229–3238
4. Yan BH, Wang AC, Huang CY, Huang FY (2002) Study of precision micro-holes in borosilicate glass using micro EDM combined with micro ultrasonic vibration machining. Int J Mach Tools Manuf 42(10):1105–1112

5. Nikumb S, Chen Q, Li C, Reshef H, Zheng HY, Qiu H, Low D (2005) Precision glass machining, drilling and profile cutting by short pulse lasers. Thin Solid Films 477(1–2):216–221
6. Arif M, Rahman M, San WY (2011) Ultraprecision ductile mode machining of glass by micromilling process. J Manuf Process 13(1):50–59
7. Daridon A, Fascio V, Lichtenberg J, Wütrich R, Langen H, Verpoorte E, de Rooij N (2001) Multi-layer microfluidic glass chips for microanalytical applications. Fresenius' J Anal Chem 371(2):261–269
8. Petersen D, Mogensen KB, Klank H (2004) Glass micromachining. Microsystem engineering of Lab-on-a-Chip devices. Wiley-VCH Verlag GmbH & Co. KGaA, Weinheim
9. Namba Y, Abe M, Kobayashi A (1993) Ultraprecision grinding of optical glasses to produce super-smooth surfaces. CIRP Ann Manuf Technol 42(1):417–420
10. Weber MJ (2002) Handbook of optical materials. CRC press, Boca Raton
11. Blackley WS, Scattergood RO (1991) Ductile-regime machining model for diamond turning of brittle materials. Prec Eng 13(2):95–103
12. Li X, Abe T, Esashi M (2001) Deep reactive ion etching of Pyrex glass using SF6 plasma. Sens Actuat A Phys 87(3):139–145
13. Plaza JA, Lopez MJ, Moreno A, Duch M, Cane C (2003) Definition of high aspect ratio glass columns. Sens Actuat A Phys 105(3):305–310
14. Slikkerveer PJ, Bouten PCP, de Haas FCM (2000) High quality mechanical etching of brittle materials by powder blasting. Sens Actuat A Phys 85(1–3):296–303
15. Dolev D (1983) A note on plasticity of glass. J Mater Sci Lett 2(11):703–704
16. Finnie I, Dolev D, Khatibloo M (1981) On the Physical Basis of Auerbach's Law. J Eng Mater Technol 103(2):183–184
17. Lawn BR, Jensen T, Arora A (1976) Brittleness as an indentation size effect. J Mater Sci 11(3):573–575
18. Yan J, Yoshino M, Kuriagawa T, Shirakashi T, Syoji K, Komanduri R (2001) On the ductile machining of silicon for micro electro-mechanical systems (MEMS), opto-electronic and optical applications. Mater Sci Eng A 297(1–2):230–234
19. Johnson KL (1970) The correlation of indentation experiments. J Mech Phys Solids 18(2):115–126
20. Giovanola JH, Finnie I (1980) On the machining of glass. J Mater Sci 15(10):2508–2514
21. Bifano TG (1988) PhD Thesis. North Caroline State University, Releigh
22. Nakasuji T, Kodera S, Hara S, Matsunaga H, Ikawa N, Shimada S (1990) Diamond turning of brittle materials for optical components. CIRP Ann Manuf Technol 39(1):89–92
23. Shimada S, Ikawa N, Inamura T, Takezawa N, Ohmori H, Sata T (1995) Brittle-ductile transition phenomena in microindentation and micromachining. CIRP Ann Manuf Technol 44(1):523–526
24. Cai MB, Li XP, Rahman M (2007) Study of the mechanism of nanoscale ductile mode cutting of silicon using molecular dynamics simulation. Int J Mach Tools Manuf 47(1):75–80
25. Sreejith PS, Ngoi BK (2001) Material removal mechanisms in precision machining of new materials. Int J Mach Tools Manuf 41:1831–1843
26. Fielder KH (1988) Precision grinding of brittle materials. In: Ultra precision in manufacturing engineering. Springer, Berlin, pp 72–77
27. Bifano TG, Dow TA, Scattergood RO (1992) Ductile-regime grinding. A new technology for machining brittle materials. J Eng for Indus 113:184–189
28. Bifano TG, Yi Y (1992) Acoustic emission as an indicator of material-removal regime in glass micro-machining. Prec Eng 14(4):219–228
29. Nakasuji T, Kodera S, Hara S, Matsunaga H, Ikawa N, Shimada S (1990) Diamond turning of brittle materials for optical components. CIRP Ann 39(1):89–92
30. Konig W, Cronjäger L, Dortmund U, Spur G, Tonshoff HK, Vigneau M, Zdeblick WJ (1990) Machining of new materials. CIRP Ann 39(2):673–681
31. Yuan YJ, Geng L, Dong S (1993) Ultra-precision machining of SiCw/Al composites. CIRP Ann 42(1):107–109

32. Yoshikawa H (1967) Brittle–ductile behaviour of crystal surface in finishing. J JSPE 35:662–667
33. Yuan ZJ, Zhou M, Dong S (1996) Effect of diamond tool sharpness on minimum cutting thickness and cutting surface integrity in ultraprecision machining. J Mater Process Technol 62(4):327–330 (2nd International Conference on Production Engineering)
34. Ueda K, Amano A, Ogawa K, Takamatsu H, Sakuta S, Murai S, Kobayashi A (1991) Machining high-precision mirrors using newly developed CNC machine. CIRP Ann Manuf Technol 40(1):555–558
35. Fang FZ, Chen LJ (2000) Ultra-precision cutting for ZKN7 glass. CIRP Ann Manuf Technol 49(1):17–20
36. Weber H, Herberger J, Pilz R (1984) Turning of machinable glass ceramics with an ultrasonically vibrated tool. CIRP Ann Manuf Technol 33(1):85–87
37. Takeyama H, Iijima N (1988) Machinability of glassfiber reinforced plastics and application of ultrasonic machining. CIRP Ann Manufac Technol 37(1):93–96
38. Moriwaki T, Shamoto E, Inoue K (1992) Ultraprecision ductile cutting of glass by applying ultrasonic vibration. CIRP Ann Manuf Technol 41(1):141–144
39. Zhou M, Wang XJ, Ngoi BKA, Gan JGK (2002) Brittle–ductile transition in the diamond cutting of glasses with the aid of ultrasonic vibration. J Mater Process Technol 121(2–3):243–251
40. Owen JD, Davies MA, Schmidt D, Urruti EH (2015) On the ultra-precision diamond machining of chalcogenide glass. CIRP Ann Manuf Technol 64(1):113–116
41. Zhou M, Ngoi BKA, Yusoff MN, Wang XJ (2006) Tool wear and surface finish in diamond cutting of optical glass. J Mater Process Technol 174(1–3):29–33
42. Bakkal M, Shih AJ, Scattergood RO (2004) Chip formation, cutting forces, and tool wear in turning of Zr-based bulk metallic glass. Int J Mach Tools Manuf 44(9):915–925
43. Malkin S, Hwang TW (1996) Grinding mechanisms for ceramics. CIRP Ann Manuf Technol 45(2):569–580
44. Chen M, Zhao Q, Dong S, Li D (2005) The critical conditions of brittle–ductile transition and the factors influencing the surface quality of brittle materials in ultra-precision grinding. J Mater Process Technol 168(1):75–82
45. Gu W, Yao Z, Li H (2011) Investigation of grinding modes in horizontal surface grinding of optical glass BK7. J Mater Process Technol 211(10):1629–1636
46. Perveen A, Wong Y, Rahman M (2011) Characterisation and online monitoring of wear behaviour of on-machine fabricated PCD micro-tool while vertical micro-grinding of BK7 glass. Int Abras Technol 4(4):304–324
47. Suratwala T, Wong L, Miller P, Feit MD, Menapace J, Steele R, Davis P, Walmer D (2006) Sub-surface mechanical damage distributions during grinding of fused silica. J Non-Cryst Solids 352(52–54):5601–5617
48. Takeuchi Y, Sawada K, Sata T (1996) Ultraprecision 3D micromachining of glass. CIRP Ann Manuf Technol 45(1):401–404
49. Matsumura T, Hiramatsu T, Shirakashi T, Muramatsu T (2005) A study on cutting force in the milling process of Glass. J Manuf Process 7(2):102–108
50. Perveen A, San WY, Rahman M (2012) Fabrication of different geometry cutting tools and their effect on the vertical micro-grinding of BK7 glass. Int J Adv Manuf Tech 61(1–4):101–115
51. Matsumura T, Ono T (2008) Cutting process of glass with inclined ball end mill. J Mater Process Technol 200(1–3):356–363
52. Ono T, Matsumura T (2008) Influence of tool inclination on brittle fracture in glass cutting with ball end mills. J Mater Process Technol 202(1–3):61–69
53. Arif M, Rahman M, San WY, Doshi N (2011) An experimental approach to study the capability of end-milling for microcutting of glass. Int J Adv Manuf Technol 53(9):1063–1073
54. Arif M, Rahman M, San WY (2012) An experimental study on the machining characteristics in ductile-mode milling of BK-7 glass. Int J Adv Manuf Technol 60(5):487–495
55. Thoe TB, Aspinwall DK, Wise MLH (1998) Review on ultrasonic machining. Int J Mach Tools Manuf 38(4):239–255

56. Lin S, Kuan C, She C, Wang W (2015) Application of ultrasonic assisted machining technique for glass-ceramic milling. World Academy of Science, Engineering and Technology. Int J Mech Aeros Indust Mechatr Manufa Eng 9(5):786–791
57. Egashira K, Masuzawa T (1999) Microultrasonic machining by the application of workpiece vibration. CIRP Ann Manuf Technol 48(1):131–134
58. Choi JP, Jeon BH, Kim BH (2007) Chemical-assisted ultrasonic machining of glass. J Mater Process Technol 191(1–3):153–156
59. D V Srikanth DMSR (2014) Abrasive jet machining-research review. Int J Adv Eng Technol V(II):18–24
60. Ramulu M, Kunaporn S, Arola D, Hashish M, Hopkins J (2000) Waterjet machining and peening of metals. J Press Vessel Technol 122(1):90–95
61. Aich U, Banerjee S, Bandyopadhyay A, Das PK (2014) Abrasive water jet cutting of borosilicate glass. Procedia Mater Sci 6:775–785
62. Azmir MA, Ahsan AK (2009) A study of abrasive water jet machining process on glass/epoxy composite laminate. J Mater Process Technol 209(20):6168–6173
63. Khan AA, Haque MM (2007) Performance of different abrasive materials during abrasive water jet machining of glass. J Mater Process Technol 191(1–3):404–407
64. Sarkar BR, Doloi B, Bhattacharyya B (2006) Parametric analysis on electrochemical discharge machining of silicon nitride ceramics. Int J Adv Manuf Technol 28(9):873–881
65. Cheng C-P, Wu K-L, Mai C-C, Yang C-K, Hsu Y-S, Yan B-H (2010) Study of gas film quality in electrochemical discharge machining. Int J Mach Tools Manuf 50(8):689–697
66. Coteață M, Slătineanu L, Dodun O, Ciofu C (2008) Electrochemical discharge machining of small diameter holes. Int J Mater Form 1(1):1327–1330
67. Jui SK, Kamaraj AB, Sundaram MM (2013) High aspect ratio micromachining of glass by electrochemical discharge machining (ECDM). J Manuf Process 15(4):460–466
68. Ohmori H, Nakagawa T (1990) Mirror surface grinding of silicon wafers with electrolytic in-process dressing. CIRP Ann Manuf Tech 39(1):329–332
69. Rahman M, Lim HS, Neo KS, Senthil Kumar A, Wong YS, Li XP (2007) Tool-based nanofinishing and micromachining. J Mater Process Technol 185(1–3):2–16
70. Han M-S, Min B-K, Lee SJ (2011) Micro-electrochemical discharge cutting of glass using a surface-textured tool. CIRP J Manuf Sci Technol 4(4):362–369
71. Cao XD, Kim BH, Chu CN (2009) Micro-structuring of glass with features less than 100 μm by electrochemical discharge machining. Precis Eng 33(4):459–465
72. Murata R, Okano K, Tsutsumi C (1985) Grinding of structural ceramics. In: Milton C Shaw, Grinding Symposium PED, vol 16, pp 261–272
73. Ohmori H, Nakagawa T (1995) Analysis of mirror surface generation of hard and brittle materials by ELID (Electronic In-Process Dressing) grinding with superfine grain metallic bond wheels. CIRP Ann Manuf Tech 44(1):287–290
74. Zhao Q, Liang Y, Stephenson D, Corbett J (2007) Surface and subsurface integrity in diamond grinding of optical glasses on Tetraform 'C'. Int J Mach Tools Manuf 47(14):2091–2097
75. Bäuerle DW (2011) Laser processing and chemistry. Springer, Berlin
76. Rihakova L, Chmelickova H (2015) Laser micromachining of glass, silicon, and ceramics. Adv Mater Sci Eng 2015:2016
77. Stuart BC, Feit MD, Herman S, Rubenchik AM, Shore BW, Perry MD (1996) Nanosecond-to-femtosecond laser-induced breakdown in dielectrics. Phys Rev B 53(4):1749–1761
78. Corbari C, Champion A, Gecevičius M, Beresna M, Bellouard Y, Kazansky PG (2013) Femtosecond versus picosecond laser machining of nano-gratings and micro-channels in silica glass. Opt Express 21(4):3946–3958
79. Gattass RR, Mazur E (2008) Femtosecond laser micromachining in transparent materials. Nat Photonics 2(4):219–225
80. Wlodarczyk KL, Brunton A, Rumsby P, Hand DP (2016) Picosecond laser cutting and drilling of thin flex glass. Opt Laser Eng 78:64–74

Chapter 3
Thermal-Assisted Machining of Titanium Alloys

O.A. Shams, A. Pramanik and T.T. Chandratilleke

Abstract Titanium alloys are used in a variety of engineering applications, especially in automotive, aerospace and nuclear fields due to their high strength and excellent corrosion resistance. Nevertheless, titanium alloys have extreme mechanical properties making them very difficult to machine with low thermal conductivity and high chemical reactivity at high temperature. Hence, titanium alloys are required to machine at low cutting speed and feed rate but that increases the cost of production of the components made by these alloys at large. Thermal-assisted machining (TAM) is an effective approach for conventional machining whereby titanium workpiece is locally softened before/during machining with external heating. Localized reduction in workpiece hardness facilitates higher material removal rate (MRR) and extended cutting tool life whilst resulting in better surface finish. This chapter compares and analyzes the merits of different heating techniques for machining of titanium alloys. The techniques under consideration are heating by laser beam, plasma torch heating and heating with the use of induction coil. The laser beam and plasma torch tend to produce more intense localized heating compare to that by induction coil. Moreover, the laser technique offers very controllable process heating compared to other two techniques. Laser-assisted machining (LAM) also largely reduces cutting forces leading to better surface finish. Thus, laser-assisted technique is recognized to be more cost-effective and productive for improving machinability of titanium alloys than rest of the heating techniques.

Keywords Titanium alloys · Thermal-assisted machining · Laser · Plasma · Induction · Surface integrity

---

O.A. Shams (✉) · A. Pramanik · T.T. Chandratilleke
Department of Mechanical Engineering, Curtin University, Bentley, Australia
e-mail: O.shams@curtin.edu.au

A. Pramanik
e-mail: alokesh.pramanik@curtin.edu.au

© Springer International Publishing AG 2017
K. Gupta (ed.), *Advanced Manufacturing Technologies*, Materials Forming, Machining and Tribology, DOI 10.1007/978-3-319-56099-1_3

List of Symbols

a_e	Radial depth of cut (mm)
a_p	Axial depth of cut (mm)
D	Tool or workpiece diameter (mm)
f_z	Feed per tooth (mm)
f_r	Feed per revolution (mm)
F	Machine linear feed, similar to the translational velocity of the plasma torch (N)
F_v	Thrust force (N)
F_t	Tangential force (N)
I	Plasma intensity (A)
L	Lens-workpiece distance on chamfer surface (mm)
L_l	Tool-laser beam distance on the surface of workpiece (mm)
P_{CO_2}	Power of CO_2 laser (kW)
RT	Room temperature (°C)
S	Rotary speed of the spindle (r.p.m)
T_{mr}	Material removal temperature
T_o	Initial workpiece bulk temperature (°C)
T_s	Surface temperature (°C)
V_c	Cutting speed (m/min)
z	Number of teeth of the milling tool
α	Angle between workpiece axis and beam axis (degree)

3.1 Introduction

In recent decades, an extensive use and increasing demand of components for titanium alloy manufacturing are witnessed due to steadily developing growth in applications requiring metal components having high strength, light-weight, good formability and excellent corrosion resistance [1–3]. Due to their extraordinary physical properties and mechanical stability at elevated temperatures, titanium alloys are primarily considered for special applications such as, automotive, aerospace, petroleum refinery and nuclear reactors, etc. [4–7]. However, titanium alloys are well recognized for being difficult-to-machine (DTM), because such alloys inherently possess low thermal conductivity and elastic modulus, and high chemical reactivity at elevated temperature [8–10]. These properties significantly lead to increase the manufacturing cost of titanium alloys components compared to other conventional alloys [5, 11, 12]. The low thermal conductivity of titanium alloys impedes dissipation of heat generated during machining, causing steep rise in tooltip temperature, hence adversely affecting tool life, resulting in poor surface finish and dimensional inaccuracies [1, 13–15]. High chemical reactivity of titanium alloys tends to rapidly dissolves and degrades the cutting tool during machining to result chipping and early tool failure. These attributes significantly impair the titanium alloy machinability with low production rate and high manufacturing cost [10, 15, 16].

Improvements in the machinability of hard and DTM materials have keenly been a subject of interest for the past decades. Several assisting techniques including thermal-assisted machining (TAM) have been proposed and attempted on "difficult-to-machine" materials to ease shear characteristics and thereby to reduce tool wear and enhance machining speed. TAM is not a new concept and originated out of applications with low-temperature heat sources [17–20]. In 1898 Tilghman [21] filed a US patent to use electrical resistance for preheating a workpiece during machining. Since then, many researchers have followed Tilghman approach and investigated many possible alternatives heat sources such as gas torches, laser beams and plasma arcs, etc.

The concept of TAM is based on the principle that the yield strength and hardness of materials decrease with the increase of temperature whilst other properties do remain relatively unaffected. In this, an external heat source is deployed to preheat the workpiece surface during machining without heating the cutting tool [22–25] as illustrated by Fig. 3.1 [26]. For any given work material, TAM requires appropriate control of workpiece temperature to keep cutting forces within an optimum regime [27], hence to achieve the best surface integrity and extended tool life [28, 29].

The basic requirements identified for TAM of DTM materials are given below [19, 24]:

- The external heat source should be localized at the shear zone;
- The heating area should be restricted to a small zone;
- Heating method is to be incorporated a temperature-regulating device;
- Overheating of machined surface is to be avoided to prevent any potential metallurgical changes to the uncut workpiece;

Fig. 3.1 Thermal- or heat-assisted machining using laser as external heating source [26] with kind permission from Elsevier

- External heat source must be of adequate intensity to produce a large heat output for imparting rapid temperature rise in advance of the cutting point.

Some recent research work has also been focused on the development of TAM techniques for improved titanium alloys machining. For example, Bermingham et al. [30] measured both tool life and wear mechanisms after preheating the workpiece in a conventional furnace prior to machining of titanium alloy (Ti-6Al-4V) at three workpiece temperatures with two carbide cutting tools. A 30% reduction in the cutting forces was achieved in addition to 7% improvement in tool life, compare to that offered by cooling technologies under identical experimental conditions. Current state-of-the art identifies high-end thermal-assisted techniques, such as induction-assisted machining [31–34], plasma-assisted machining [35–40] and laser-assisted machining (LAM) [41–45] are more promising contenders that significantly improve the overall machinability of titanium alloys with recognized benefits of high MRR and economics of machining.

This book chapter carries out an extensive review of recent advances in methodology and appraises preheating to the TAM techniques that are being used for titanium alloys. Specific emphasis is applied on the reduction of cutting forces, extension of cutting tool life and enhancement in surface finish quality. This study is intended to deliver a clearer view of merits and capture the capabilities of each preheating technique in machining titanium alloys.

3.2 Thermal-Assisted Machining Techniques

3.2.1 Laser-Assisted Machining

Recently, LAM has been used as an auxiliary process for machining various DTM materials and alloys as well as alloys possess high melting point, and composites. LAM deploys a high-power laser focussed at a certain spot on the workpiece to preheat upstream of tool path surface during traditional machining [46–49]. The laser heating point is confined to less than 3 mm diameter [50, 51], as illustrated in Fig. 3.2. This localized heating lowers the yield strength of DTM at elevated temperature, thus reducing cutting forces and tool wear to improve surface finish quality and cut down the machining time [48, 52, 53]. However, the intense laser heating causes oxidation, melting and/or vaporization of the workpiece surface while the transient thermal response of the workpiece would potentially induce uneven thermal expansion of the material [21, 49, 54]. The performance of laser preheating technique depends on the laser attributes such as laser power, beam diameter, scan speed and approach angle (tool-beam distance), as well as the machining parameters such as cutting speed, feed rate and depth of cut. Therefore, LAM is rather complex to control for as it is governed by many parameters and their mutual interactions [13, 50, 55]. Nonetheless, this technique has been widely extended to various machining operations such as milling, grinding and turning, etc.

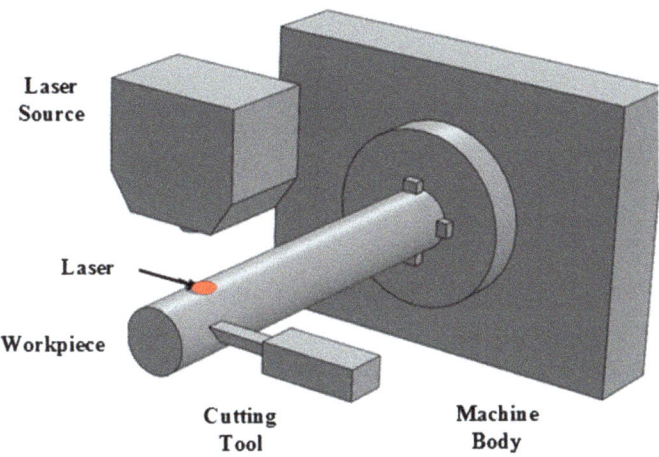

Fig. 3.2 Schematic representation of laser assisted turning

[56], recognizing its many advantages: (i) adaptability to different types of machining processes, (ii) localization of high temperature and (iii) process stability [52], etc. This technique offers flexibility in machining of DTM material as it allows adjusting of input parameters to reduce machining time, thus yielding a significant reduction in production cost, as much as 60–80% [19, 27, 57, 58].

Laser-assisted turning (LAT) centre consists of three main components, where the first component contains a high-power continuous-wave laser to supply strong, localized and controlled heating to the workpiece; the second part is a system of optics for guiding the laser beam from the lasing chamber to the workpiece; and the third component contains machine tool, i.e. lathe itself [45]. The position of a laser beam is arranged in such a way that the temperature distribution remains uniform into the cutting zone [59]. It uses two control systems namely, revolver-kind optics and auto focusing to control the laser beam size [56]. The revolver-kind optic system is equipped with various lenses that process selectively, while in auto focusing, the laser beam diameter and the focal length are set by automatic focusing system. The laser beam can be transferred from its module to a specific machining zone by using optical fibres or mirrors [56].

In LAT, the workpiece rotates at high speed and while it is being subjected to cyclic heating at a specific rotating point via a laser beam focused on an area ahead of the cutting edge. The workpiece temperature gets progressively elevated at the heated point and shows slight cooling when moving away along the cutting path [60–62]. Generally, a continuous-wave beam with a Gaussian distribution is useful to minimize the thermal shock. To the contrary, with a pulsed-wave mode laser, heating and cooling tend to be more rapid at the workpiece, leading to a workpiece surface hardening process that adversely affects the machining performance [62]. The convection and conduction heat transfers processes impact considerably on the temperature distribution in the cutting zone [60, 62, 63].

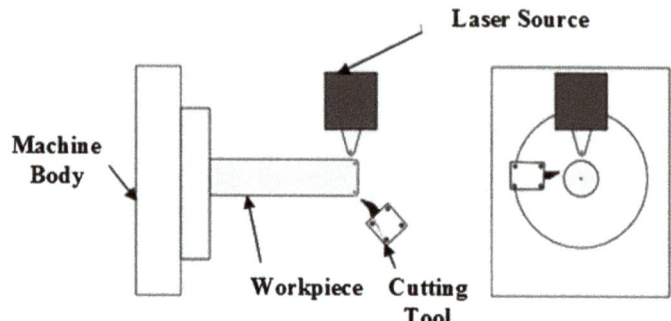

Fig. 3.3 Illustration of laser beam position (perpendicular to the feed direction) in turning

It has been found that arranging the laser beam perpendicular to the feed direction is more effective for not heating the machined surface and makes machining easier [46], see Fig. 3.3. In addition, in this method, the laser output is regulated by a pyrometer, allowing a constant temperature to be achieved in the component [64]. However, the temperature at a depth of machining may not be enough for a deeper cut and the measurements do not warrant recording such temperature for precise determination of effective cutting depth [42, 64–66]. Whilst LAM can be practically used for machining titanium alloys, this technique is observed to be difficult for cutting straight holes or pear-shaped holes [48, 52].

Another possible arrangement of the laser beam is to set it vertical orientation to the workpiece chamfer surface during turning operation [45, 59], as illustrate in Fig. 3.4 [67]. This reduces the components of cutting forces significantly [22, 67, 68], hence minimizing the chances of any mechanical and/or thermal issues. For uniform reduction of cutting forces, it is essential that the laser spot size should completely cover the chamfer surface, so that the softening of plastic deformation

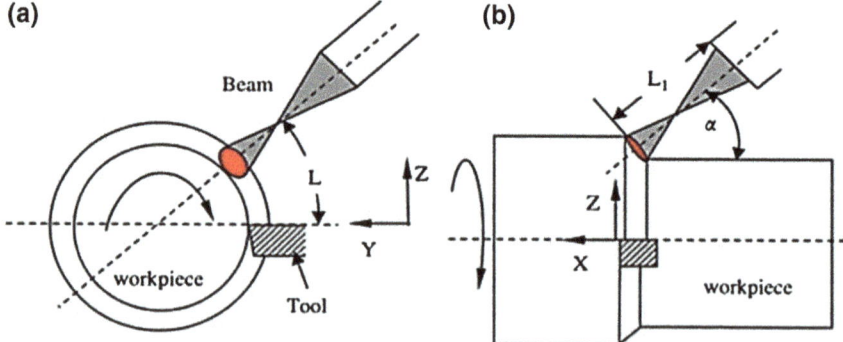

Fig. 3.4 The relative position of the laser beam during turning operation: **a** end-view and **b** side-view [67] with kind permission from John Wiley and Sons

zone by the laser becomes easy. This will reduce the shear-generated heat to level below conventional machining [42].

In laser-assisted milling operation, multiple laser beams can be used in various orientations over the cutting area, because a large zone is covered by cutting tool [43, 69]. The temperature at the laser spot depends on (i) the heating duration, (ii) the laser power density distribution and (iii) the number of heating/cooling cycles [42]. Due to the ability to control spot size and laser power density, the thermal distortion and heat-affected area are generally small with LAT methods. Therefore, intense thermal gradient is confined to a very thin surface layer at the workpiece, enabling machining without interfering on the integrity of workpiece subsurface. According to many studies [54, 70], the temperature gradient at the cutting zone is important to understand the mechanism of chip formation, investigate the thermo-mechanical characteristics in LAM, and to determine the reduction values of cutting forces [71].

The effectiveness and benefits of LAM for various DTM materials are reported in literature, identifying a significant reduction in cutting forces, tool wear, chatter and producing better machined surface and increased MRR, such as for steels [42, 72–75], inconel 718 [72, 76–79], nickel alloys [80, 81], waspaloy alloy [82] ceramics [51, 54, 61, 83–85], magnesium alloys [86, 87] and metal matrix composites [60, 71, 88–90]. There have been many investigations in recent years examining the feasibility of LAM for titanium alloys.

Hedberg et al. [91] experimentally studied the improvement in the machinability of titanium alloy (Ti-6Al-4V) by using LAM technique. It was determined from the study that the flank wear and compressive residual stresses were decreased by 10% and cutting speed was increased around 35% in LAM compared with conventional process to machine Ti-6Al-4V. Furthermore, increasing cutting speed showed a 33% reduction in cost in spite of the additional cost of laser equipment. Germain et al. [92] investigated the effect of LAM on residual stresses of machined bearing steel (100Cr6) and titanium alloy (Ti-6Al-4V). This study indicated that the higher laser power reduces the cutting forces and increases the residual stresses towards tensile or positive stresses at the surface layer. Braham-Bouchnak et al. [57] studied the improvement in productivity of titanium alloy (Ti555-3) through LAM to improve chip formation by a thermal softening phenomenon. It was identified that a reduction of surface temperature and the change of chip formation mechanisms are achieved by increasing the frequency of sawtooth. Furthermore, the surface integrity of workpiece was modified in terms of strain hardening and residual stresses, as shown in Fig. 3.5 [57]. Dandekar et al. [13] investigated the effect of LAM with hybrid machining on titanium alloy (Ti-6Al-4V) machinability. The results show that it is possible to increase the cutting speed during machining (Ti-6Al-4V). Moreover, the cutting forces were reduced with reasonably higher MRR.

Rashid et al. [93] studied the effects of LAM on machining of the beta titanium alloy (Ti-6Cr-5Mo-5V-4Al), by comparing cutting temperatures and cutting forces during LAT and conventional machining at different values of feed rates and cutting speeds. This comparison showed that the cutting forces reduced by 15% in LAM.

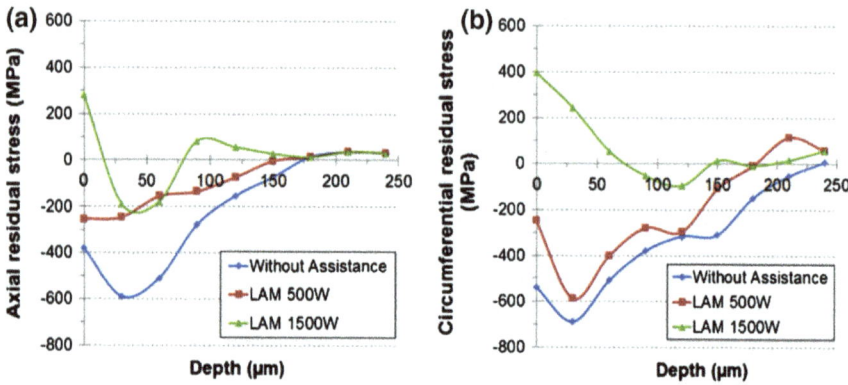

Fig. 3.5 Residual stresses as a function of depth during machining with and without laser heating, **a** axial residual stresses and **b** circumferential residual stresses [57]

The results show that the optimum cutting speed range of 25–100 m/min is suitable to avoid chip/tool welding, and leads to unacceptable characteristics of surface integrity. Furthermore, the optimum range of cutting temperatures was found to be 1050–1250 °C causing a moderate reduction in cutting forces. Sun et al. [94] reported that local heating by a laser beam in front of the cutting tool during laser-assisted milling of titanium alloy (Ti-6Al-4V) caused a dramatic reduction in cutting forces, especially feed force at 200–450 °C. This higher temperature at tool/chip interface zone may accelerate the dissolution/diffusion and adhesion wear, thereby leads to softening the cutting tool [94, 95]. Therefore, the maximum tool life achieved between 230 and 350 °C for LAM with compressed air delivered through the spindle during laser-assisted milling of Ti-6Al-4V [94]. It was noted that both the cutting tool life and tool failure modes depend on the temperature at cutting zone.

LAM of DTM materials produces a dense layer with fine grains which improves the hardness of the surface during the phase change and fast solidification in the surface area [96, 97]. The most widely used laser sources that have been investigated as preheating media in LAM experiments are a neodymium-doped yttrium aluminium garnet laser (Nd:YAG), carbon dioxide laser (CO_2) and high-power diode laser (HPDL) [56, 98, 99]. All types of lasers generate a concentrated beam of light. Fundamentally, Nd:YAG laser is a solid-state laser that emits a wavelength of 1.064 μ [100, 101]. However, CO_2 laser is a gas laser that emits a wavelength of 10.64 μ, which is ideal for optimum absorption. The CO_2 laser produces spot size 10 times greater than that produced in Nd:YAG laser, when both types are used in the same machine set-up. HPDL laser has broader wavelength bands than others and emits a wavelength from 0.808 to 0.980 μ [102, 103]. Nd:YAG laser is ideally used for metals and coated metals, while CO_2 laser is best suited for organic materials for example wood, paper, plastics, glass, textiles and rubber. The CO_2 laser has limitations, where it requires a beam transfer technique using a mirror compared to other types of the laser using fibre optic cables [62, 99, 102]. In HPDL

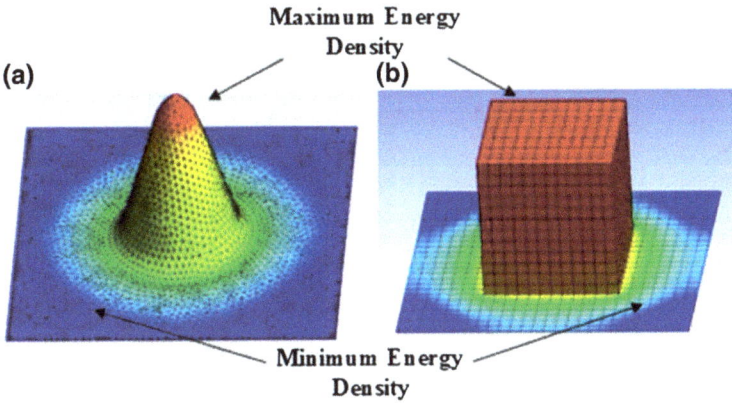

Fig. 3.6 The distribution shape of laser energy density, **a** Gaussian distribution and **b** Top-Hat distribution [56]

laser, the laser energy density has uniform geometry in Top-Hat shaped, which represents much better results than the Nd:YAG and CO_2 lasers that have a Gaussian distribution [56, 102, 104], see Fig. 3.6 [56]. The Nd:YAG laser is most useful on most of the "hard-to-machine" materials like hardened steel and titanium alloys due to the shorter wavelength [4, 50, 75]. On the other hand, HPDL laser shows an absorption rate of about 40% at the surface of metals, thus making this type being largely suited for most metallic materials and better beam stability in LAT [103–105].

Rashid et al. [106] have studied the effects of laser power on the beta titanium alloy (Ti-6Cr-5Mo-5V-4Al) machining and noted that the optimum range of laser power is 1.2–1.6 kW to reduce cutting forces significantly at cutting speed ranging from 25 to 125 m/min. The influence of laser power on residual stresses during LAT of AISI 4340 steel via finite element analysis have examined by Nasr and Balbaa [95]. The results showed that surface tensile residual stresses decrease with the increase of laser power. At lower laser power, limited preheating in front of the cutting tool causes easier compressive deformation. Higher laser power produces mostly thermal deformation, which leads to minimizing the material resistance of workpiece at cutting zone and causes tensile residual stress at the machined surface. The effect of laser power and cutting speed on the machining of Inconel 718 superalloy have been investigated [107]. It was noted that the surface temperature increases with higher laser power and falls when the approach angle increases. Kannan et al. [108] investigated the LAM feasibility of machining alumina ceramics over a range of laser scan speeds on cutting tool wear. The results showed that the surface temperature fluctuates around 1250 °C when the laser scan speed is around 35–55 mm/min at a laser power of 0.35 kW. The cutting forces and flank wear decreased significantly at this temperature. However, above this range of laser

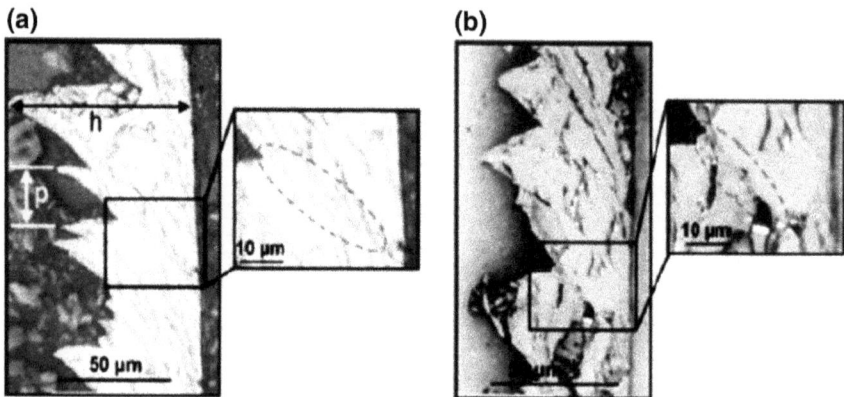

Fig. 3.7 Chip morphologies of titanium metal matrix composite: **a** with LAM, $V_c = 100$ m/min, $T_s = 500$ °C, and, **b** conventional machining, $V_c = 100$ m/min, RT [71] with kind permission from Elsevier

scan speeds, the cutting forces and flank wear were observed to increase. This is because of the reduction of laser-material interaction time at the higher speed of laser scan. Subsequently, the temperature at the depth of cut would be much less than the softening temperature. Sun et al. [109] analyzed the influence of the laser beam diameter on chip formation during machining titanium alloy (Ti6Al4V) at various machining speeds. This investigation showed segmented chip formation at low and high speeds. Continuous chips are formed at intermediate speeds because of the thermal softening at the shear zone, which causes ductile deformation.

The surface layer of workpiece needs to be heated to a temperature that will cause ductile deformation of material during the machining operation. Due to high temperature during LAM, the continuous chip is formed, indicating plastic deformation rather than brittle fracture [44, 64]. Chips thus formed have a highly irregular segmentation pattern, that is distinguished by larger fluctuations in segment height (h) and pitch (p) compared to the one that formed in conventional machining, as illustrated in Fig. 3.7 [71]. As a principle requirement, the heating temperature must not be high enough to change the bulk material properties of workpiece and/or softening of the cutting tool [64]. Therefore, laser heating needs to be confined to the workpiece surface and should be readily removed before heat propagates into the bulk of workpiece [16]. This results in surface reduced roughness, as shown in Fig. 3.8 [72], and lessen tool wear by about 90% when compared with traditional machining [64, 72].

The distance from cutting edge to laser beam is a critical factor during preheating by laser. This determines the time interval between heating and machining processes, therefore the temperature distribution at the machining zone [34]. During machining commercially pure titanium, Sun et al. [67] noted that the highest reduction of cutting forces occur when the laser spot position is nearest to the cutting edge. However, the distance between tool and beam must not be very close to prevent cutting tool damage by overheating [63]. High laser heating temperatures

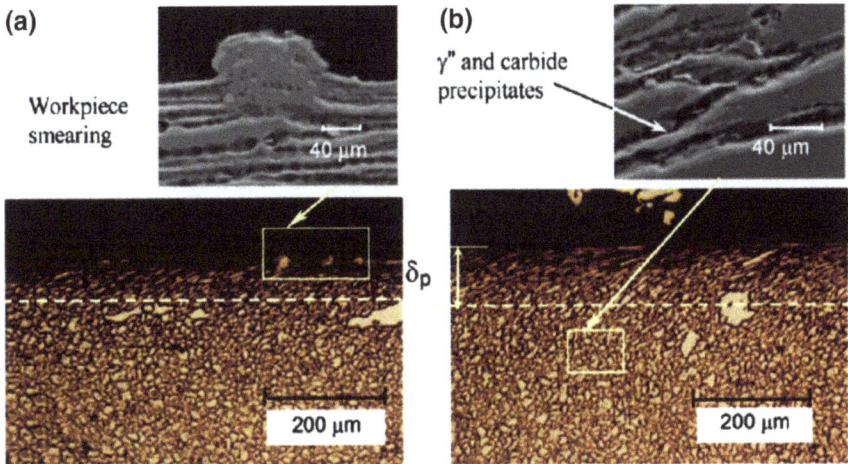

Fig. 3.8 The micrograph of the machined surface (300 m/min, 0.4 mm/rev) under: **a** conventional and **b** LAM machining conditions [72] with kind permission from Elsevier

may prematurely degrade the cutting edge, leading to subsurface damage of workpiece [59].

The material removal temperature is an important variable that requires to be controlled to prevent metal melting and/or undesirable microstructure change during machining [42]. The material removal temperature can be obtained empirically for turning Ti-6Al-4 V with single CO_2 laser from Eq. (3.1) [13]:

$$T_{mr} = \frac{e^{3.4} P_{CO_2}^{0.66}}{f_r^{0.31} D^{0.34} V_c^{0.31}}. \tag{3.1}$$

Equation (3.1) shows that the material removal temperature increases with higher laser power. However, this temperature decreases with increasing cutting speed and workpiece diameter, owing to the reduced interaction time between the laser beam and the workpiece. Therefore, increasing rotating speed leads to a reduction in both the cutting force and cutting tool life [42, 67]. Attia et al. [72] stated that MRR and the surface finish improve approximately by 800% and 25%, respectively, during LAT compare to that with conventional turning. The past work on the effect of LAM on the machinability of titanium alloys is summarized in Table 3.1.

This section has reviewed the set-up for LAM and all laser parameters, such as laser power, laser spot size, laser scanning speed, laser beam/cutting tool lead distance, laser focusing lens/workpiece distance and induction beam/workpiece axis angle. It has been reported from many studies that there are positive influences of LAM parameters on the machining of titanium alloys. Rising preheating temperature with laser power causes a decrease the microhardness at machined zone, thereby reducing the cutting tool pressure on the workpiece surface. Subsequently,

Table 3.1 Summary of LAM of titanium alloys

Author	Type of study	Workpiece material	Machining process	Process parameters	Objectives	Results and conclusions
Rajagopal et al. [45]	Experimental	Inconel 718 and Ti-6Al-4V	Turning	Laser power, feed rate	To study the influence of laser beam/cutting tool distance on the machining efficiency and tool wear	Significant reduction in tool wear and cutting forces
Germain et al. [92]	Experimental	Bearing steel (100 Cr6) and Ti-6Al-4V	Turning	Cutting speed, feed rate, depth of cut, laser power	To analyze the effect of process parameters on fatigue strength in LAM, and to further optimize them	- Significant reduction in cutting forces and surface roughness - Improvements in residual stresses - Slight reduction in the fatigue strength
Sun et al. [67]	Experimental	Pure titanium	Turning	Laser beam-tool lead distance, laser focusing lens-workpiece lead distance, incident beam/workpiece axis angle, laser power, cutting speed, feed rate	To investigate the effect of LAM on the cutting forces and chip formation	- Low cutting forces - Smoother surface finish - Continuous chip formation
Sun et al. [109]	Experimental	Ti-6Al-4V	Turning	Cutting speed, laser power	To investigate the influence of laser beam on chip formation	- Continuous chip formation at higher laser power and cutting speed

(continued)

Table 3.1 (continued)

Author	Type of study	Workpiece material	Machining process	Process parameters	Objectives	Results and conclusions
Dandekar et al. [13]	Numerical and experimental	Ti-6Al-4V	Turning	Tool material, cutting speed, material removal temperature	To study the enhancement in MRR and cutting tool life	- Significant enhancement in MRR and cutting tool life - Low machining costs
Yang et al. [110]	Numerical and experimental	Ti-6Al-4V	Without machining	Laser power, laser scanning speed, laser spot size, angle of incidence	Characterization of heat-affected area produced by laser heating	- Increasing width and depth of the heat affected area with laser power increases and laser spot size decreases
Sun et al. [94]	Experimental	Ti-6Al-4V	Milling	Cutting speed, feed rate, depth of cut, laser power, cutter axis/laser spot centre distance	Analyzing the influence of LAM on cutting forces and cutting tool wear.	- Dramatic decrease in feed force - Significant reduction in tool chipping
Rashid et al. [93, 106]	Experimental	Ti-6Cr-5Mo-5V-4Al	Turning	Cutting speed, feed rate	To analyze the effects of cutting speed and feed rate on cutting forces and cutting temperatures	- Reduction in cutting forces. - Cutting temperature decreased with increasing cutting speed
Zamani et al. [111]	Numerical and experimental	Ti-6Al-4V	Milling	Cutting speed, laser power, feed rate	Investigating the effect of machining and	- Significant cutting forces reduction in three

(continued)

Table 3.1 (continued)

Author	Type of study	Workpiece material	Machining process	Process parameters	Objectives	Results and conclusions
					laser parameters on cutting forces	directions (X, Y and Z) - Increasing cutting tool life
Rashid et al. [93, 106]	Experimental	Ti-6Cr-5Mo-5V-4Al	Turning	Cutting speed, laser power	To analyze the influence of laser power on cutting forces and cutting temperature	- Reduction in cutting forces at high laser power and moderate to high cutting speeds
Zamani et al. [112]	Numerical	Ti-6Al-4V	Milling	Feed rate, laser power	To analyze the influence of feed rate and laser power on the cutting speed and cutting tool	- Appreciable reduction in cutting forces and tool wear
Rashid et al. [86]	Experimental	Ti-10V-2Fe-3Al	Turning	Cutting speed, feed rate	To study the effect of cutting speed and feed rate on cutting forces, cutting temperature and chip formation	- Low cutting forces and temperature - Cutting temperature is greatly dependent on the cutting speed and feed rate - Continuous chip was formed at moderate cutting

(continued)

3 Thermal-Assisted Machining of Titanium Alloys

Table 3.1 (continued)

Author	Type of study	Workpiece material	Machining process	Process parameters	Objectives	Results and conclusions
						speed and segmented chip at high cutting speed
Braham-Bouchnak et al. [57]	Experimental	Ti555-3	Turning	Cutting speed, feed rate, depth of cut, laser power	Investigating the effect of LAM on cutting forces, chip formation and machining productivity	- LAM has small effect on surface roughness criteria - Increasing frequency of discontinuous chip with increase in laser power
Joshi et al. [113]	Numerical	Ti-6Al-4V	Without machining	Laser power, laser spot diameter, rotation speed	To analyze the temperature distribution inside the workpiece heated in LAM	- The surface temperature increases when laser power increases and laser spot diameter decreases - The surface temperature decreases with increase in scanning speed
Rashid et al. [114]	Experimental	Ti-10V-2Fe-3Al	Turning	Cutting speed, laser power	To investigate the effect of laser power on machining	- Optimum value of laser power resulted in reduction of cutting forces and

(continued)

Table 3.1 (continued)

Author	Type of study	Workpiece material	Machining process	Process parameters	Objectives	Results and conclusions
					parameters during LAM	improvement in surface finish - Optimal range of cutting speed helps to avoid force fluctuations during LAM
Xi et al. [115]	Numerical	Ti-6Cr-5Mo-5V-4Al	Turning	Laser scanning speed, rotation speed	Simulation of LAM using traditional finite element method and Split Hopkinson Pressure (SPH) method	- Significant reduction in temperature is achieved with increase in laser scanning speed
Ayed et al. [116]	Numerical and experimental	Ti-6Al-4V	Turning	Laser power, cutting speed, feed rate, laser beam/cutting tool distance	To study the effects of machining and laser parameters on cutting forces and chip formation	- Increase in laser power minimizes the shear angle and reduces the chip thickness - Significant reduction in the cutting forces
Hedberg et al. [91]	Experimental	Ti-6Al-4V	Milling	Laser power	To study the machinability of Ti alloy	Significant reduction in flank wear and cutting forces

this leads to a significant reduction in cutting force and flank wear, thereby increasing cutting tool life. Moreover, the chip formation during LAM is continuous because of preheating temperature improves the deformability of workpiece surface layer near to cutting zone. Furthermore, lowering down dynamic cutting forces and hardness near the cutting surface leads to much smoother machined surface with very few defects such as material build-up and grain pull-out. Thus, LAM effectively improves the machinability of titanium alloys even at high cutting speed.

3.2.2 Plasma-Assisted Machining

Plasma-assisted machining (PAM) is another technique that is recommended for machining of DTM materials such as Ti-6Al-4V that undergo change in mechanical properties at high temperature [36, 38, 117]. PAM is an economical option for heating work material because its cost of heat generation is much lower than the laser technique [38]. PAM has the ability to generate and transfer required amount of heat to the workpiece by a plasma torch to improve machinability [38, 39, 118]. This technique elevates temperature locally ahead of the cutting tool edge and can raise workpiece temperature to 400–1000 °C by convective plasma energy transformation [37, 119]. Although, plasma-heating operates with the same principle as laser heating, but the power density obtained from plasma is less than that of laser [24, 62].

A PAM set-up consists of a conventional machine tools (i.e. lathe and milling machines), a plasma-heating platform with a torch and a control unit. The layout of PAM and plasma torch is illustrate in Figs. 3.9 and 3.10, respectively [38]. In this, the thermal arc generator has two parts such as electrode and nozzle, as shown from Fig. 3.10. A critical issue with PAM is the difficulty to keep the localized heating temperature constant in the allocated cutting zone. Direct current is used to generate arcs from the reaction between plasma gas and electrode sparking to produce equilibrium or thermal plasma [38, 62]. In this preheating technique, the heat spot diameter is around (4–5 mm) and can be located at machining zone near the cutting tool. The electrode is generally made of tungsten and act as a cathode. The plasma gas flows through the nozzle that works as an anode when machine nonconductive materials. However, the workpiece serves as an anode for conducting materials machining [38, 119]. Extremely localized energy is obtainable at low gas flow rates when PAM of materials that have a good electrical conductivity [118, 119]. When cutting is performed nearer to the chuck, a special enclosure is attached to the chuck to minimize the turbulent air flow effects generated by the chuck rotation [38, 119].

There are many technical considerations affecting the selection of cutting parameters [37], where PAM technique has inherent difficulties in controlling the hotspot diameter. The size of heating spot influences the process parameters f_z and a_e. The surface temperature of workpiece can be obtained from the empirical Eq. (3.2) [120], with fixed values of the cathode setback, the plasma gas flow rate

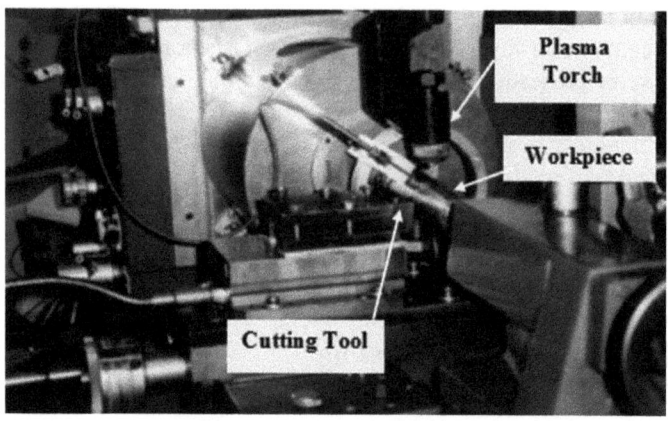

Fig. 3.9 Illustration set-up of PAM [38] with kind permission from Elsevier

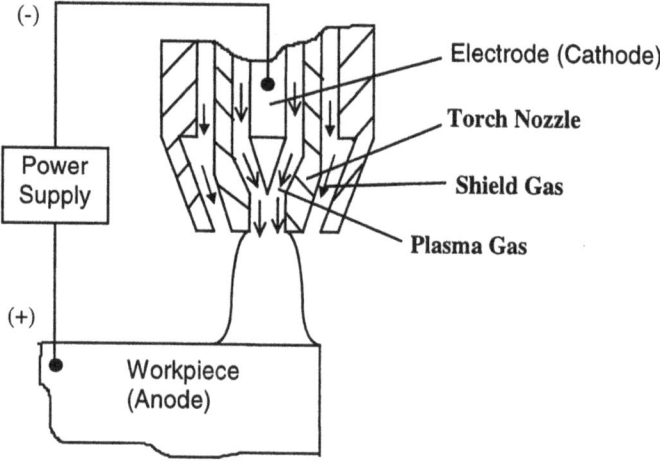

Fig. 3.10 A schematic drawing of plasma torch [38] with kind permission from Elsevier

and the shield gas flow rate. The axial depth of cut, a_p is directly related to the temperature distribution under the surface layer of the workpiece. The machining parameters namely, f_z, V_c, a_p and a_e have direct effect on the operation performance and cutting tool behaviour. Equation (3.3) [37], shows a cross relationship between heating parameter (F) and machine parameters (f_z, V_c, z, D):

$$T_s = 80.3275 \frac{I^{0.584} T_o^{0.06}}{V_c^{2.206} D^{0.405} f_r^{0.2026}} \qquad (3.2)$$

3 Thermal-Assisted Machining of Titanium Alloys

$$F = \frac{f_z 1000\, V_c z}{\pi D}. \qquad (3.3)$$

Applying plasma technique enables high MRR and increased cutting tool life [38]. PAM is widely used only for turning operations because of the difficulties with plasma heating in end-milling operation [40, 62].

During PAM, the formed chip morphology transformed from segmented chip (brittle fracture type) to continuous chip (plastic flow type) due to increased heating temperature [121]. Additional, the chip temperature that formed in this technique tends to be higher than in traditional machining, leading to higher flank wear rates [37, 122]. The cutting forces in this method are substantially lower around 20–40% with the elevated surface layer temperature until a critical temperature is reached at the chip deformation zone [37–39]. Besides the reduction in cutting tool wear, the PAM technique also improves the surface finish of the workpiece without occurring much defects [24, 39, 40].

3.2.3 Induction-Assisted Machining

Induction-assisted machining (IAM) is a new approach in thermal-assisted machining, where a high-frequency induction coil is used as an external heat source to preheat workpiece at localized areas or surface zones adjacent to the coil. Unlike other methods, the coil generates heating along a line on the workpiece, not at a localized point. Thus, the heating region tends to be much wider in IAM compare to other methods. Nonetheless, this method is capable of quickly achieving high temperatures that are controllable and stable [62, 123]. Depending on the machining application of turning or milling processes, IAM induction coil is configured to match the operational requirement, as illustrated by Fig. 3.11. In these, the heating device consists of three main components, namely (i) matching transformer and condenser, (ii) high-frequency transformer (invertors) and (iii) cooling unit. The induction coil is energized by high-frequency alternating electric current to create a magnetic field, thereby generating intense heating. This heating is achieved without any contact between the induction coil and the workpiece while the rate of heating is determined by parameters such as, material magnetic permeability, frequency and intensity of electric current and, material electrical resistance and specific heat [31–34].

Induction workpiece preheating reduces tensile strength and strain hardening [62, 124], and softens the work surface layers [18, 31, 125], hence decreasing the cutting forces. A past work conducted on induction-assisted milling of (Ti-6Al-4V) at 650 °C, thermal softening resulted in cutting force reduction from 338.2 N to 265.6 N [32]. Baili et al. [124] studied the effect of preheating by induction coil on machinability of titanium alloy (Ti-5553). They reported that the components of cutting forces remain constant at a temperature lower than 100 °C while for a

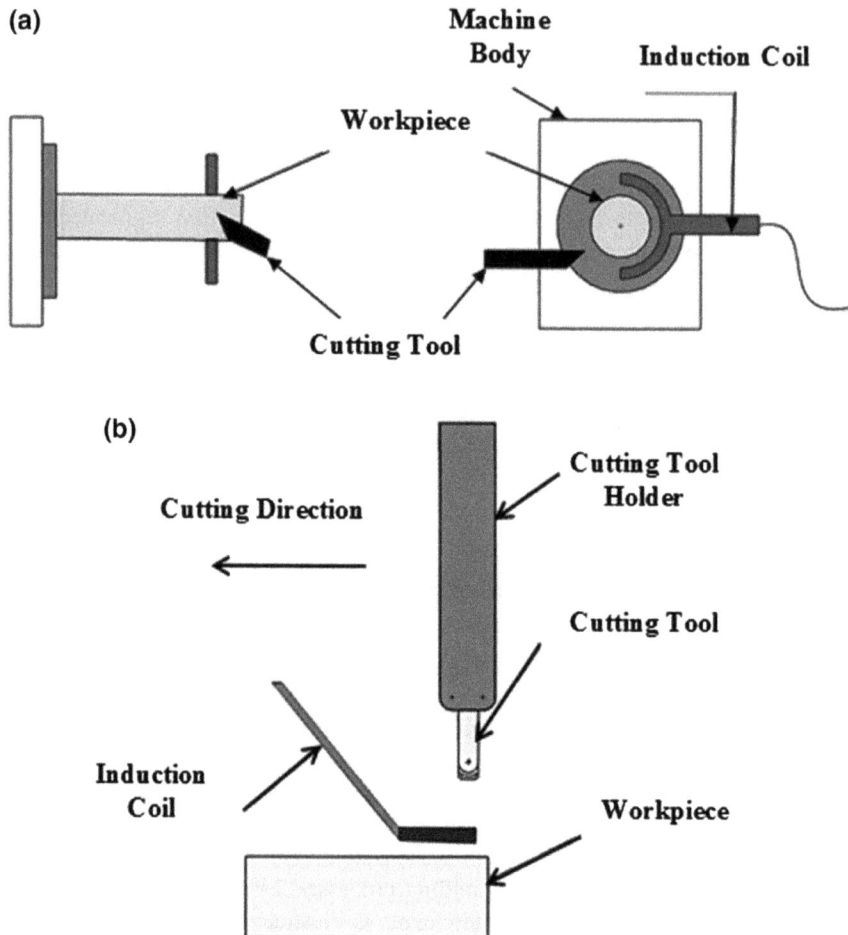

Fig. 3.11 Experimental set-up of **a** induction-assisted turning, and **b** induction-assisted milling

temperature between 100 and 500 °C, about 10% force reduction was observed. Then this reduction increases significantly when the temperature exceeds 500 °C [124]. However, the thermal stresses encountered in induction coil heating are higher than that for laser technique [18].

With IAM, reduced tool wear dramatically extends the cutting tool life, much similar to other workpiece preheating techniques [18, 31, 32, 123]. At moderate cutting speeds in end-milling process, titanium alloy preheating by IAM increased the cutting tool life around 80% [123]. In end-milling of Inconel 718, tool life improvement of 83% is noted at the lowest cutting feed while around 28% is observed at the highest feed during [126]. With high-frequency induction heating, tool vibration and chatter are also reduced to acceptable levels [18, 123]. Ginta et al.

[32] have observed 67% reduction in vibration amplitude when the titanium workpiece was heated to 420 °C led to enhance machinability. Additionally, their results show [18] 88% less amplitude acceleration for Ti-6Al-4V with IAM at workpiece temperature of 650 °C. Furthermore, the vibration suppression minimizes dynamic loads on the cutting edge, thereby reduces flank wear [32–34, 123].

During machining titanium alloys, the Built-up-edge (BUE) is considered as a major factor that affects surface quality. The formation of BUE depends on the chemical reactivity between chip and cutting tool materials, which in turn depend

Table 3.2 Summary of past work on IAM of titanium alloys

Author	Workpiece material	Machining process	Variable process parameters	Objectives	Results and conclusions
Hossain et al. [123]	Ti-6Al-4V	End-milling	Cutting speed, feed rate	To study the influence of induction assistance on the surface integrity in milling process	- Tremendous increment in tool life - Significant reduction in cutting forces - Reduced chatter and vibrations
Ginta and Amin [32]	Ti-6Al-4V	End-milling	Preheating temperature	Investigating the effect of workpiece heating on the tool life	- Increased cutting tool life - Significant reduction in cutting forces - Reduced machining vibrations
Ginta and Amin [18]	Ti-6Al-4V	End-milling	Preheating temperature	To study the effect of workpiece heating by induction source on the machinability in milling process	- Significant reduction in cutting forces - Improvements in the tool life - Reduced chatter and vibrations
Baili et al. [124]	Ti-5553	Turning	Preheating temperature	To enhance the machinability of Titanium	- Low cutting forces - Improvements in surface integrity
Ginta et al. [127]	Ti-6Al-4V	End-milling	Depth of cut, preheating temperature	To study the influence of IAM on the surface integrity during end-milling process	- Decrease in microhardness with increase in preheating temperature

Table 3.3 Comparison of preheating techniques used in machining of titanium alloys

	LAM	PAM	IAM
Heating source	Laser	Plasma	Electric current
Heating tool	Laser beam	Plasma torch	Induction coil
Heating tool movement	Unlimited movement of heating tool	Unlimited movement of heating tool	Limited movement of heating tool
Complexity of the components	Complex components	Less complex components	Simple components
Heating area	Localized point	Localized point	Line
Degree of heat concentration	High degree	High degree	Low degree
Controlling heat source	Easy control of heat source	Impossible to precisely control	Easy to use
Cost of equipment	High cost	Medium cost	Low cost

on the preheating temperature [126]. The size of BUE increases with increase in temperature during IAM, until it reached a certain size, then pass off with a chip [32]. In end-milling of Ti-6Al-4V, the temperature generated is basically responsible for the built-up of chip fragments on the machined zone that finally depreciates the surface finish [127]. The temperature generated during cutting process is rationally adequate as a driving force to enhance annealing arrangements during IAM. The driving force increases with the increase of preheating temperature, thereby leading to lower microhardness [127]. With this condition, initiation and establishment of grain growth happen, causing hardness reduction of materials [127]. IAM produces lower surface roughness compared to that achieved in machining process at room temperature [124, 125]. Previous studies on the effect of IAM on the machinability of titanium alloys are summarized in Table 3.2.

A comparative evaluation of compares all the heating techniques applied in machining of titanium alloys is given in Table 3.3.

3.3 Summary

The accelerated demands of special alloys such as titanium alloys in manufacture of engineered components have substantially increased the innovations and developments in machining techniques. Preheating of workpiece before initiating cutting process has indicated much more effectiveness in increasing the machinability than other techniques used for DTM materials. Thermal-Assisted Machining (TAM) has proven to be practically useful for a board range of DTM materials, including titanium alloys. In this book chapter, a detailed description (working principle, mechanism and research progress and development) of three important TAM techniques(laser-, plasma- and IAM) used for titanium alloys is given.

The chapter can be summarized as follows:

(a) All three preheating techniques lead to a cutting forces reduction, extended tool life and improved surface quality at different levels during machining of titanium alloys;
(b) LAM is capable to reduce the generation of cutting forces up to 60%, decrease tool wears about 20–30%. Furthermore, the surface quality obtained by LAM is equivalent to those obtained by grinding process;
(c) LAM and PAM offer the advantages of rapid and localized heating with sharp temperature rise in workpiece compared to induction heating techniques;
(d) Laser or plasma beam offers the best controlling possibility and flexibility of heat distribution in the workpiece.
(e) PAM is capable of decreasing cutting force up to 25%, with up to 40% extension in tool life with reasonable improvement in surface finish when compared with conventional machining;
(f) Induction-assisted machining produces high capacity preheating, but unable to provide concentrated and localized workpiece preheating;
(g) To Control preheat rate is quite easier in LAM than PAM and IAM.

TAM techniques have proven viability with improved machinability for Titanium alloys and a board range of other DTM metals and polymers. LAM is noted to offer the "best option" as a heat source for achieving highly localized and intense power density for workpiece preheating requirements with the advantage of output controllability. In essence, it can be concluded that the heating ability of LAM with its focused beam is ideally suitable for machining Titanium alloys and other hard materials. This technique allows a significant improvement in the machinability with low cutting forces, increased MRR, less heat-affected zone, increased tool life and superior workpiece surface integrity.

References

1. Ezugwu E, Wang Z (1997) Titanium alloys and their machinability—a review. J Mater Process Technol 68(3):262–274
2. Mantle A, Aspinwall D (1998) Tool life and surface roughness when high speed machining a gamma titanium aluminide, progress of cutting and grinding. In: Fourth international conference on progress of cutting and grinding, Urumqi and Turpan. International Academic Publishers, China, pp 89–94
3. Abele E, Hölscher R (2014) New technology for high speed cutting of titanium alloys. In: New production technologies in aerospace industry. Springer, New York, pp 75–81
4. Ezugwu E (2005) Key improvements in the machining of difficult-to-cut aerospace superalloys. Int J Mach Tools Manuf 45(12):1353–1367
5. Veiga C, Davim J, Loureiro A (2013) Review on machinability of titanium alloys: the process perspective. Rev Adv Mater Sci 34(2):148–164
6. Pramanik A, Littlefair G (2015) Machining of titanium alloy (Ti-6Al-4V)—theory to application. Mach Sci Technol 19(1):1–49
7. Boyer R (1996) An overview on the use of titanium in the aerospace industry. Mater Sci Eng A 213(1):103–114

8. Verma DRSV et al (2003) Effect of pre-drilled holes on tool life in turning of aerospace titanium alloys. In: Proceedings of the national conference on advances in manufacturing system, Production Engineering Department, Jadavpur University, Kolkata, India, pp 42–47
9. Dornfeld D et al (1999) Drilling burr formation in titanium alloy, Ti-6AI-4V. CIRP Ann Manuf Technol 48(1):73–76
10. Oosthuizen GA et al (2010) A review of the machinability of titanium alloys. R&D J S Afr Inst Mech Eng 26:43–52
11. Ezugwu E, Bonney J, Yamane Y (2003) An overview of the machinability of aeroengine alloys. J Mater Process Technol 134(2):233–253
12. Gupta K, Laubscher RF (2016) Sustainable machining of titanium alloys: a critical review. Proc Inst Mech Eng Part B J Eng Manuf, p 0954405416634278
13. Dandekar CR, Shin YC, Barnes J (2010) Machinability improvement of titanium alloy (Ti-6Al-4V) via LAM and hybrid machining. Int J Mach Tools Manuf 50(2):174–182
14. Khanna N et al (2012) Effect of heat treatment conditions on the machinability of Ti64 and Ti54M alloys. Procedia CIRP 1:477–482
15. Pramanik A (2014) Problems and solutions in machining of titanium alloys. Int J Adv Manuf Technol 70(5–8):919–928
16. Brecher C et al (2011) Laser-assisted milling of advanced materials. Phys Procedia 12:599–606
17. Przestacki D, Jankowiak M (2014) Surface roughness analysis after laser assisted machining of hard to cut materials. J Phy Conf Ser (IOP Publishing)
18. Ginta TL, Amin AN (2010) Machinability improvement in end milling titanium alloy TI-6AL-4V, vol 3, pp 25–33
19. Krabacher EJ, Merchant ME (1951) Basic factor of hot machining of metals. J Eng Ind 73:761–776
20. Pfefferkorn FE et al (2009) A metric for defining the energy efficiency of thermally assisted machining. Int J Mach Tools Manuf 49(5):357–365
21. Radovanovic MR, Dašić PV (2006) Laser assisted turning. In: Research and development in mechanical industry, RaDMI 2006, Budva, Montenegro, pp 312–316
22. Shin YC (2011) LAM benefits a wide range of difficult-to-machine materials. J Ind Laser Solut Manuf
23. Pentland W, Mehl C, Wennbery J (1960) Hot machining. Am Mach Metalwork Manuf 1:117–132
24. Madhavulu G, Ahmed B (1994) Hot machining process for improved metal removal rates in turning operations. J Mater Process Technol 44(3):199–206
25. Çakır O, Altan E (2008) Hot machining of high manganese steel: a review. In: Trends in the development of machinery and associated technology, Istanbul, Turkey, pp 105–108
26. Özler L, Inan A, Özel C (2001) Theoretical and experimental determination of tool life in hot machining of austenitic manganese steel. Int J Mach Tools Manuf 41(2):163–172
27. Rajopadhye RD, Telsang MT, Dhole NS (2009) Experimental setup for hot machining process to increase tool life with torch flame. In: Second international conference on emerging trends in engineering (SICETE), Nagpur, Maharashtra, India, pp 58–62
28. Tosun N, Ozler L (2004) Optimisation for hot turning operations with multiple performance characteristics. Int J Adv Manuf Technol 23(11–12):777–782
29. Rebro PA et al (2002) Comparative assessment of laser-assisted machining for various ceramics, vol 30. Transactions of North American Manufacturing Research Institution, pp 153–160
30. Bermingham M, Palanisamy S, Dargusch M (2012) Understanding the tool wear mechanism during thermally assisted machining Ti-6Al-4V. Int J Mach Tools Manuf 62:76–87
31. Amin A, Abdelgadir M (2003) The effect of preheating of work material on chatter during end milling of medium carbon steel performed on a vertical machining center (VMC). J Manuf Sci Eng 125(4):674–680
32. Ginta TL et al (2009) Improved tool life in end milling Ti-6Al-4V through workpiece preheating. Eur J Sci Res 27(3):384–391

33. Lajis MA et al (2009) Hot machining of hardened steels with coated carbide inserts. Am J Eng Appl Sci 2(2):421–427
34. Amin AN et al (2008) Effects of workpiece preheating on surface roughness, chatter and tool performance during end milling of hardened steel D2. J Mater Process Technol 201(1):466–470
35. Kttagawa T, Maekawa K (1990) Plasma hot machining for new engineering materials. Wear 139(2):251–267
36. Popa L (2012) Complex study of plasma hot machining (PMP). Revista de Tehnologii Neconventionale 16(1):26
37. De Lacalle LNL et al (2004) Plasma assisted milling of heat-resistant superalloys. J Manuf Sci Eng 126(2):274–285
38. Leshock CE, Kim J-N, Shin YC (2001) Plasma enhanced machining of Inconel 718: modeling of workpiece temperature with plasma heating and experimental results. Int J Mach Tools Manuf 41(6):877–897
39. Novak J, Shin Y, Incropera F (1997) Assessment of plasma enhanced machining for improved machinability of Inconel 718. J Manuf Sci Eng 119(1):125–129
40. Shin YC, Kim J-N (1996) Plasma enhanced machining of Inconel 718. Manuf Sci Eng ASME MED 4:243–249
41. Jau BM, Copley SM, Bass M (1981) Laser assisted machining. In: Proceedings of the ninth north american manufacturing research conference, University Park, Pennsylvania, pp 12–15
42. Dumitrescu P et al (2006) High-power diode laser assisted hard turning of AISI D2 tool steel. Int J Mach Tools Manuf 46(15):2009–2016
43. Thomas T, Vigneau JO (1999) Laser-assisted milling process. Google Patents
44. Chryssolouris G, Anifantis N, Karagiannis S (1997) Laser assisted machining: an overview. J Manuf Sci Eng 119(4B):766–769
45. Rajagopal S, Plankenhorn D, Hill V (1982) Machining aerospace alloys with the aid of a 15kW laser. J Appl Metalwork 2(3):170–184
46. Dubey AK, Yadava V (2008) Laser beam machining—a review. Int J Mach Tools Manuf 48(6):609–628
47. Venkatesan K, Ramanujam R, Kuppan P (2014) Laser assisted machining of difficult to cut materials: research opportunities and future directions-a comprehensive review. Procedia Eng 97:1626–1636
48. Shin YC (2000) Laser assisted machining. Mach Technol 11(3):1–6
49. Velayudham A (2007) Modern manufacturing processes: a review. J Des Manuf Technol 1(1):30–40
50. Jeon Y, Park HW, Lee CM (2013) Current research trends in external energy assisted machining. Int J Precis Eng Manuf 14(2):337–342
51. Rozzi JC et al (2000) Experimental evaluation of the laser assisted machining of silicon nitride ceramics. J Manuf Sci Eng 122(4):666–670
52. Warap N, Mohid Z, Rahim EA (2013) Laser assisted machining of titanium alloys. In: Materials science forum. Trans Tech Publications, Switzerland
53. Rebro PA, Shin YC, Incropera FP (2004) Design of operating conditions for crackfree laser-assisted machining of mullite. Int J Mach Tools Manuf 44(7):677–694
54. Lei S, Shin YC, Incropera FP (2001) Experimental investigation of thermo-mechanical characteristics in laser-assisted machining of silicon nitride ceramics. J Manuf Sci Eng 123(4):639–646
55. Wu J-F, Guu Y-B (2006) Laser assisted machining method and device. Google Patents
56. Kim K-S et al (2011) A review on research and development of laser assisted turning. Int J Precis Eng Manuf 12(4):753–759
57. Braham-Bouchnak T et al (2013) The influence of laser assistance on the machinability of the titanium alloy Ti555-3. Int J Adv Manuf Technol 68(9–12):2471–2481
58. Shi B, Attia H (2013) Integrated process of laser-assisted machining and laser surface heat treatment. J Manuf Sci Eng 135(6):061021

59. Anderson M, Patwa R, Shin YC (2006) Laser-assisted machining of Inconel 718 with an economic analysis. Int J Mach Tools Manuf 46(14):1879–1891
60. Kannan V, Radhakrishnan R, Palaniyandi K (2014) A review on conventional and laser assisted machining of Aluminium based metal matrix composites. Eng Rev 34(2):75–84
61. Klocke F, Bergs T (1997) Laser-assisted turning of advanced ceramics. In: Lasers and optics in manufacturing III. International Society for Optics and Photonics
62. Rahim E, Warap N, Mohid Z (2015) Thermal-assisted machining of nickel-based alloy. Superalloys
63. Kong XJ et al (2014) Laser-assisted machining of advanced materials. In: Materials science forum. Trans Tech Publications, Switzerland
64. Pfefferkorn FE, Incropera FP, Shin YC (2005) Heat transfer model of semi-transparent ceramics undergoing laser-assisted machining. Int J Heat Mass Transf 48(10):1999–2012
65. Xuefeng W, Hongzhi Z, Yang W (2009) Three-dimensional thermal analysis for laser assisted machining of ceramics using FEA. In: Proceedings of SPIE
66. Rozzi JC et al (1998) Transient thermal response of a rotating cylindrical silicon nitride workpiece subjected to a translating laser heat source, part I: comparison of surface temperature measurements with theoretical results. J Heat Transfer 120(4):899–906
67. Sun S, Harris J, Brandt M (2008) Parametric investigation of laser-assisted machining of commercially pure titanium. Adv Eng Mater 10(6):565–572
68. Gratias J et al (1993) Proposition of a method to optimize the machining of XC42 steel with laser assistance. CIRP Ann Manuf Technol 42(1):115–118
69. Yang B, Lei S (2008) Laser-assisted milling of silicon nitride ceramic: a machinability study. Int J Mechatron Manuf Syst 1(1):116–130
70. Lei S, Shin YC, Incropera FP (2000) Deformation mechanisms and constitutive modeling for silicon nitride undergoing laser-assisted machining. Int J Mach Tools Manuf 40 (15):2213–2233
71. Bejjani R et al (2011) Laser assisted turning of titanium metal matrix composite. CIRP Ann Manuf Technol 60(1):61–64
72. Attia H et al (2010) Laser-assisted high-speed finish turning of superalloy Inconel 718 under dry conditions. CIRP Ann Manuf Technol 59(1):83–88
73. Anderson M, Shin Y (2006) Laser-assisted machining of an austenitic stainless steel: P550. Proc Inst Mech Eng Part B J Eng Manuf 220(12):2055–2067
74. Ding H, Shin YC (2010) Laser-assisted machining of hardened steel parts with surface integrity analysis. Int J Mach Tools Manuf 50(1):106–114
75. Germain G, Dal Santo P, Lebrun JL (2011) Comprehension of chip formation in laser assisted machining. Int J Mach Tools Manuf 51(3):230–238
76. Garcí V et al (2013) Mechanisms involved in the improvement of Inconel 718 machinability by laser assisted machining (LAM). Int J Mach Tools Manuf 74:19–28
77. Kim D-H, Lee C-M (2014) A study of cutting force and preheating-temperature prediction for laser-assisted milling of Inconel 718 and AISI 1045 steel. Int J Heat Mass Transf 71:264–274
78. Venkatesan K, Ramanujam R, Kuppan P (2014) Analysis of cutting forces and temperature in laser assisted machining of inconel 718 using Taguchi method. Procedia Eng 97:1637–1646
79. Dong-Gyu A, Kyung-Won B (2009) Influence of cutting parameters on surface characteristics of cut section in cutting of Inconel 718 sheet using CW Nd: YAG laser. Trans Nonferr Metals Soc China 19:s32–s39
80. Kong X et al (2015) Cutting performance and coated tool wear mechanisms in laser-assisted milling K24 nickel-based superalloy. Int J Adv Manuf Technol 77(9–12):2151–2163
81. Thawari G et al (2005) Influence of process parameters during pulsed Nd: YAG laser cutting of nickel-base superalloys. J Mater Process Technol 170(1):229–239
82. Ding H, Shin YC (2013) Improvement of machinability of Waspaloy via laser-assisted machining. Int J Adv Manuf Technol 64(1–4):475–486

83. Rebro PA, Shin YC, Incropera FP (2002) Laser-assisted machining of reaction sintered mullite ceramics. J Manuf Sci Eng 124(4):875–885
84. Lee S-J, Kim J-D, Suh J (2014) Microstructural variations and machining characteristics of silicon nitride ceramics from increasing the temperature in laser assisted machining. Int J Precis Eng Manuf 15(7):1269–1274
85. Kim J-D, Lee S-J, Suh J (2011) Characteristics of laser assisted machining for silicon nitride ceramic according to machining parameters. J Mech Sci Technol 25(4):995–1001
86. Rashid RAR et al (2013) Experimental investigation of laser assisted machining of AZ91 magnesium alloy. Int J Precis Eng Manuf 14(7):1263–1265
87. Pfefferkorn FE et al (2004) Laser-assisted machining of magnesia-partially-stabilized zirconia. J Manuf Sci Eng 126(1):42–51
88. Wang Y, Yang L, Wang N (2002) An investigation of laser-assisted machining of Al_2O_3 particle reinforced aluminum matrix composite. J Mater Process Technol 129(1):268–272
89. Chang C-W, Kuo C-P (2007) Evaluation of surface roughness in laser-assisted machining of aluminum oxide ceramics with Taguchi method. Int J Mach Tools Manuf 47(1):141–147
90. Dandekar CR, Shin YC (2010) Laser-assisted machining of a fiber reinforced metal matrix composite. J Manuf Sci Eng 132(6):061004
91. Hedberg G, Shin Y, Xu L (2015) Laser-assisted milling of Ti-6Al-4V with the consideration of surface integrity. Int J Adv Manuf Technol 79(9–12):1645–1658
92. Germain G et al (2006) Effect of laser assistance machining on residual stress and fatigue strength for a bearing steel (100Cr6) and a titanium alloy (Ti 6Al 4V). In: Materials science forum. Trans Tech Publications, Switzerland
93. Rashid RR et al (2012) An investigation of cutting forces and cutting temperatures during laser-assisted machining of the Ti–6Cr–5Mo–5V–4Al beta titanium alloy. Int J Mach Tools Manuf 63:58–69
94. Sun S et al (2011) Experimental investigation of cutting forces and tool wear during laser-assisted milling of Ti-6Al-4V alloy. Proc Inst Mech Eng Part B J Eng Manuf 225(9):1512–1527
95. Nasr MNA, Balbaa M (2014) Effect of laser power on residual stresses when laser-assisted turning of AISI 4340 steel. In: Proceedings of the canadian society for mechanical engineering international congress, Toronto, Ontario, Canada
96. Germain G et al (2008) Laser-assisted machining of Inconel 718 with carbide and ceramic inserts. Int J Mater Form 1(1):523–526
97. Lee J-H et al (2008) Trends of laser integrated machine. J Korean Soc Precis Eng 25(9):20–26
98. Dahotre NB, Harimkar SP (2008) Laser fabrication and machining of materials. Springer, New York
99. Nath A (2013) High power lasers in material processing applications: an overview of recent developments. In: Laser-assisted fabrication of materials. Springer, New York, pp 69–111
100. König W, Zaboklicki AK (1993) Laser assisted hot machining of ceramics and composite materials, vol 847. National Institute of Science and Technology, NIST Special Publication
101. König W, Wageman A (1991) Fine machining of advanced ceramics. In: Vincenzini P (ed) Ceramics today—tomorrow's ceramics, Montecatini Terme, Italy, pp 2769–2784
102. Kennedy E, Byrne G, Collins D (2004) A review of the use of high power diode lasers in surface hardening. J Mater Process Technol 155:1855–1860
103. Bachmann F (2003) Industrial applications of high power diode lasers in materials processing. Appl Surf Sci 208:125–136
104. Li L (2000) The advances and characteristics of high-power diode laser materials processing. Opt Lasers Eng 34(4):231–253
105. Choi S et al (2007) Characteristics of metal surface heat treatment by diode laser. J Korean Soc Manuf Process Eng 6(3):16–23
106. Rashid RR et al (2012) The effect of laser power on the machinability of the Ti-6Cr-5Mo-5V-4Al beta titanium alloy during laser assisted machining. Int J Mach Tools Manuf 63:41–43

107. Venkatesan K, Ramanujam R, Kuppan P (2016) Parametric modeling and optimization of laser scanning parameters during laser assisted machining of Inconel 718. Opt Laser Technol 78:10–18
108. Kannan MV et al (2014) Effect of laser scan speed on surface temperature, cutting forces and tool wear during laser assisted machining of alumina. Procedia Eng 97:1647–1656
109. Sun S, Brandt M, Dargusch M (2010) The effect of a laser beam on chip formation during machining of Ti6Al4V alloy. Metall Mater Trans A 41(6):1573–1581
110. Yang J et al (2010) Experimental investigation and 3D finite element prediction of the heat affected zone during laser assisted machining of Ti6Al4V alloy. J Mater Process Technol 210(15):2215–2222
111. Zamani H et al (2012) Numerical and experimental investigation of laser assisted side milling of Ti6Al4V alloy. In: Proceedings of materials science & technology conference and exhibition
112. Zamani H et al (2013) 3D simulation and process optimization of laser assisted milling of Ti6Al4V. Procedia CIRP 8:75–80
113. Joshi A et al (2014) A study of temperature distribution for laser assisted machining of Ti-6Al-4V alloy. Procedia Eng 97:1466–1473
114. Rashid RR et al (2014) A study on laser assisted machining of Ti10V2Fe3Al alloy with varying laser power. Int J Adv Manuf Technol 74(1–4):219–224
115. Xi Y et al (2014) Numerical modeling of laser assisted machining of a beta titanium alloy. Comput Mater Sci 92:149–156
116. Ayed Y et al (2014) Experimental and numerical study of laser-assisted machining of Ti6Al4V titanium alloy. Finite Elem Anal Des 92:72–79
117. Pérez J, Llorente J, Sanchez J (2000) Advanced cutting conditions for the milling of aeronautical alloys. J Mater Process Technol 100(1):1–11
118. Pfender E, Spores R, Chen WLT (1995) A new look at the thermal and gas dynamic characteristics of a plasma jet. Int J Mater Prod Technol 10(3–6):548–565
119. Pfender E, Fincke J, Spores R (1991) Entrainment of cold gas into thermal plasma jets. Plasma Chem Plasma Process 11(4):529–543
120. Wang Z et al (2003) Hybrid machining of Inconel 718. Int J Mach Tools Manuf 43(13):1391–1396
121. Kitagawa T, Maekawa K, Kubo A (1988) Plasma hot machining for high hardness metals. Bull Jpn Soc Precis Eng 22(2):145–151
122. Armendia M et al (2010) Comparison of the machinabilities of Ti6Al4V and TIMETAL® 54M using uncoated WC-Co tools. J Mater Process Technol 210(2):197–203
123. Hossain MI et al (2008) Enhancement of machinability by workpiece preheating in end milling of Ti-6Al-4V. J Achiev Mater Manuf Eng 31(2):320–326
124. Baili M et al (2011) An experimental investigation of hot machining with induction to improve Ti-5553 machinability. In: Applied mechanics and Materials. Trans Tech Publications, Switzerland
125. Amin AN et al (2007) Influence of preheating on performance of circular carbide inserts in end milling of carbon steel. J Mater Process Technol 185(1):97–105
126. Amin A, Hossain MI, Patwari AU (2011) Enhancement of Machinability of Inconel 718 in End Milling through Online Induction Heating of Workpiece. In: Advanced materials research. Trans Tech Publications, Switzerland
127. Ginta TL, Amin AN (2013) Surface integrity in end milling titanium alloy Ti-6Al-4V under heat assisted machining. Asian J Sci Res 6(3):609

Chapter 4
Abrasive Water Jet Machining of Composite Materials

Sumit Bhowmik, Jagadish and Amitava Ray

Abstract In order to fulfil various advanced technological requirements in defence, aerospace, automotive and other industrial sectors, the use of composite materials have rapidly been increased. Machining of composite material is one of the important operations while manufacturing different engineered components. Machining of composite materials using conventional processes is extremely difficult and involves high cost due to their inherent material properties that often causes various types of damages during machining. Abrasive water jet machining (AWJM), an important non-conventional-type machining, is capable to overcome the limitations as regards to the machining of composites and therefore widely employed to machine them. This chapter provides an overview of various issues related to the machining of composites in traditional techniques followed by detailed fundamental scientific aspects of the abrasive water jet machining including material erosion mechanisms, influencing parameters that affect the abrasive water jet machining performance, etc. An experimental case study on machining of wood dust-based filler-reinforced polymer composites in abrasive water jet machining process and optimization of its process parameters via parametric analysis is presented.

Keywords Composites · Abrasive water jet machining · Polymer · Optimization · MRR · Surface integrity

---

S. Bhowmik (✉) · Jagadish
Department of Mechanical Engineering,
National Institute of Technology, Silchar, India
e-mail: bhowmiksumit04@yahoo.co.in

A. Ray
Training and Placement Department,
Jalpaiguri Government Engineering College,
Jalpaiguri, West Bengal, India

© Springer International Publishing AG 2017
K. Gupta (ed.), *Advanced Manufacturing Technologies*, Materials Forming, Machining and Tribology, DOI 10.1007/978-3-319-56099-1_4

4.1 Introduction

In order to meet the demands of new technologies in different industrial applications, the materials science community is continuously engaged to develop new materials with improved properties. In the recent days, composite materials (such as polymers, metals matrix and ceramics composite) have extensively being used in defence, aircraft, automotive industries, etc. [1]. Composite materials offer some exceptional merits such as minimal cost, high quality, high strength-to-weight ratio, high modulus-to-weight ratio, good damage tolerance, and excellent fatigue tolerance. Moreover, these materials have specific limitations such as low working temperatures, high thermal and moisture coefficient and low elastic property in transverse direction [2, 3]. With the increasing demand and various applications, many types of composite material have been developed, including metal matrix composites (MMC), polymer matrix composites (PMC) and ceramic matrix composites (CMC). These composites are subjected to extensive machining operations before actual use. Due to the intrinsic properties and material behaviour, the machining of composite materials is extremely difficult using conventional machining process, i.e. drilling, milling, sawing and grinding. Conventional machining of composite materials results in various types of damages such as delamination, fibre pull-out, burning of composites due to the poor thermal conductivity, poor dimensional accuracy, etc. It also results in excessive tool wear and thus shorter tool life and poor work surface quality [4, 5]. To overcome the machining difficulties of composite materials, non-conventional machining is the best alternative option. Non-conventional machining (NCM) process such as electrical discharge machining (EDM), ultrasonic machining (USM), laser machining (LM) and abrasive water jet machining (AWJM) have been used for machining of composites [5–9]. Furthermore, selection of machining process for machining composites is not simple, not all the non-conventional machining processes are suitable. In some of the cases, electrical discharge machining has been used for machining of composites specifically metal matrix composites, because they are hard, possess high strength, temperature-resistant materials and conductive in nature [8]. However, the fibre-reinforced polymers, natural filler/fibre-reinforced polymers and ceramic matrix composites are cannot be machined by EDM due to hard, brittle and non-conductive nature. While laser machining process can be used for machining of metal matrix composites, fibre-reinforced polymers, natural filler/fibre-reinforced polymers and ceramic matrix composites cannot be machined because this process suffers from the problem of a large heat-affected zone which results in melting of these composites due to lesser thermal conductivity [9]. In order to overcome the difficulties, AWJM is used in most of the cases, as this process offers many advantages such as reduced waste materials, less heat-affected zone, higher flexibility and versatility, and does not generate any fumes. Overall, AWJM is a green process [10]. Hence, abrasive water jet machining process has been known to machine composite materials with greater ease, providing unique capabilities to machine with the desired shape, size and dimensional accuracy [11]. The remainder of this chapter is as follows.

In Sect. 4.2, brief introduction of composite materials along with their properties is given. In Sect. 4.3, machining problems with composites followed by exhaustive reviews on improvement in the machinability of composites by the previous researchers is discussed. In Sect. 4.4, detailed fundamental scientific aspects of abrasive water jet machining including working principle, material erosion mechanisms, important process parameters, advantages, limitation and its applications are presented. At last, before ending up the chapter with conclusions and future scope, an experimental case study on machining of wood dust-based filler-reinforced polymer composites in abrasive water jet machining process and its parametric optimization is discussed in Sect. 4.5.

4.2 Composite Materials

Composites are the advanced materials that overcome the conflicting properties of their constituent materials. These are formed by combining two or more constituent materials with significantly different physical or chemical properties, when combined; the characteristics of the material obtained differ from the individual components [12]. It consists of a discontinuous phase known as reinforcement and a continuous phase known as matrix. Generally, fibres are used as reinforcements in order to harden the matrix for better strength and stiffness. The properties of the composites can be affected due to reinforcements which can be cut, placed and aligned in different ways. Therefore, the matrix is used in order to keep the reinforcements in their desired orientations. The matrix bonds the reinforcements for transferring the applied loads effectively. It also protects the reinforcements from chemical and environmental attacks.

Composites find plenty of general and industrial applications in aerospace, marine, and automotive industries due to their superior properties. The detailed classification of composite materials is shown in Fig. 4.1 [13].

Composite materials are widely used in several engineering areas due to the following properties [1, 3, 14, 15]:

- Composites are lighter than wood and metals which makes them suitable to be used in aircraft and automobiles;
- High strength-to-weight ratio;
- Great resistance to corrosion degradation and chemical and weathering;
- Good dimensional stability, design flexibility and close tolerances;
- Enhanced magnetic properties and electrical insulation;
- Composites have lower aerodynamic resistance and better translucency and radiation characteristics
- Any shade of any colour can be incorporated into the composite product during manufacture by pigmenting the gel coat used.
- Great flexibility, fire resistant can be used to make complicated shapes;

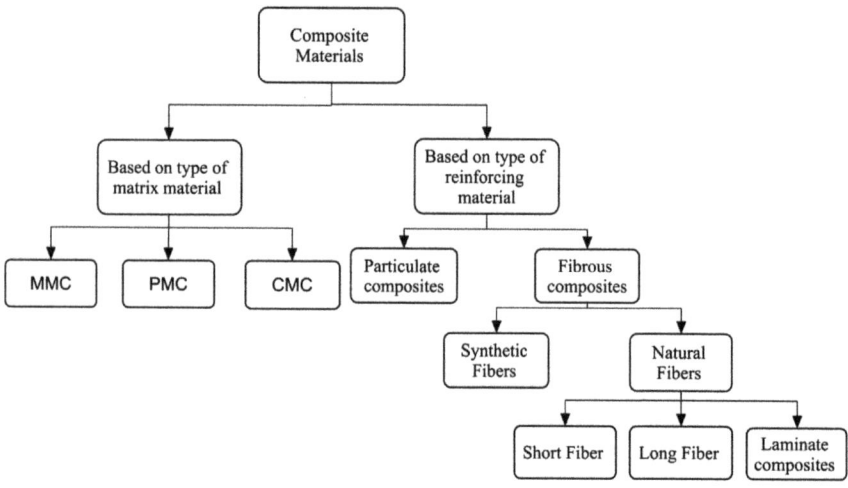

Fig. 4.1 Classification of composite materials

4.3 Problems in Machining of Composites

An extensive review of the available literature is done before writing this chapter. These studies reveal that machining of composite materials found to be differed in many aspects from the machining of traditional materials and also found to be extremely difficult and expensive [1, 13, 15]. Damages that commonly occur during machining of composites are burrs, interlaminar cracks, delamination and thermal damage. The amount of defect depends on the selected process parameters. Thermal damages, burr generation and dimensional accuracy are very important aspects to be taken care of. Burrs are small portion of broken material at the edge of the surface or at the corner of the hole. The presence of the burr results in disturbance in the cutting. Further, thermal damage occurs because of localized heat generated during the machining and thus results in thermal damage like burning or melting [14]. Amongst other damages caused during the machining of composites, the most problem faced during machining is the delamination. The delamination is the separation of layers in laminate and fibre pull-out and is defined as the ratio of damaged area around the hole to the actual hole area, as shown in Fig. 4.2.

Especially, in machining, the problem of delamination is widely observed around a hole. This creates many hurdles such as reduction of structural integrity of the machined surface, poor assembly tolerance, and addition a potential for long-term deterioration of machined surface. Normally, the delamination occurs at both the entrance and exit plane [15]. In the entrance side, the peel-up delamination will observe while in exist side push-out delamination. The graphical representation of peel-up delamination and push-out delamination is shown in Fig. 4.3. The push-out delamination occurs at a certain point where interlaminar bond strength exceeds as shown in Fig. 4.3a, whereas peel-up delamination occurs, because the

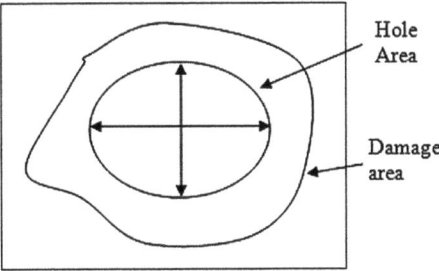

Fig. 4.2 Schematic layout of the damage area and actual hole area

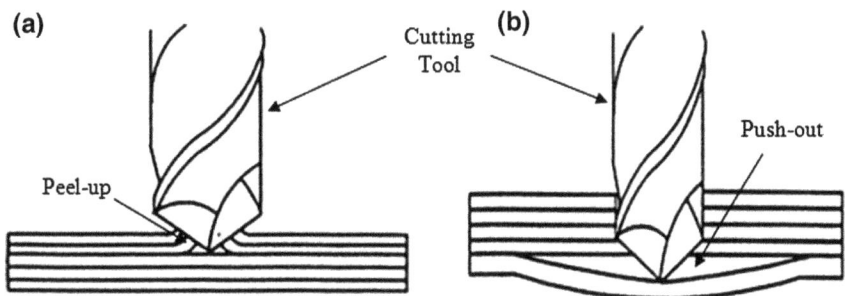

Fig. 4.3 a Peel-up delamination at entrance **b** Push-out delamination at exit

drill first abraded the laminate and then pulled the abraded material away along the flute, causing the material to spiral up before being machined completely, as shown in Fig. 4.3b [14].

The problems related to the machining of composite materials in conventional technique such as exposure of fibres to environmental attack, material degradation due to localized heating, poor dimensional accuracy, shorter tool life, delamination, and fibre pull-out due to anisotropic, nonhomogeneous and abrasive nature of such composite materials have been reported in several studies [14–18]. Some studies have also been explored the effects of tool wear [13, 19] and optimal processing variables [20, 21]. However, machining of composite by conventional machining is associated with inherent problems such as rapid tool wear, significant damage of work piece subsurface in the form of chipping, cracking and delamination and poor surface finish [13, 14]. Apart from these defects, the surface integrity of machined composite is also hard to control, including surface roughness, residual stresses and subsurface damages due to varying mechanical properties of the fibre and the matrix [19]. This difference in the material properties and the degree of anisotropy causes difficulty in defining the cutting parameters to deal with characteristics of all materials, individually within a composite material and the whole composite together and also prediction of the material behaviour while being machined [19].

In spite of these difficulties in machining for composite materials, the conventional technique requires excessive time to complete the job due to the usage of frequent and expensive tools. On the other hand, the conventional technique generates fine and powder chips particles along with a high proportion of dust released into the air resulting in serious occupational health and environmental issues. Also, these toxic substances can enter the body of the operator through ingestion, inhalation and skin contact [22].

In order to overcome these difficulties, now industries have started to utilize the non-conventional technique such as electrical discharge machining (EDM), ultrasonic machining (USM), laser machining (LM) and abrasive water jet machining (AWJM) [6, 7]. The selection of suitable non-conventional machining technique for the composite materials depends on various constraints. As the metal matrix composites are in hard, high strength, complex geometry shape, and temperature-resistant materials and conductive in nature, the electrical discharge machining process can be selected for machining of these materials. Recent advances in the electrical discharge machining technology made it a valuable and viable process in the manufacturing sector due to their unique advantages such as machining of any type of conductive materials (metals, alloys, graphite, ceramics, etc.) of any hardness, allowing for programming of complex and intricate profiles using simple electrode, machining of very small work pieces, attaining high accuracy, providing good surface roughness, etc. Despite its advantages, the electrical discharge machining process is considered as a hazardous process because it discharges large amount of toxic substances in the form of solid, liquid, and gases wastes, resulting in serious occupational health and environmental issues [22, 23]. On the other hand, the fibre-reinforced polymers, natural filler/fibre-reinforced polymers and ceramic matrix composites are hard, brittle and non-conductive in nature, and electrical discharge machining process cannot be used for them.

While laser cutting machining process can be used for machining of polymer and ceramic composites, this process suffers from the problem of a large heat-affected zone, which results in melting of these composites due to lesser thermal conductivity and change in the material dimensions [9]. Several studies indicate that good quality of cut surface can be attained, if thermal conductivity of fibres is nearer to that of the matrix [9, 24–27].

Furthermore, ultrasonic machining (USM) can be implemented for machining composites but limited to metal matrix composites and ceramic matrix composites. This process also possesses workpiece and tool size limitations [13] and it is a slow process as well [14].

Another advanced technique for machining all kind of composites is the abrasive water jet machining (AWJM) process. This process uses erosive effect of high-velocity water jet mixed with abrasive to remove work piece material for cutting two- and three-dimensional profiles [5]. AWJM results in very less heat-affected zone, higher flexibility and versatility, and minimal force during machining [7]. Since high-velocity stream of water flushes the eroded materials form the work piece, possibilities of environmental contamination are also less [8]. Additionally, this process requires no cooling or lubricating oils, and no toxic fumes generate during the cutting process. Hence, this process inherently called as an environmentally friendly (green) process. As a result, abrasive water jet machining process has been a foundation for the manufacturing process, providing unique capabilities to machine any kind of composite materials with desired shape, size and required dimensional accuracy [7, 10]. Due to the various advantages of the abrasive water jet machining, it was widely applied in the different metal matrix composites and ceramic composites. Considerable amount of literature is readily available on use of AWJM for improving the machinability of MMC and CMC [15, 17, 19], with very limited work on AWJM of fibre-reinforced polymers and natural filler/fibre-reinforced polymer composites.

Quality (geometry of the cut and surface integrity of the part) and productivity are the two important criteria to determine the performance of AWJM. The dominant mechanism of material removal is determined by the properties of impacting abrasive particles in terms of shape, size and hardness, the jet impact angle, abrasive feed rate and the properties of the work piece material. The variation of these process parameters may result in errors which subsequently affect the performance as well as efficiency of the machining process. This compels the need of optimization of process parameters of AWJM.

4.4 Abrasive Water Jet Machining

Abrasive water jet machining is one of most commonly used non-traditional machining processes. This process works on the principle of mechanical erosion where a fine jet of high pressure water is used to accelerate the abrasive slurry to cut the target material [5]. The schematic diagram of abrasive water jet machining process illustrating its cutting head assembly and the basic machining principle is shown in Fig. 4.4.

The velocity of abrasive mixed water jet is used about 300 m/s and for the special cases the velocity may be used as high as 900 m/s depending on the requirements. Such a high velocity of the jet results in high kinetic energy, and thereby leads to the erosion of target work surface. The operating pressure for water jet is about 400 MPa, which is adequate to produce maximum jet velocity (i.e. 900 m/s). When the water pressure is increased, the increase in the jet kinetic energy results in a high momentum which is transferred to the abrasive particles such that the impact and change in momentum of the abrasive material leads to cut the target [5]. This high-velocity jet is able to process the variety of materials such as ceramics, composites, rocks and metals [6]. The material removal takes place by erosion wear in the upper surface while it follows by deformation wear at the lower portion of the specimen being cut. Currently, this process is superior to many other machining techniques in processing materials like all kinds of metals, ceramics, marbles and composites etc., [5].

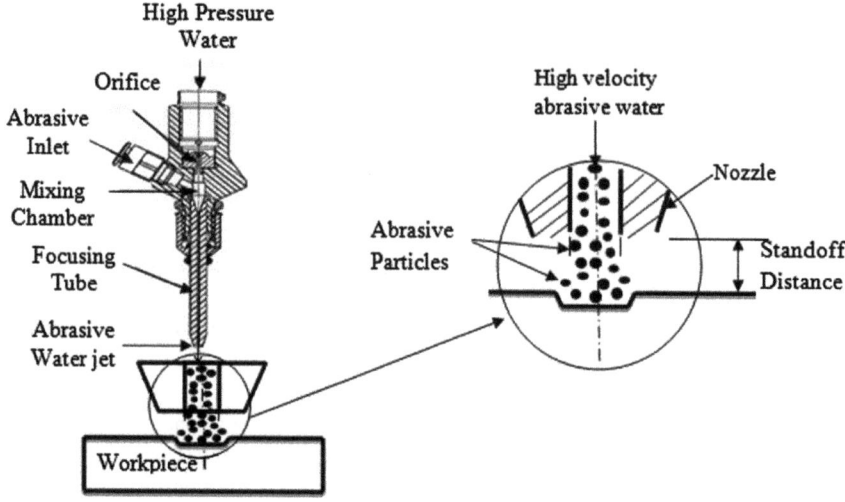

Fig. 4.4 Schematic of AWJM head assembly

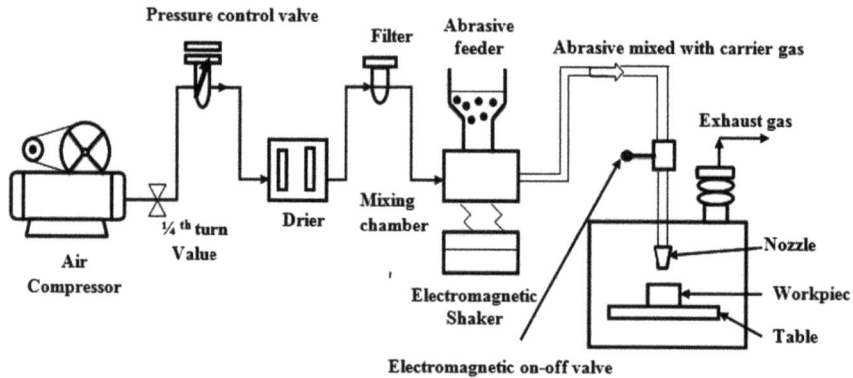

Fig. 4.5 Schematic diagram of abrasive water jet machining system

4.4.1 Abrasive Water Jet Machine

In the late 1980s, the first abrasive water jet machining system was introduced and commercialized for machining various engineering materials. The schematic diagram of a typical abrasive water jet machining setup is shown in Fig. 4.5. The machining setup consists of pressure unit, water supply system, abrasive feed unit, cutting head (nozzle) unit, operator control unit, and catcher unit [28].

Pressure unit is used for pumping the pressurized water from the reservoir to the abrasive cutting head using a radial displacement pump and air compressor. The different pressure control values are used to regulate the pressure. The range of pressure used for pumping the water is 150–450 MPa. Then, the pressurized water is delivered to the abrasive cutting head via. the water supply system. In the water supply system, various forms of values, joints, and pressure pipes are employed for controlling the pressurized water. Then, the stream of water which emerges from the nozzle is mixed with abrasive particles. The mixing of water with abrasive particles is done in the abrasive feed system. Usually, the abrasive feed system consists of an abrasive hopper, metering value, and delivery hose. The abrasive particles are stored in the abrasive hopper while control of abrasive mass flow rate and pumping of abrasive particles from the hopper to the cutting head (i.e. mixing chamber nozzle) are done using metering value and delivery nose, respectively [28].

The nozzle unit is used to increase the velocity of the mixer (water + abrasive particles) by passing the mixer through a small orifice of diameter ranging from 0.1 to 0.5 mm. The most common material used for the nozzle is sapphire, carbide or boron carbide, tungsten, diamond, etc. Basically, nozzle unit consists of three major components such as orifice mixing chamber and nozzle jet. To get the optimum cutting performance, the appropriate diameter and length of nozzle is recommended. Generally, the nozzle diameter depends on the orifice diameter and it is in the range of 2–2.5 times the orifice diameter. Similarly, appropriate size of nozzle length is also recommended because the too short nozzles create insufficient acceleration of abrasive particles, on the other hand, too long nozzles result in high wear of the nozzle and loss of momentum of the abrasive particles [5]. Further, position of cutting head is controlled by the position control system. The control system usually consists of various controlling devices such as computer design or manufacturing software with CNC controller, mechanical manipulator, etc. Another element of the abrasive water jet is the catcher. The catcher system is used as a reservoir or a point catcher depending upon the configuration of the machine and their motion. The catcher system collects the abrasive particles and debris from the cut surface, collecting the water as well as depleting the residual energy of the jet.

4.4.2 Cutting Mechanism and Parameters

The basic mechanism of abrasive water jet machining (AWJM) is based on the erosion principle, where a very high-velocity stream of water jet mixed with abrasive particles is allowed to hit a given specimen resulting in removal of material [5]. During this process kinetic energy is converted into pressure energy that results in stress on the specimen. When the stresses induced in the specimen reaches to high enough, the particles of the specimen are inevitably removed. As the composites are inhomogeneous and brittle in nature, the machining behaviour differs from the other engineering materials. This causes damage of composite material due to the force exerted by the water jet in abrasive water jet machining. The research of Hashish [29] has justified the physics of the material removal process and the striation or waviness formation mechanisms in ductile and homogenous materials. An experiment was performed on abrasive water jet machining process using a high-speed camera to record the material removal process in a Plexiglas sample. The study shows that the material removal process in abrasive water jet cutting is a cyclic penetration process which consists of two cutting zones such as "cutting wear zone" and "deformation wear zone". The graphical representation of these two zones and the abrasive particle trajectory path in these two zones is shown in Fig. 4.6.

In the cutting wear zone, material removal occurs primarily by micro-cutting with particle impact at shallow angles of attack. This process is in a steady cutting state with characteristics that the material removal rate is equal to the jet material displacement rate by traversal. While the material removal in the deformation wear zone is achieved by the impinging of the abrasive particles at large angles of attack, in this zone, the material is removed in a different erosion mode that is associated with multiple particle bombardment, surface hardening by plastic deformation and crack formation. The geometrical expression of this deformation zone is jet-induced waviness "striation" which is regularly distributed on lower portion of kerf and waviness is increased with increase in the depth of cut. Hashish [29] also suggested that the cutting process in abrasive water jet machining (AWJM) process is performed in three stages; the first stage is the entry stage, second is the cyclic cutting stage and third is the exit stage as shown in Fig. 4.7.

In the entry stage, the different cutting mechanisms develop until the maximum depth of cut (h_c) reached. At this stage, jet enters the work piece and cutting is occurred at shallow angle of attack with increase in depth of cut. As the depth of cut (h_c) increases further, the penetration occurs gradually due to erosion at larger angle of impact until its depth of cut reaches at maximum. When the depth of cut (h_c) reaches it maximum level, the penetration process is fully developed and controlled by erosion wear at larger angle of impact associated with jet upward deflection

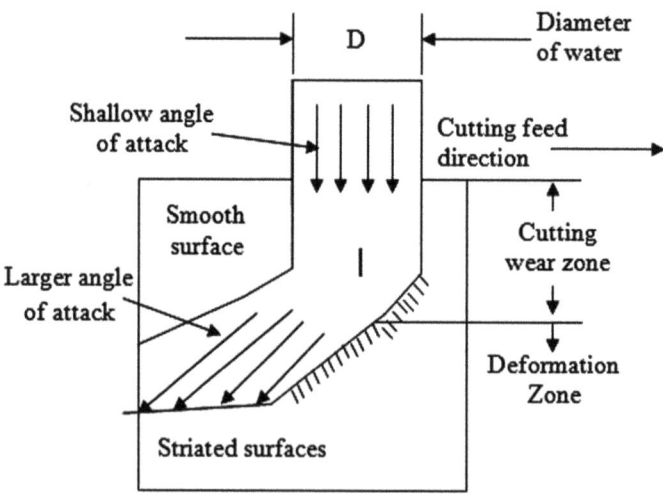

Fig. 4.6 Cutting and deformation wear zones in AWJ cutting

Fig. 4.7 Stages in the abrasive water jet (AWJ) cutting process

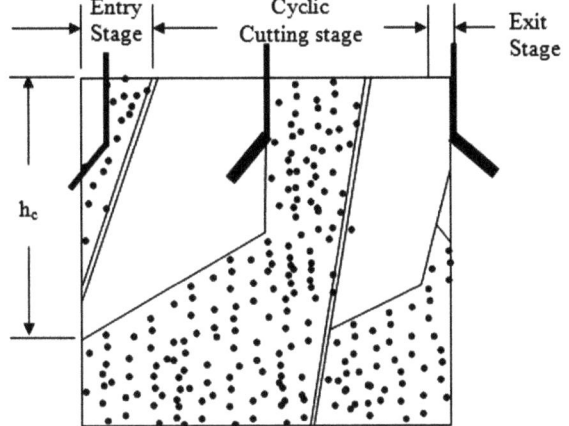

followed by the entry stage cutting. This stage is cyclic cutting process that continues until the jet reaches the end of the work material. During this stage, a steady-stage interface to a depth of h_c exists at the top portion of the kerf. Below h_c, there exists a step(s) and it will under the impact of the jet until it reaches the final depth. The kerf curvature at depth h_c changes suddenly and this makes a transition from one material removal model to another. In the exit stage, cutting process comes to an end. During this stage, an uncut triangle is observed which is associated with a jet sideward deflection. This phenomenon shows that over a certain depth (h_c) the cutting process is steady, which approximately marks the location of the top of the triangle. This important observation indicates that for effective, separation of two cut sides without triangle requires (h_c) to be greater than or equal to the materials thickness. In general, cut quality obtained by abrasive water jet cutting shows a smooth upper zone followed by a lower striated zone [30].

4.4.3 Important Process Parameters

There are numerous parameters of abrasive water jet machining (AWJM) process that can influence the performance, efficiency and effectiveness of the abrasive water jet machining [5]. Generally, these process parameters are categorized into two types: input (independent) parameters and output (dependent) parameters.

The details of some of the important parameters are explained as follows:

- **Abrasive Type**: It is the most important machining parameter in AWJM process because it directly relates to the material removal rate (MRR) and machining accuracy. The selection of abrasive type is usually determined by the hardness of the material that is being cut. The harder the material, the harder the abrasive particle should be. It also depends on the MRR and machining accuracy required. Basically, garnet, silica, silicon carbide (SiC), glass beads, crushed glass, aluminium oxide (Al_2O_3) and sodium bicarbonate are some of commonly used abrasive types in AWJM process. So most commonly used abrasives for machining include granite, aluminium oxide and then silicon carbide. In cleaning, etching and polishing operations, glass beads and dolomites are generally used. The use of the harder abrasive substantially reduces the surface roughness (R_a) but on the other hand increases the MRR which subsequently reduces the process times (PT).

- **Abrasive Grain Size**: The size of the abrasive particles or grain size also plays an important role in type of machining operations in abrasive jet machining. Coarse grains particles are recommended for cutting operations while fine grains are recommended for finishing or polishing operations. In this, an abrasive with smaller mesh number has a larger average value of particle size and fewer particles per unit. The lower surface roughness is obtained by an abrasive with larger mesh number, while the higher roughness is achieved by an abrasive with smaller mesh number.
- **Water Jet Pressure**: Water pressure refers to the pressure required to flow of abrasive with water to cut the material in AWJM. In case of water pressure, a higher water pressure increases the kinetic energy of the abrasive particles and enhances their ability for material removal. As a result the surface roughness decreases [5] and process takes the lesser time. This is because the increase of water pressure results in an increase in particle velocity and particle fragmentation inside the abrasive nozzle, which causes positive effect on surface quality. But when water pressure is higher, it will generate negative effect. High water pressure makes the abrasive particles to loose cutting ability when they become too fragmented. The higher value of water pressure enhances the material removal rate (MRR) and reduces the process times (PT) but increases the surface roughness. The water jet pressure affects the process by determining the abrasive particle velocity. Pressure can be high as 415 MPa, but more commonly it remains in between 172 and 275 MPa [12].
- **Stand-off Distance**: Further, the stand-off-distance (SoD) is one of the most important parameters in abrasive water jet machining (AWJM) process. It is defined as the distance between the tip of the nozzle and the work surface. Generally, higher stand-off distance allows the jet to expand before impingement which may increase vulnerability to external drag from the surrounding environment [5]. Therefore, increase in the stand-off distance results in an increased jet diameter as cutting is initiated and in turn, reduces the kinetic energy density of the jet at impingement. Densities of abrasive particles in the outer perimeter of the expanding jet are low due to this expansion. This decrease of density in the external rim generates more random peaks and valleys on the machined surface created by singular particles and thus results in a higher surface roughness [10]. The influence of stand-off distance decreases with the depth of cut to the deformation wear zone of material removal. It is desirable to have a lower stand-off distance which may produce a smoother surface due to increased kinetic energy which subsequently increases the processing efficiency of the cutting process.

- **Abrasive Mass Flow Rate**: Abrasive mass flow rate plays an important role in AWJM and it is directly relates to the cutting efficiency. It is defined as the rate at which the flow of abrasive particle involved in mixing and cutting process. An increase in abrasive flow rate leads to a proportional increase in the depth of cut [29]. When the abrasive flow rate is increased, the jet can cut through the material easily and as a result, the cut surface becomes smoother and higher material removal rate (MRR) is achieved.
- **Traverse Speed**: It is the speed at which nozzle head moves over the work/job surface to perform machining operation. High traverse rate allows less overlap machining action and fewer abrasive particles to impinge the surface which therefore increases the MRR and thus reduces the cutting time but on the other hand deteriorates the surface quality. Low traverse rate is desirable to produce high-quality surface finish.

4.4.4 Advantages, Limitations and Applications

4.4.4.1 Advantages of Abrasive Water Jet Machining Process

The abrasive water jet machining process offers various advantages that are listed below [31, 32]:

- Extremely versatile process,
- Minimum material waste due to cutting,
- Less heat effected zone,
- Higher flexibility and versatility,
- Require minimal force during machining,
- Less possibilities of environmental contamination,
- Process requires no cooling or lubricating oils, no toxic fumes will generate during the cutting process. Hence, this process is inherently called as an environmentally friendly process or green process.

4.4.4.2 Limitations of Abrasive Water Jet Machining Process

Despite having significant advantages, the AWJM process also possesses the following limitations [31, 32]:

- An inappropriate selection of the cutting velocity may produce surface roughness values and kerf taper angles out of normal. It may also cause the burr, which would require secondary finishing.
- The short life of some parts, like nozzle and orifice, adds replacement costs and overheads to abrasive water jet machining (AWJM) operation.

- Another disadvantage is the fact that the cutting material is placed on top of support bars. The support bars may represent a problem in the final presentation of the work pieces, due to jet deflection.
- The capital cost is high.
- High noise during operation.

4.4.4.3 Applications of Abrasive Water Jet Machining

The potential applications of AWJM are numerous. Because of the technical and economic performance of abrasive water jet, many industries could immediately benefit from this new technology. Over the last decades, abrasive water jet machining (AWJM) has been found to be widely used in various industries, including manufacturing industry, civil and construction industry, coal mining industry, food processing industry, oil and gas industry, electronic industry and cleaning industry. Abrasive water jet machining is highly used in aerospace, automotive, electronics, construction, mining and food industries [31, 32]. Some specific applications are as follows:

- In aerospace industries, parts such as titanium bodies for military aircrafts, engine components (aluminium, titanium, heat resistant alloys, and stainless steel), aluminium body parts and interior cabin parts are made using abrasive water jet cutting. Abrasive water jet machining has been particularly used in cutting "difficult-to-cut" materials.
- In automotive industries, parts like interior trim (head liners, trunk liners, door panels) and fibre glass body components and bumpers are made by this process. Similarly, in electronics industries, circuit boards and cable stripping are made by abrasive water jet cutting. These materials were cut by a variety of other technologies before the development of abrasive water jet machining including diamond saws, plasmas and lasers. However, the thermal devices may produce undesirable changes in material characteristics and, in some cases, may not be able to cut as effectively as an abrasive water jet.
- In the construction industry, in addition to cutting and scarifying for reinforced concretes, abrasive water jet machining could perform several other useful functions. It could be used to sandblast and cut corroded rebar, and it can drill holes for bolting posts. It is also often used in road and bridge repair, underground work and pile cutting.
- In the oil and gas industry, a ship can be equipped with an abrasive water jet machining (AWJM) cutting system for offshore work. Some examples are casing cutting for decommissioning of oil wells, rescue operations, platform cutting and repair and pipe cutting.
- The coal mining industry can benefit from abrasive water jet machining cutting by being able to safely cut metal structures in the potentially explosive environment underground. Coal picks can be replaced with abrasive jets for high productivity.

- In electronic industry, water jet cutting is mostly used to cut out smaller circuit boards from a large piece of stock. With water jet cutting, a very small kerf, or cutting width, can be generated, and there is no much waste of materials. Because the jet is so concentrated, it can also cut very close to the given tolerances for parts mounted on the circuit board without damaging them.
- In food industry, water jet cutting can be used for food preparation. The cutting of certain foods such as bread and trimming fat from meats can also be easily done with it.

4.5 Machinability Study of Natural Filler/Fibre-Reinforced Polymer Composite Using Abrasive Water Jet Machining Process

4.5.1 Background

In this section, an experimental study on machinability of nature filler/fibre-reinforced polymer composites using abrasive water jet machining is given. The natural filler-reinforced polymers (NFRP) composite used in this study is wood dust-based filler-reinforced polymer composite. The size of the wood filler is of 2 μm and density 0.799 gm/cc after cleaning [10, 33–35]. Due to the anisotropic and inhomogeneous properties, machining of these composites by conventional processes is extremely difficult and expensive [15].

In the experimental work discussed, the abrasive water jet machining is employed to improve the machinability of wood dust-based filler-reinforced polymer composites. The working principle of this process has been already described in Sect. 4.4. Further, a parametric analysis of the experimental data has been performed to examine the effect of process parameters on one of the important performance measures, i.e. surface roughness (mainly average roughness 'R_a'). Finally, recommendations are made on selection of the optimum combination of the process parameters for practical applications.

4.5.2 Experimental Procedure

The experiments were conducted on abrasive water jet (AWJ) machine manufactured by KMT water jet systems (Fig. 4.8) to cut the wood dust-based filler-reinforced polymer composite specimen. The specimen was prepaid by hand layup process. The reinforcement material is used as wood dust with size of 2 μm and density 0.799 gm/cc after cleaning and matrix material is used as low-temperature epoxy resin (Araldite LY 556) and corresponding hardening agent (HY 951). A size of 200 mm × 120 mm × 5 mm has been prepaid for the

Fig. 4.8 The AWJM experimental setup

Table 4.1 Machining parameters and their levels

Input parameters	Symbol	Units	Level 1	Level 2	Level 3
Abrasive material grain size	AMGS	mesh	60	80	100
Stand-off distance	SoD	mm	1.5	2.5	3.5
Working Pressure	WP	MPa	150	225	300
Abrasive mass flow rate	AMFR	g/s	3	5	7
Nozzle speed	NS	mm/min	125	175	225

experimental work. Even though the abrasive water jet machining (AWJM) includes a large number of process variables as stated by Hashish [29] and all these variables affect the cutting performance, only major and easy-to-adjustable process variables were selected in the study. The process variables and their levels are presented in Table 4.1.

During machining, a square hole of 20 mm × 20 mm was drilled for each of the experimental settings according to the Taguchi Design (L_{27}) orthogonal array. The corresponding performance characteristics, i.e. surface roughness (R_a), has been measured at the machined surface using surface texture measuring instrument (SURFCOM 130A-Monochrome) manufactured by TOKYO SEIMITSU CO.LTD in microns and average value of R_a has been taken for the analysis.

4.5.3 Results and Discussions

The parametric analysis in terms of studying the effects of abrasive material grain size (AMGS), stand-off distance (SoD), working pressure (WP), abrasive mass flow rate (AMFR) and nozzle speed (NS) on surface roughness (R_a) was done and is presented as the main effect plot shown in Figs. 4.9, 4.10, 4.11, 4.12 and 4.13.

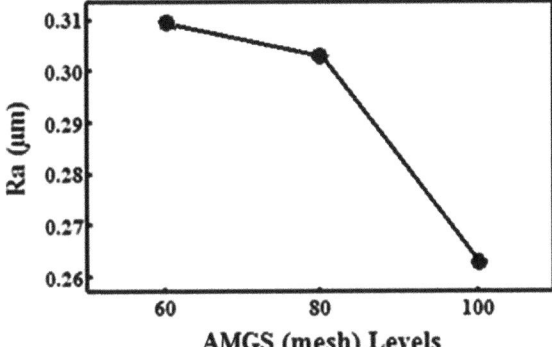

Fig. 4.9 Main effect plot for AMGS

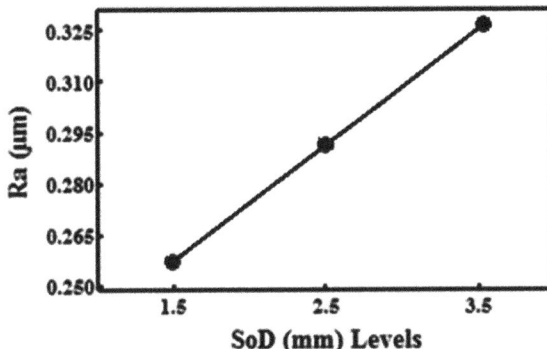

Fig. 4.10 Main effect plot for SoD

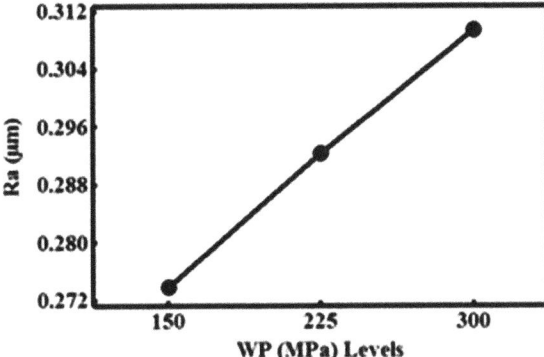

Fig. 4.11 Main effect plot for WP

In case of abrasive material grain size, an abrasive with smaller mesh size has a larger average value of particle size and fewer particles per unit weight. It can be seen from the Fig. 4.9 that lower surface roughness is obtained by an abrasive with a higher mesh size (i.e. at 100 mesh) while the higher roughness are achieved by an abrasive with smaller mesh size (i.e. 60 mesh) (see Fig. 4.9).

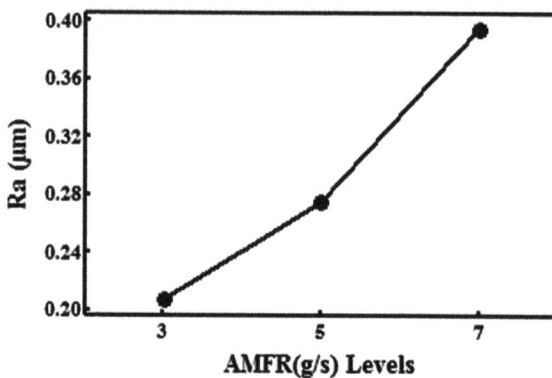

Fig. 4.12 Main effect plot for AMFR

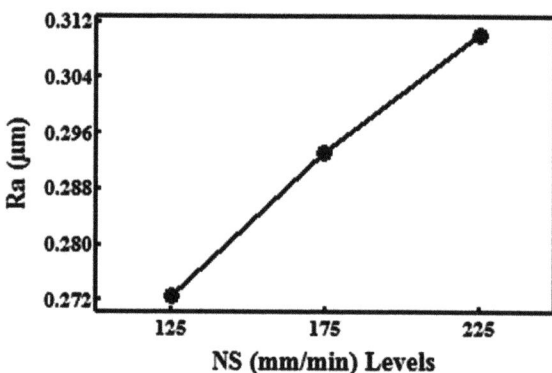

Fig. 4.13 Main effect plot for NS

In case of stand-off distance (SoD), the surface roughness (Ra) rate is significantly higher when the stand-off distances increased from 1.5 to 3.5 mm (Fig. 4.10). This is because higher stand-off distance allows the jet to expand before impingement which may increase susceptibility to external drag from the surrounding environment [5, 36]. Therefore, increase in the stand-off distance results in an increased jet diameter as cutting is initiated and in turn, reduces the kinetic energy density of the jet at impingement results in rougher surface. It is desirable to have a lower stand-off distance which may produce a smoother surface (Fig. 4.10).

In case of higher water pressure (WP), the surface roughness is found to be maximum. This is due to the fact that an increase in water pressure increases the kinetic energy of the individual particles inside the jet and enhances their capability for the material removal. However, higher water pressure may also result in number of particle collisions on work piece due to the acceleration; rough surfaces can be obtained [36]. When the working pressure increased from 150 to 300 MPa, the surface roughness (Ra) rate is gradually increased during the machining which is shown in Fig. 4.11.

Abrasive mass flow rate (AMFR) has direct effect on the surface roughness (R_a). The general explanation of higher surface roughness (R_a) is that the higher the abrasive mass flow rate, the higher the number of particles involved in the mixing and cutting processes results in proportional increase in the cut depth and obtains higher surface roughness, i.e. when abrasive mass flow rates increased from 3 to 7 g/s the surface roughness (R_a) rate is significantly increased which is shown in Fig. 4.12. In case of nozzle speed, as the water jet nozzle speed moves faster, less number of particles is available which pass through a unit area. Thus, less number of impacts and cutting edges will be available per unit area that results in rougher surfaces (as depicted in Fig. 4.13) [5, 36].

Furthermore, the parameters abrasive mesh size (AMGS) and stand-off distance (SoD) are found to be the major influential parameter while working pressure (WP), abrasive mass flow rate (AMFR) and nozzle speed (NS) are found to be minor influential parameters.

The optimal combination of process parameters (for high surface finish) recommended via parametric analysis is abrasive material grain size (100 mesh, Level 3), stand-off distance (1.5 mm, Level 1), working pressure (WP) (150 MPa, Level 1), abrasive mass flow rate (3 g/s, Level 1) and nozzle speed (125 mm/min, Level 1).

4.6 Summary

This chapter provides an insight into the abrasive eater jet machining (AWJM) of composites. An experimental investigation on abrasive water jet cutting of wood dust-based filler-reinforced polymer composite composites is discussed along with the fundamental discussion on composites, their machining and aspects of AWJM. It is summarized that AWJM process is a viable and effective alternate to the conventional cutting processes for machining of wood dust-based filler-reinforced polymer composite with good quality and high productivity. The parametric analysis of the experimental data reveals that abrasive material grain size (AMGS) and stand-off distance (SoD) are found to be the major influential parameter while working pressure (WP), abrasive mass flow rate (AMFR) and nozzle speed (NS) are found to be minor influential parameters for surface roughness (R_a). The optimal combinations of process parameters obtained are abrasive material grain size (100 mesh), stand-off distance (1.5 mm), working pressure (300 MPa), abrasive mass flow rate (7 g/s) and nozzle speed (225 mm/min) for producing the best surface finish while machining wood dust-based filler-reinforced polymer composite by AWJM. The machinability of wood dust-based filler-reinforced polymer composites in abrasive water jet machining process is proven satisfactory regarding the surface roughness (R_a) and the process is worth further development for industrial applications.

The possible avenues of future research include the investigating the effects on abrasive water jet machining on other composites such as fibre-reinforced polymer composites, and natural filler-reinforced polymer composites, etc.

References

1. Chawla KK (2013) Chapter-1 Introduction, Chawla KK (Ed.) Composite materials science and engineering. Springer, New York
2. Gachter R, Muller H (1990) Plastics additives, 3rd edn. Hanser Publishers, New York
3. La-Mantia FP, Morreale M (2011) Green composites: a brief review. Compos Part A 42:579–588
4. Surppa MK (2003) Aluminum matrix composites: challenges and opportunities. Sadhana 28:319–334
5. Momber AW, Kovacevic R (1998) Principle of abrasive waterjet machining. Spinger, London
6. Bradford JD, Richardson DB (1980) Production engineering technology, 3rd edn. Macmillan, London, pp 74–93
7. Hashish M (1991) Advances in composite machining with abrasive-waterjets. Process Manuf Comp Mater 49(27):93–111
8. Bisaria H, Shandilya P (2015) Machining of Metal Matrix Composites by EDM and its variants: a review. Daaam International, Vienna, pp 267–282
9. Tagliaferri V, DiIlio A, Crivelli VI (1985) Laser cutting of fiber reinforced polyesters. Compos 16(4):317–325
10. Jagadish, Bhowmik S, Ray A (2015) Prediction of surface roughness quality of green abrasive water jet machining: a soft computing approach. J Intell Manuf. doi:10.1007/s10845-015-1169-7
11. Hocheng H, Tsai HY, Chang KR (2013) Advanced analysis of nontraditional machining. Springer, New York
12. Schwartz MM (1992) Composite materials handbook, 2nd edn. Mc Graw Hill Inc., USA, pp 34–35
13. Teti R (2002) Machining of composite materials. CIRP Ann Manuf Technol 51(2):611–634
14. Arifur Rahman M et al (2015) Chapter-2, Introduction to manufacturing of natural fibre-reinforced polymer composites. In: Salit MS, Jawaid M, Yusoff NB, Hoque ME (eds) Manufacturing of natural fibre reinforced polymer composites. Springer, Switzerland
15. Abrate S (1997) Machining of composite materials. In: Mallick PK (ed) Composites engineering handbook. Marcel Dekker, New York, pp 777–809
16. Abrao AM, Faria PE, Rubio JCC, Reis P, Davim JP (2007) Drilling of fiber reinforced plastics: a review. J Mater Process Technol 186(1–3):1–7
17. Monaghan JM, Reilly PO (1992) The drilling of an Al/SiC metal matrix composites. J Mater Process Technol 33:469–480
18. Ramulu M (1988) EDM sinker cutting of ceramic particulate composites. Adv Ceram Mater 3(4):324–327
19. Gordon S, Hillery MT (2003) A reviews of the cutting of composite materials. Proc Inst Mech Eng J Mater Des Appl, 217(1):35–45
20. Miller JA (1987) Drilling graphite/epoxy at Lockheed. Am Mach Autom Manufact 131(10):70–71
21. Koenig W, Wulf CH, Grass P, Willerscheid H (1985) Machining of fibre reinforced plastics. Manuf Technol CIRP Ann 34:537–548
22. Konig W, Rummenholler S (1993) Technological and industrial safety aspects in milling FRP. ASME Mach Adv Comp 45(66):1–14
23. Sivapirakasam SP, Mathew J, Surianarayanan M (2011) Multi-attribute decision making for green electrical discharge machining. Expert Syst Appl 38:8370–8374
24. Negarestani R, Li L (2012) Fibre laser cutting of carbon fibre–reinforced polymeric composites. J Engg Manuf 227(12):1755–1766
25. Emmrich M, Levsen K, Trasser FJ (1992) The behavior of fiber-reinforced plastics during laser cutting. Zentralbl Hyg Umweltmed 193(1):67–77
26. Caprino G, Tagliaferri V (1988) Maximum cutting speed in laser cutting of fiber reinforced plastics. Int J Mach Tools Manulact 28(4):381–398

27. Ibraheem HMA (2013) Numerical optimization for cutting process in glass fiber reinforced plastic using conventional and non - conventional methods. Dissertation, Institute of Graduate Studies and Research Eastern Mediterranean University, Gazimagusa, North Cyprus
28. Jain VK (2008) Advanced (non-traditional) machining processes. Springer, London, pp 300–327
29. Hashish M (1984) A modeling study of metal cutting with abrasive water jets. J Eng Mater Techno 106:88–100
30. Hashish M (1991) Characteristics of surfaces machined with abrasive water jets. J Eng Mater Techno 113:354–362
31. Dixit A, Dave V, Baid MR (2015) Water jet machining: an advance manufacturing process. Int J Eng Res Gen Sci 3(2):288–292
32. Patel D, Thakkar J, Bhatt T, Patel MC (2014) Review on current investigation and enlargement of abrasive water jet machining. Int J Technol Res Eng 3(3):2347–4718
33. Markarian J (2002) Additive developments aid growth in wood-plastic composites. Plast Addit Compou 4:18–21
34. Pritchard G (2004) Two technologies merge: wood-plastic composites. Plast Addit Compou 6:18–21
35. Jagadish Bhowmik S, Ray A (2015) Prediction and optimization of process parameters of green composites in AWJM process using response surface methodology. Int J Adv Manuf Technol. doi:10.1007/s00170-015-8281-x
36. Arola D, Ramulu M (1993) A study of kerf characteristics in abrasive waterjet machining of graphite/epoxy composite. ASME J Eng Mater Technol 45(66):125–151

Part II
Advanced Repair and Joining

Chapter 5
Advanced Joining and Welding Techniques: An Overview

Kush Mehta

Abstract Joining and welding is an essential component of manufacturing technology. New developments in joining and welding are evolved in order to acquire extraordinary benefits such as unique joint properties, synergistic mix of materials, cost reduction of component, increase productivity and quality, complex geometrical configurations, suitability and selection of material to manufacture new products. This chapter provides an update on recent developments of welding and joining to showcase above benefits. Theoretical background, process parameters, novel aspects, process capabilities, and process variants along with its application are presented in this chapter. Advanced welding and joining techniques are addressed under different headings of fastening and bonding processes, developments of arc welding processes, advanced beam welding techniques, sustainable welding processes, micro-nano joining and hybrid welding.

Keywords Arc · Bonding · Fastening · Heat · Joining · Productivity · Quality · Welding

5.1 Introduction

Joining and welding technologies play a crucial role in the area of manufacturing. There are many conventional joining and welding technologies available and that can be, and are applied successfully, the scope of their modification and development always exists and is tremendously increasing day by day with the accelerated commercial requirements. Modifications and development in conventional welding and joining techniques and innovations in this field are being made to attain the tangible benefits such as favorable joint properties, synergistic mix of materials, significant reduction in the component cost, tremendous enhancement in produc-

K. Mehta (✉)
Mechanical Engineering Department, School of Technology (SOT), Pandit Deendayal Petroleum University (PDPU), Raisan, Gandhinagar 382007, Gujarat, India
e-mail: kush_2312@yahoo.com; kush.mehta@sot.pdpu.ac.in

© Springer International Publishing AG 2017
K. Gupta (ed.), *Advanced Manufacturing Technologies*, Materials Forming, Machining and Tribology, DOI 10.1007/978-3-319-56099-1_5

tivity, excellent quality, attaining complex geometrical configuration, and ability to deal with variety of materials to manufacture new products. Innovative approaches of joining and welding technology have provided a new window through which the variety of materials such as reactive metals, composites, plastics, non-metallic materials and dissimilar materials can be joined. These advancements of welding and joining techniques are discussed in the present chapter.

The main objective of this chapter is to provide an introductory knowledge of recent developments in the area of welding and joining processes along with its basic concepts and special features of different techniques, assuming the basic understanding of the readers with regard to conventional welding and joining.

Over the past few years, sustainable welding processes, hybrid mechanical fastening, micro-nano joining, adhesive joining, activated arc welding processes, pulsed arc welding processes, narrow groove arc welding and hybrid welding and joining processes have been the prime topics of research in this field. Figure 5.1 presents the classification (based on recent advancements) of advanced welding and joining processes that can be classified in five subsections such as advanced fastening and bonding, advanced fusion welding, micro and nano-joining, advanced non fusion welding and hybrid welding processes.

5.2 Advanced Fastening and Bonding Processes

There are so many conventional fastening and bonding processes available which are capable to join variety of materials. The main objective behind the developments in conventional fastening and bonding processes is to successfully join special materials with excellent joint properties. Recently developed advanced fastening and bonding processes such as hybrid bonded fastened joining, clinching and electrostatic bonding are discussed here.

5.2.1 Hybrid Bonded Fastened Joints

Hybrid bonded fastened joints are produced by simultaneous actions of adhesive bonding and mechanical fastening. There are two types of hybrid bonded fastened joints, such as, the joints that use bolts/rivets are known as hybrid bonded–bolted joints (see Fig. 5.2a), and the joints that use of pins with adhesive are known as hybrid bonded pinned joints (see Fig. 5.2b). Individual approaches of mechanical fastening and adhesive bonding are successfully applied to metallic materials and plastics. However, an individual approach of mechanical fastening and adhesive joining is difficult for joining of dissimilar materials, composite materials and brittle metallic materials [1, 2]. Considering the aforementioned limitations, hybrid approach of bonded fastened joint is developed in order to avoid individual limitations of fastening and adhesive joining.

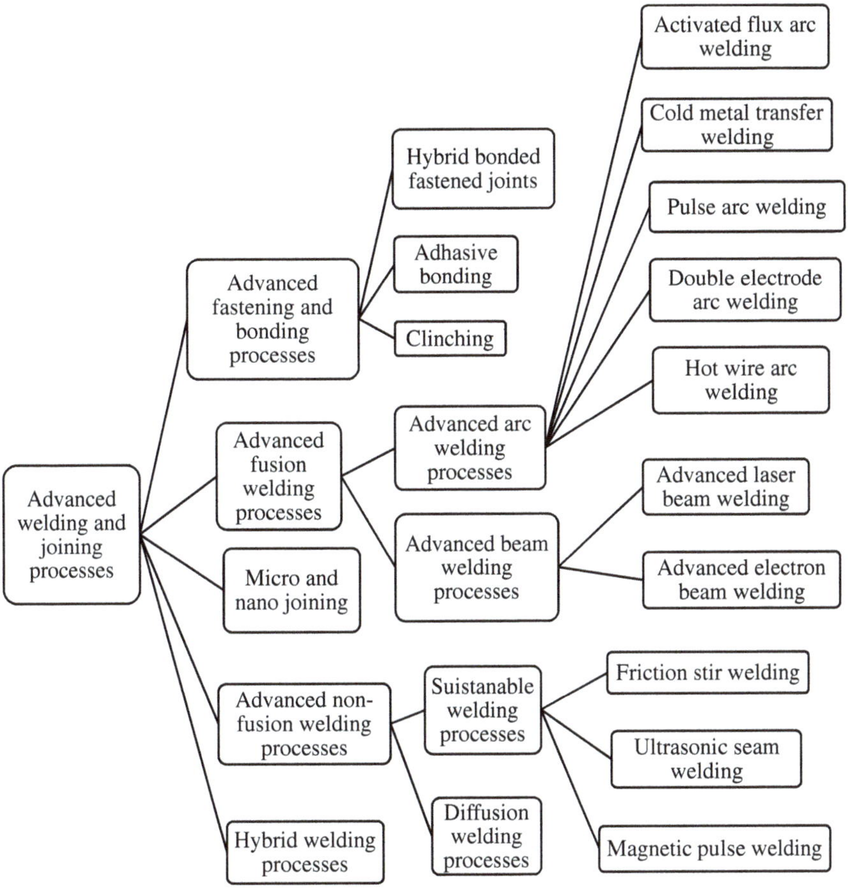

5 Advanced Joining and Welding Techniques: An Overview

Fig. 5.1 Classification of advanced welding and joining processes

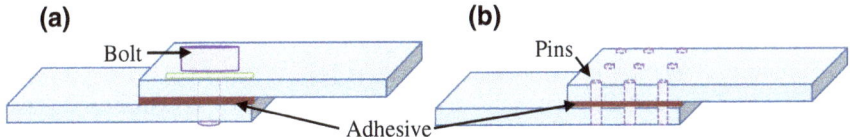

Fig. 5.2 Hybrid bonded fastened joints **a** hybrid bonded bolted joint and **b** hybrid bonded pinned joint

Selection of adhesives, strength of adhesive in service, ability to sustain fatigue load are common difficulties that exist, when joining of composite–composite, composite-metal and dissimilar materials systems. In case of mechanical fastening, load distribution capacity and mechanical strength are issues associated with joining of composite-composite, composite-metal, and brittle metallic materials. In order to

Table 5.1 Process parameters of hybrid bonded fastened joints [1, 2]

Parameters of workpiece materials	Geometrical parameters	Different types of joint configuration	Different types of adhesives	Different types of fasteners[a]
Type of workpiece material	Adhered thickness	Single-lap and double lap joints	Hysol Shell 951 Epoxy	Sphere type pin fastener
Thickness	Overlap length	Scarf joint	Pliogrip 7400/7410 Polyurethane	Wedge type pin fastener
Strength of material	Width	Stepped lap joint	3 M 2216 epoxy	Protruding type fastener head
Young's modulus of material	Bond-line thickness	T-joint	Montagefix-PU Polyurethane	Countersunk type fastener head
Similar or Dissimilar system of materials	Fastener-hole clearance	Flush and triangular fillet joints		Rounded shaped head type fastener
		Butt joint		Polygonal shaped head type fastener

[a]Fasteners: pin mean diameter less than 1 mm without top head, bolt and rivet are denoted as fasteners

address these issues, hybrid approach of bonded fastened joining is developed. Fastening system can help to sustain axial loads while its distribution can be handled by adhesives. Most important parameters such as type of joint configuration, type of fastener, adhesive material and its thickness, loading condition, type of material being used and fastener-hole clearance of hybrid bonded fastened joints are required to take into the considerations [1]. Table 5.1 shows process parameters reported under the literature of hybrid bonded fastened joints. Combinations of these parameters lead to the formation of metallurgically sound joints.

Applications

- Joining of materials such as polymers, aluminum composites, steel-composite, aluminum–aluminum, titanium-composite, and magnesium-composite materials, in order to obtain light weight structure for aerospace applications [1, 2].
- Joining of automobile components, especially body work assembly made out of aluminum and its alloys and composites [1].
- Joining of fiberglass in the nautical industries [1].
- Joining of non-metallic materials such as plastics, ceramics, polymers and glass, etc. [1].

5.2.2 Adhesive Injection Fastening

Adhesive injection fastening is advanced hybrid bonded fastened joining technique, invented at the welding institute (TWI) for joining of plastics [3]. Later, this technique is applied to materials such as aluminum, composites, and aerospace

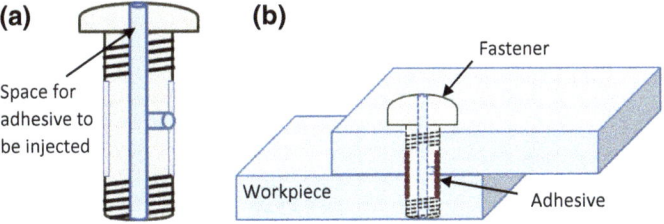

Fig. 5.3 Adhesive injection fastening, **a** internal mechanism of fastener and **b** Adhesive injection fastening joint

metals. Process principle of adhesive injection fastening is illustrated in Fig. 5.3. A novel mechanism of adhesive injection through fastener is introduced at the interface between fastener and workpiece in this technique (Fig. 5.3a). Use of this technique provides rapidity of process, excellent sealing ability, and fatigue resistance of adhesive. The use of adhesive can be saved with this simplified hybrid bonded fastened joint technique [3]. Stiffening and strengthening can be further improved by this technique as the adhesive is exactly located at the interface between fastener and workpiece as shown in Fig. 5.3b.

5.2.3 Clinching

Clinching is known as press-joining technique, aimed to join thin sheets using specially designed tools or without application of the tool, through interlock formed by plastic deformation of base materials (refer Fig. 5.4 for process description). Clinching is alternative joining method of riveting, screwing, adhesive joining, and spot welding. The clinching process is found advantageous compared to other joining methods in terms of flexibility, repeatability, cleanness, processing time, cost, energy consumption, eco-friendliness, temperature, mechanical and thermal stresses generated, and fatigue characteristics of joints [4].

The most common types of clinch are cylindrical, beam, star, flat and self-piercing elements. The round, flat, square and star type of clinching joints do not require tooling as a filler material while self-piercing needs an external consumable tool as shown in Fig. 5.4. Clinching can be applied to most of the metals, plastics, polymers, wood, composites, and dissimilar materials. Super hard and brittle materials are difficult to join by clinching process [4]. Process parameters of clinching are presented in Table 5.2.

Recent developments of clinching
In order to improve clinching process, further developments such as heat assisted clinching [5] and clinching with adhesives [6] have been reported. Heat assisted clinching helps to increase the material ductility of workpiece that subsequently extend the clinching joint ability. Heat assisted clinching also leads benefits in terms

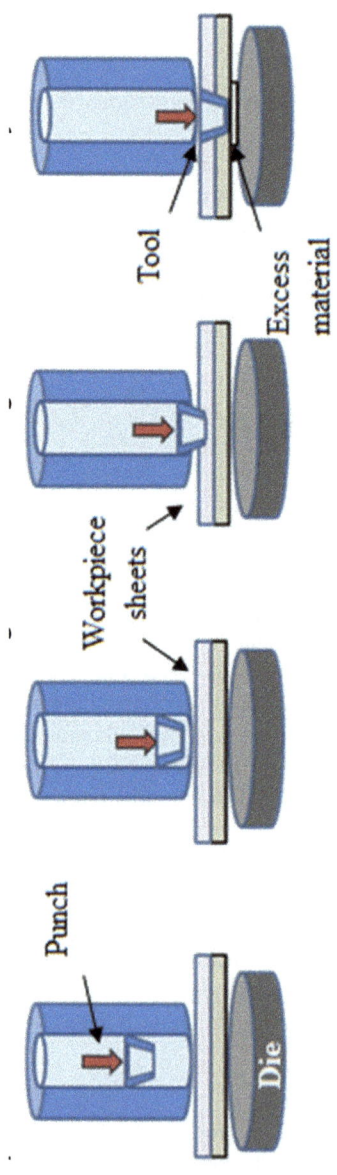

Fig. 5.4 Basic process description of clinching, **a**, **b** offsetting, **c** upsetting, and **d** flow pressing

Table 5.2 Process parameters of clinching [4, 6]

Parameters related to workpiece materials	Parameters related to die and punch	Working process parameters
Material of workpiece	Shape and dimensions of the punch and die	Punch force
Thickness	Tool material and its geometry	Length of interlock
Similar or dissimilar system of materials	Die groove dimensions	Loading direction
Thinning of workpiece sheet	Punch corner radius	
Mechanical properties of workpiece material	Draft radius	

of required punch force as the yield stress of the material is reduced when heat is applied externally. However, the amount of heat needed for preheating is a critical process parameter of this technique. Clinch with adhesives is called "clinchbonding" wherein, suitable adhesive is applied between workpiece sheets. Clinchbonding can improve fatigue and tensile properties of joints [6].

Applications of clinching

- Joining of metal to metal for the application of automobile chassis (automotive, trucks, buses, railways) [6].
- Joining of aerospace materials such as composites, aluminum, titanium, composite-metal dissimilar joints [4, 6].
- Manufacturing applications of ropeways and different nautical equipments [4, 6].
- Joining of railways and automobile bodies [6].
- Joining of dissimilar materials [6].

5.3 Advanced Arc Welding Processes

Arc welding falls under the classification of fusion joining techniques used to join a variety of materials through melting and solidification of the base material by an arc with or without application of filler material. Developments of arc welding processes such as activated flux arc welding, cold metal transfer arc welding, narrow gap arc welding, pulse arc welding, tabular wire arc welding and double electrode arc welding are discussed in the following subsections.

5.3.1 Activated Flux Arc Welding Processes

Activated flux arc welding processes are advanced methods of joining in which the activating layer of flux containing of metallic oxides, fluorides and chlorides are applied on the weld surface of a workpiece as shown in Fig. 5.5a through liquid paste or spray. Liquid paste of activated flux is prepared with the help of solvent like acetone/ethanol. Activated arc welding increases the weld bead penetration and reduces bead width of weld as shown in Fig. 5.5c, while conventional arc welding provides weld bead with low penetration and broad width as shown in Fig. 5.5b. Concept of activated flux arc welding process is observed for processes such as activated tungsten inert gas welding (A-TIG), activated flux gas metal arc welding (A-GMAW), activated flux plasma arc welding (A-Plasma welding) and activated flux laser welding [7–10].

5.3.1.1 Activated Flux Tungsten Inert Arc Welding

The first study on activated flux arc welding is applied on A-TIG of titanium and its alloys by Paton Electric welding institute. A-TIG is also known as flux assisted-TIG welding process. The maximum literatures for activated flux arc welding are found for A-TIG process hitherto [7]. A-TIG can significantly improve the penetration level of weld up to 8–12 mm which is impossible to obtain through conventional TIG technique. Improvement of penetration is achieved by the effect of *"reverse*

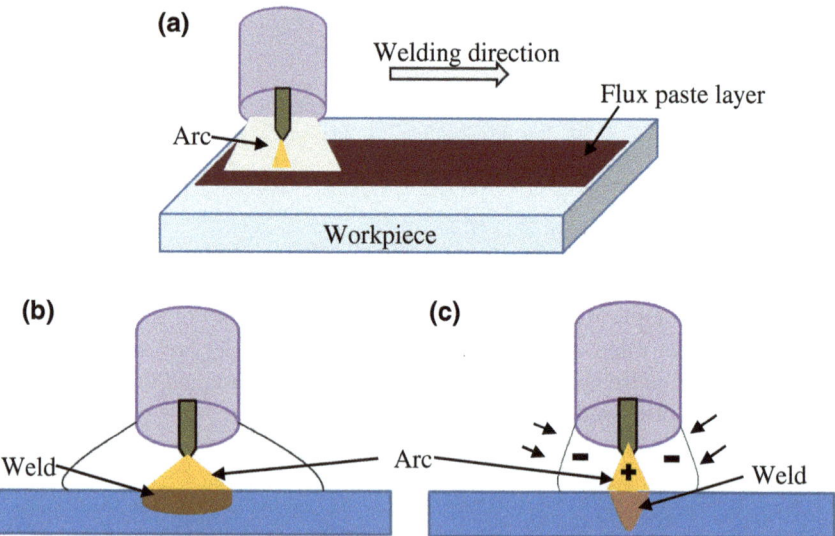

Fig. 5.5 a Process principle of activating flux arc welding, b weld characteristics of conventional welding, and c weld characteristics of activated flux welding

Marangoni convection" and *"arc constriction mechanisms."* Marangoni convection is known as surface-tension driven convection or thermo-capillary convection, which is having a considerable effect on the penetration level of the weld. The reverse Marangoni effect comes into the effect when activating flux is pasted on the workpiece, in which the direction of surface tension becomes inward towards the center, which consequently affects the fluid flow of the weld pool. This phenomenon ultimately increases the penetration level and reduces the width of the weld. In case of arc constriction mechanism, the arc generated on the surface of the flux pasted workpiece is reduced its width as shown in Fig. 5.5c. This arc constriction provides focused heat source by an arc with its higher power and density that subsequently results in higher penetration. Phenomenon of electronegativity of flux is found responsible for arc constriction in the literatures [7].

A-TIG for repair application

A-TIG can also be applied for the repair work of any mechanical component. A-TIG is suitable for repair work due to its advantages such as no need to prepare V-groove, filler wire is not required, weld shrinkage and distortion is reduced, and single pass of welding up to 10–12 mm is achievable by its autogenous mode. Development of A-TIG for nuclear power plant is reported in the article published by TWI team [11].

Process Parameters affecting A-TIG performance

The process parameters that affect A-TIG performance are summarized in Table 5.3. Type of flux, forces acting on process, input parameters and workpiece material are the major factors which affects A-TIG performances. Selection of fluxes depends on the type of workpiece to be welded. Compatibility of different fluxes with chemical compositions of workpiece is most important parameter.

Table 5.3 Summary of process parameters of A-TIG [7]

Major parameters	Sub-parameters	Types
Flux	Mixtures of acetone or ethanol with metal oxides, fluorides or chlorides	CuO, HgO, ZnO, MgO_2, TiO_2, Cr_2O_3, SiO_2, ZrO_2, TiO_2, $MoO3$, Al_2O_3, Fe_2O_3, Co_3O_4, MoS_2, NiF_2, CaF_2, AlF_3
Forces acting in A-TIG	Marangoni force, electromagnetic or Lorenz force, bouncy force, aerodynamic drag force	Surface tension, density caused forces, magnetic effect caused forces, current caused forces
Process input parameters of welding	Welding current, welding speed, arc length, electrode geometry, shielding gas composition	Conical shaped electrode tip, frustum and wedge shaped electrode
		Argon, Helium, mixture of argon and helium, mixture of argon and nitrogen
Properties of workpiece material to be welded	Thickness of the workpiece, chemical composition of workpiece material	Ferrous and non-ferrous workpiece materials, surface active and reactive elements of workpiece such as sulfur, oxygen, tellurium, selenium, manganese, aluminum, silicon

Fluxes reported in literatures for different workpiece materials are presented in Table 5.3. In addition to this, Marangoni forces, electromagnetic force, buoyancy force and aerodynamic drag force are acting on A-TIG process. These forces are affected by the properties of workpiece, type of flux used, welding current, and magnetic effect. Welding current and welding speed affects heat input mainly while arc length, electrode geometry, shielding gas composition governs operational performances, which ultimately affects weld bead shape of the joint [7].

It is reported that, the A-TIG technique has been successfully applied on different metallic materials such as carbon steel, stainless steel, aluminum, magnesium, copper, zirconium, nickel-based alloys and low activation ferritic/martensitic (LAFM) steels [7].

5.3.1.2 Other Activated Arc Welding Processes

A-GMAW technique is carried out for AISI 1020 carbon steel with three oxide fluxes such as Fe_2O_3, SiO_2, and $MgCO_3$. The flux of $MgCO_3$ has performed excellent in terms of obtaining optimum weld bead profile, improved tensile strength, and hardness [10]. A-plasma arc welding is analyzed with flux elements of Ti, Cr, and Fe. Concentrated arc with narrow temperature field can improve weld properties if A-plasma arc welding is applied [8]. A-laser welding can improve penetration level significantly at low power. A unique mixture of flux such as 50% ZrO_2, 12.09% $CaCO_3$, 10.43% CaO, and 27.48% MgO has provided 2.23% higher penetration for ferritic stainless steel than the normal laser welding [12]. The conduction mode of laser welding is required to achieve improved penetration [9]. However, A-GMAW, A-plasma arc welding and A-laser welding are not much applied for research and practical applications so far.

5.3.2 Cold Metal Transfer Arc Welding

Cold metal transfer (CMT) arc welding is developed for metal inert gas welding (MIG)/metal active gas welding (MAG) by Fronius company. CMT is fully mechanized automatic welding process in which arcing and wire feeding are well controlled during operation. In this advance process, the metal transfer occurs in cold condition. Since this process is a cold arc process, it leads with advantages such as low heat input, spatter free metal transfer and extremely stable arc [13].

The process principle of CMT is illustrated in Fig. 5.6. CMT is a cold welding process in which the controlled short circuit type arc is generated between workpiece and filler type electrode of MIG/MAG welding. Initially, the filler wire is melted and then after it is transferred to the workpiece through controlled short circuit mode. The filler wire moves forward during the welding (see Fig. 5.6a, b) and that is pulled back again as soon as the short circuit occurs (see Fig. 5.6c, d). This way the metal transfer to the workpiece happens in cold condition. In this

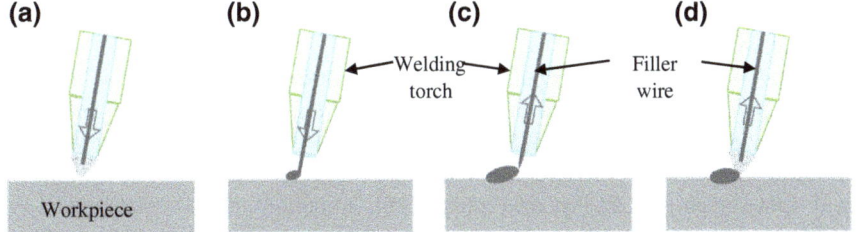

Fig. 5.6 Process principle of CMT, **a** arc generation with forward wire motion, **b** metal transfer with short circuit mode and forward wire motion, **c** metal transferred and reverse wire motion, and **d** wire is back to its position

process, metal transfer occurs drop by drop as presented in Fig. 5.6b–d. Re-ignition of arc again happens and this cycle is repeated [13, 14]. Process parameters such as forward and reverse motion of electrode and electrical characteristics are major factors that affect performance of CMT technique. Low voltage and low current are recommended parameters in order to obtain drop by drop metal transfer phenomena of CMT [14].

CMT pulse
The CMT pulse technique combines a CMT cycle and pulsed cycle that leads to the more heat input. The performance and flexibility of process can be significantly improved by combining pulse cycle with CMT cycle [14, 15].

Advanced CMT
Advanced CMT is even cooler relative to CMT. In this case, process control is carried out through polarity of the welding current. At the phase of short circuit, the polarity reversal is done that in turn provides results of extremely controlled heat input and intensely gap bridge-ability that allows up to 60% higher deposition rate [14, 15].

Applications of CMT

- CMT can weld ultra- thin sheets (0.3 mm) of metallic materials [15].
- Welding of dissimilar materials such as aluminum–steel, aluminum–magnesium, and aluminum–titanium, etc. [14–17].
- A well-controlled and precise cladding can be done by CMT [14, 16, 18].
- Repair of mechanical components (for example: cracks generated in the steam turbine case are efficiently repaired by CMT process [18]).

5.3.3 Pulse Arc Welding

Pulse gas tungsten arc welding and pulse gas metal arc welding (P-GMAW) process are the most commonly used important types of pulse arc welding techniques reported in literatures for different metallic materials under the classification of pulse arc welding.

P-GMAW operates on the pulsed current power source which is developed to improve process stability, weld penetration, deposition rate of filler wire and weld deposition properties [19]. Weld deposition can be improved by two steps of advanced solidification: one at the time of pulse off period and second at the time of weld spot developed during next pulse [20]. Important process parameters of P-GMAW are pulse duration, pulse frequency and three different currents such as mean current, peak current and base current that subsequently affects the velocity of molten droplet [21]. Additionally, process parameters such as welding speed, filler wire size, rate of filler wire feed and shielding gas affects performance of P-GMAW [20]. Weld defects of bad weld surface, lack of fusion, undercut, burn-backs, and stubbing-in may have caused due to improper selection of these process parameters [21].

P-GMAW provides spray transfer with overall reduced heat input. This process is able to result in excellent weld bead appearance due to tiny molten droplet caused through spray transfer mode. Formation of spatter is eliminated by P-GMAW. Heat affected zone is reduced to a great extent through this process [20, 21].

P-GMAW is applied to join different materials such as aluminum and steels and alloy steels, etc. [19, 20].

5.3.4 Double Electrode Arc Welding

Double electrode arc welding process is a novel welding technique in which consumable or non-consumable type electrodes are brought into the action in order to bypass the wire current. This bypass current of the second electrode reduces the heat input on main set-up while increases the deposition rate of first consumable electrode [22].

The process principle of double electrode arc welding concept is demonstrated in Fig. 5.7. It shows two different electrodes such as consumable type electrode of gas metal arc welding (GMAW) and non-consumable type electrode of plasma torch. The non-consumable electrode of plasma torch is added as bypass torch while GMAW electrode is main electrode. Both of these bypass torch and GMAW torch are operated with same power source as presented in Fig. 5.7. Because of bypass arc, the heat input by current of base material is reduced with an increase in deposition rate of consumable wire. Additionally, deposition efficiency is also increased for the reduced heat input at the same or increased or controllable deposition rate [22]. There are different technologies reported in the literatures for double electrode arc welding, such as gas tungsten arc welding (GTAW)-GMAW process, plasma welding-GMAW process and submerged arc welding-GMAW

5 Advanced Joining and Welding Techniques: An Overview

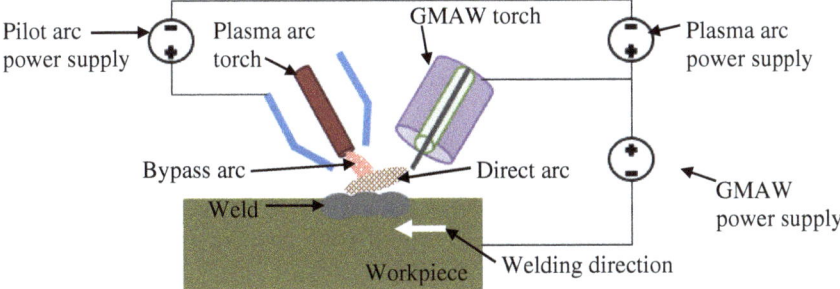

Fig. 5.7 Concept of bypass current of plasma arc welding and gas metal arc welding [22], with kind permission from Elsevier

process [22]. The aforementioned double electrode arc welding techniques are discussed in the subsequent sections.

5.3.4.1 Gas Tungsten Arc Welding

GTAW torch is provided for the purpose of bypass current through the non-consumable electrode for GTAW–GMAW double electrode arc welding. Two torches are added to the GMAW system in order to provide two bypass loops for the melting current. The main current flows through the base material while bypass current flows through two GTAW bypass loops. This process is able to reduce the heat input without affecting the deposition rate of filler that subsequently leads to the reduction of heat affected zone and distortion of the workpiece without any effect on productivity [22]. However, it is difficult to make configuration of dual bypass GMAW with GTAW system due to its setup complexity.

5.3.4.2 Pulse Gas Metal Arc Welding

Pulse GMAW process is developed at Lanzhou University of Technology in which two pulsed type power is supplied to control base and bypass current at different required levels with two GMAW processes. The main and bypass currents are supplied to two different consumable electrodes of GMAW. This process results in stabilized arc even at low current with required low heat input in order to obtain dissimilar aluminum-galvanized steel joints. The spray transfer mode can be able to achieve at low heat with the help of pulsed bypass current that consequently result in quality joints [22, 23]. The deposition rate of this system can be claimed higher than GTAW-GMAW double electrode welding due to the consumable electrodes. However, further investigations are being carried out to prove the conceptual benefits.

5.3.4.3 Plasma-GMAW and Plasma-GTAW

The plasma arc welding torch is used as a bypass arc with its pilot arc advantage while GTAW/GMAW is used as a main arc system which is developed at the University of Kentucky. Pilot arc can avoid the delay time for establishment of the main arc to the bypass arc, which is required in another double electrode arc welding. Hence, burned through effect of the workpiece can be eliminated by this method. Significant reduction in heat input and arc pressure is reported without reducing melting speed in plasma-GMAW process [22].

5.3.4.4 Submerged Arc Welding-GMAW

Submerged arc welding-GMAW double electrode arc welding is developed by Adaptive Intelligent Systems LLC for the shipbuilding manufacturing. Submerged arc welding is used as main wire and GMAW as bypass filler wire in order to obtain benefits such as use of extra high current at high speed with higher deposition rate at low heat input. Submerged arc welding usually causes high heat input that results in heat affected zone and workpiece distortion, which can be remarkable improved by the application of this double electrode system. Weld bead geometry and heat input can be controlled easily with submerge arc welding-GMAW double electrode arc welding process [22].

5.3.5 Hot Wire Arc Welding

Hot wire gas tungsten arc welding is the advanced joining technique in which filler wire is preheated initially, and then added to the arc, in order to melt faster than the conventional method [24]. Separate resistive heating source is provided to preheat the filler wire. The heating current of this system is settled ideally in such a way that the filler wire reaches to its melting point as soon as it comes to the weld pool. The deposition rate of filler wire is improved through this method, which increases the productivity. Independent wire feeding allows flexibility of process control and its feed [24]. Hot wire GTAW can be applied to most of the metallic materials. Hot wire arcing GTAW is further development of hot wire GTAW process in which the filler is added in such a way that it creates a side arcing as shown in Fig. 5.8a. In this process, the filler wire is completely melted before it goes to the weld pool while in hot wire GTAW, filler wire melts at the weld pool. Side arc helps to increase the deposition rate significantly even than the hot wire GTAW as the wire is heated not only from wire heating power source. For higher conductive filler wires such as copper and aluminum, the excellent melting efficiency is reported

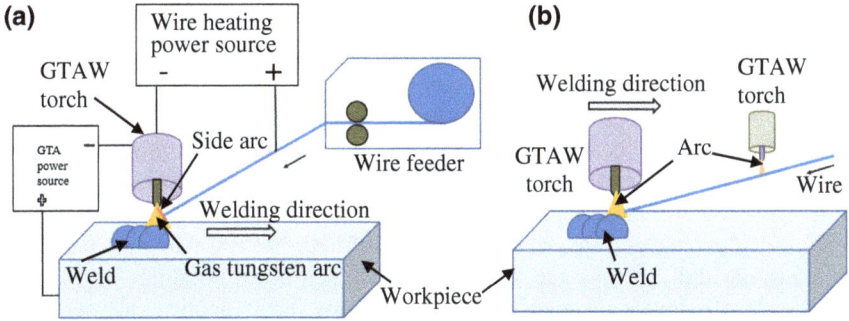

Fig. 5.8 a Hot wire GTAW with arcing and **b** arc assisted hot wire GTAW [22], with kind permission from Elsevier

through hot wire GTAW with arcing. A study conducted at the Beijing University of Technology has found that, there are three types of metal transfer such as free transfer, touching transfer and bridging transfer involved with hot wire GTAW with arcing. In free transfer, the metal separates from the filler wire before it makes contact with the weld pool. In touching transfer, the melted metal droplet initiates to form with a gap to the weld pool surface and travels into the weld pool after it makes contact to the surface periodically. In bridging transfer, the wire goes to the weld pool, even if it is solid [22]. However, a proper combination of melting current and wire position is required to obtain these modes.

Arc assisted hot wire GTAW is developed by researchers of Harbin Institute of Technology in order to obtain higher deposition rate than the hot wire GTAW. Figure 5.8b represents the arc assisted hot wire GTAW technique, in which pre-heating of wire is done through secondary gas tungsten arc. Separate parameters are able to control the temperature of wire that in turn leads to the enhancement of the deposition rate [22].

Other welding processes such as submerged arc, gas metal arc, plasma arc, laser and electron beam are also utilized for hot wire, but applications of the same are found to be limited [19, 24].

5.4 Advanced Beam Welding Processes

Beam welding processes are type of fusion welding techniques in which workpiece is being melted with the help of intense, focused heat provided through a beam of photons or electrons or plasma. Laser beam welding, electron beam welding and plasma beam welding are examples of beam welding processes. Advances of laser beam welding and electron beam welding are discussed here.

5.4.1 Laser Beam Welding

The laser is an extraordinary candidate, and widely applied in many fields of engineering for different applications. Laser beam welding is considered as one of the most versatile process in the area of manufacturing. Laser beam welding is developed for its different advantages such as precise process, excellent depth to width ratio, low weld heat input, small heat affected zone, rapid cooling, applied for complicated geometries. Recent developments of laser beam welding are presented in terms of ultra narrow gap laser welding, laser beam welding for plastics, non-metals, dissimilar materials, laser welding for repair applications and laser powder welding, under the subsections.

5.4.1.1 Ultra Narrow Gap Laser Welding

The ultra narrow gap is a type of advanced narrow gap welding used to join thick sections in a more economic way. In this welding procedure, the joint preparations are being prepared with one half of width of those in the conventional narrow gap between two workpieces which is to be joined with small included angles [25]. It is reported that, <2 mm of narrow width between workpieces to be joined can be considered as ultra narrow gap configuration [26]. This configuration requires less weld metal and less welding time to fill the cavity, which in turn increases productivity with material cost saving. The laser beam is most capable process to go into such small width groove as the laser beam diameter is very small. Additionally, laser beam welding is having a high power density and high precision control of process that perfectly suits for ultra narrow gap welding of thick sections. Ultra narrow gap laser beam welding offers low heat inputs, high welding speeds, lower levels of residual stresses and distortion, while consuming less filler material and power [27]. Defects such as lack of sidewall fusion, hot cracking, and porosity can be eliminated from the multiple-pass narrow gap laser beam welding of thick-section [26]. Process parameters such as joint design configuration, material transfer mode, current, laser power, laser-wire distance, welding speed, defocus of laser and laser beam diameter affects the output of the process [28].

5.4.1.2 Laser Beam Welding for Plastics

Plastic is a mixture of polymers that is either having homogeneous structure or heterogeneous structure. Properties of plastics such as reflection, absorptivity, melting point, thermal conductivity, and coefficient of thermal expansion affect its weldability [1]. Laser beam welding process is applied successfully to obtain sound joints of plastics. Since the evolution of diode lasers, plastics are actively investigated for its laser beam welding applications. Different types of lasers such as YAG, fiber and diode can be utilized under its low power defocused conditions in order to weld transparent materials of plastics, as the absorption and transparency can be controlled through concentration of substances such as carbon black [19, 29]. Suitable absorption additive has to be chosen for transparent plastic material that again depends on different conditions such as laser wavelength, demands on the color and transparency and economic aspects of the application. It is reported that, the transparent thermoplastics can be welded without the use of additional absorber in which diode lasers with wavelengths of 1500 nm or 2000 nm are recommended. However, such laser sources can be expensive [29].

5.4.1.3 Laser Beam Welding for Dissimilar Materials

Dissimilar materials are difficult to weld due to differences in its thermal, physical, chemical, metallurgical, and mechanical properties. Laser beam welding is one of the most feasible process for dissimilar joints because of precise control of the heat input, focused beam, flexible process, low distortion, quick heat dissipation, limited formation of brittle intermetallic compounds, and so on [19, 30–32]. Novel parameters of laser beam welding of dissimilar materials such as laser beam offset, dual beam laser, surface modification of workpiece, compatible filler wire, special edge preparation, wavelength and laser modulation are reported as the latest developments. Laser beam offset means displacement of laser beam towards particular base material according to its favorable properties, which can help to distribute the heat to both dissimilar base materials equally. Similarly, dual beam can provide equal distribution of heat. Filler material to laser welding of dissimilar materials can provide compatibility for excellent bonding with third suitable materials. Special joint preparation such as scarf joint configuration, close butt joint, and lap joint also provide favorable conditions for sound bonding. Control of laser modulation and wavelength provides suitable heat input that subsequently gives good conditions for dissimilar joints [31]. Dissimilar combinations like aluminum-steel, aluminum–copper, aluminum–magnesium, copper–steel, steel–Kovar, steel–nickel, dissimilar metal grades, and composite-metal joint can be successfully obtained through laser beam welding [31, 32].

5.4.1.4 Laser Welding for Repair Applications

Laser beam welding is suitable for different repair applications such as cladding of different material, repairing of damaged components, mold repair, repair through laser peening, under water laser welding and so on. Molds for plastic products and die casting of magnesium and aluminum alloys are subjected to strong thermo-mechanical loads that may lead to damage the mold surface in the form of fatigue cracks or wear. These defects can be repaired by laser welding using the ND-YAG lasers due to its characteristics like flexible method, less change of metal composition around repairing zone, accurate deposition of small volume of filler material, small heat affected zone, no distortion and able to operate at small thickness [33]. Extremely difficult material of nitride coated steel and chrome plated steel can be repaired by laser welding using heat pre-treating, re-melting and low alloyed filler wire [34]. In order to avoid replacement of steam circuit components of thermal power station, laser powder welding is employed through which coating can be applied with less processing time [35]. The details about laser powder welding are discussed in next Sect. 5.4.1.5.

5.4.1.5 Repair Using Laser Powder Welding

Laser powder welding is a variant of laser beam welding in which powder is fed to the shielded laser as shown in Fig. 5.9. The powder is being melted with the use of laser onto a substrate. This results in solid deposit as molten powder gets solidified [35]. Major advantages of this process are: (I) Liquation cracking is eliminated as the process draws low heat input and (II) Laser spot of very small size can produce highly accurate deposit onto a substrate. Besides these, the process is suitable for original part manufacturing. The functionally graded component can also be developed using varying powder composition. Dissimilar materials can be welded using laser powder welding (as for example, aluminum–steel [31]).

Another interesting development is hot wire laser welding, in which hot wire is fed to the laser instead of powder. Control of wire transfer and the initial temperature of the wire are two most important parameters of hot wire laser welding. Hot wire laser welding is employed in repairing purpose due to flexibility and low heat input technique [36].

5.4.2 Electron Beam Welding

Electron beam welding is the technique in which stream of electron is impinged on workpiece that transmits heat and causes joining due to melting and solidification.

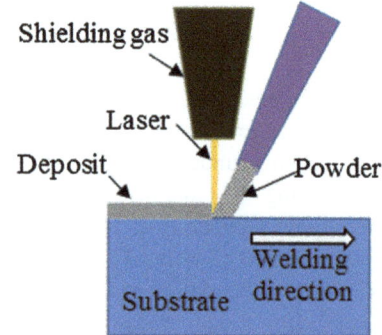

Fig. 5.9 Laser powder welding

This process requires a vacuum chamber to handle stream of the electron beam. Electron beam welding is capable to operate on ultra-thin thickness to a higher thickness for different materials such as plastics, aluminum, magnesium, steel, tungsten, nickel, molybdenum, zirconium, beryllium, dissimilar materials, composites, etc. [19]. The electron beam is most focused heat source among all welding processes, even compare to the laser beam welding, which is the biggest advantage of this process [37]. The electron beam welding is developed for repair applications, surface modification, non-vacuum process, high speed welding, and improved control system welding.

5.4.2.1 Electron Beam Welding for Repair Applications

Electron beam welding is adopted for repair applications of mechanical components due to its characteristics such as low distortion, high density heat input, and ability to weld dissimilar materials. Additionally, repairing thickness of deposition and fusion penetration can obtain, according to the requirements of the component due to good controllability of the process. Repair of large components such as bearing boxes, the rotor of a gas turbine, compressor blades and engines chamber can also be possible by electron beam welding [37].

Similar to laser powder welding and laser hot wire welding (as discussed in Sect. 5.4.1.4), external application of powder and filler wire is applied for the purpose of repair through deposition layer for different mechanical components such as roller bearing seats shafts, turbine, blades, die blocks and splines. Electron beam welding is having special advantage of minimizing dilution between base material and filler that in turn results in reduction of the thickness of deposition [19, 37].

5.4.2.2 Surface Modification Through Electron Beam Welding

Surface modification is conducted using a deflection system of single gun or lead multi-beam process which allows modifications of large surfaces. For the single gun set-up, contour path needs to be followed while for multi-beam type set-up having multi beams operates at the same time on the whole surface. Electron beam diameter, movement of beam and power density sophisticatedly controlled for the purpose of surface modification that subsequently affects the heating rate, surface microstructure, and modification depth [19, 37].

Surface modification through deposit layer can be achieved through the addition of filler material or powder (as mentioned in Sect. 5.4.2.1). Surface hardening and alloying are some other applications of surface electron beam welding [37].

5.4.2.3 Non-vacuum Electron Beam Welding

Non-vacuum electron beam welding is a technique in which beam is generated in vacuum and impinged to the workpiece which is kept in the atmosphere. Beam generator is excited with 175 kV voltage and transmitted to the workpiece by multi stage orifice assembly and special nozzle system. The pressure of 10^{-2} up to 1 mbar is kept for the pressure chamber which is connected to the beam generator chamber having vacuum 10^{-4} mbar that is evacuated and separated by pressure nozzles [19, 37]. This is high speed electron beam welding process in which productivity is increased significantly. However, one has to refer and apply the radiation protection guidelines before using this technique [37].

5.5 Sustainable Welding Processes

Sustainable welding processes are techniques in which workpiece materials are joined at minimum energy consumption, minimum material wastage, with minimum resources, at highest efficiency with maximum cost saving, highest quality with maximum environmental benefits (represented in Fig. 5.10). Solid state welding processes such as friction stir welding, magnetic pulse welding and ultrasonic welding fall under the categories of sustainable welding processes.

Fig. 5.10 Sustainable welding process aspects

5.5.1 Friction Stir Welding

Friction stir welding (FSW) is a type of solid state welding process, invented at The Welding Institute (TWI) initially for aluminum and its alloys. Developments of FSW for different materials such as plastics, composites, magnesium, titanium, nickel-based alloys, steel alloys and dissimilar materials are carried out on later stage because of its solid state nature. FSW is also known as "green process" due to many advantages such as no fumes generated, shielding gas is not required, and melting of material is not involved. Furthermore, energy benefits in terms of less power requirement (for example, 2.5% of less power required than laser welding), reduction in weight through enhancement of weldability of light weight material and saving of consumable materials are the other advantages [38, 39].

The FSW process principle is represented in Fig. 5.11. The non-consumable rotating tool made-up from hardened material, consist of pin and shoulder inserted into the workpiece, in such a way that the pin is totally inserted into the workpiece and shoulder generates friction through rubbing with the top surfaces of the workpiece. Frictional heat causes plastic deformation and stirring leads material flow, which consequently produces joint after transverse movement of the tool without the addition of external material [38–40]. Process parameters of FSW are summarized in Table 5.4.

Friction stir welding is further developed for its different variants such as friction stir spot welding, stationary shoulder FSW, friction bit joining, filling FSW, friction stir extrusion, under water FSW and friction crush welding. Some of these advances are discussed here as under.

5.5.1.1 Friction Stir Spot Welding

Friction stir spot welding (FSSW) is applied as an alternative of resistance spot welding in which tool is being kept in same position instead of its transverse speed at suitable rotational speed, like in FSW. FSSW is also one of the sustainable welding process through which energy consumption can be reduced up to nearly 99% compare to conventional spot welding (claimed by Mazda reported in [38]), while cost of installation can be saved up to approx. 40% relative to resistance spot welding [38, 39]. FSSW is capable to join highly conductive materials, non-metallic materials such as plastics and polymers as well as dissimilar materials. This can be operated with

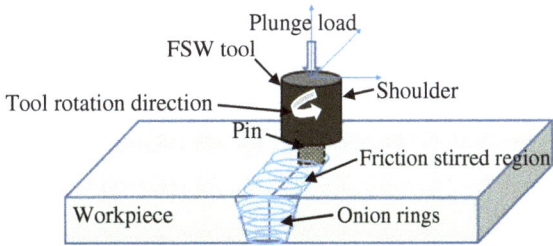

Fig. 5.11 Process principle of friction stir welding

Table 5.4 The important process parameters of FSW [38–44]

Tool design and material	Process parameters	FSW joint configurations	Workpiece material and thickness
Pin diameter, shoulder diameter, shoulder to pin diameter ratio	Rotational speed	Butt joint	Dissimilar materials
Pin shape and features (cylindrical, polygonal shapes, cone shaped, whorl pin, trifluet pin, trivex pin, thread-less pin, threaded pin,)	Welding speed	Lap joint	Metallic materials
Shoulder features (Concave, flat, convex, spiral, scoops, concentric circles)	Forces acting during process (axial force, transverse force, translational force)	T-joint	Refractory materials
Tool material (tool steel, tungsten carbide, nickel-cobalt alloys, tungsten-cobalt alloys, poly crystalline boron nitride)	Tilt angle	Multiple lap joint	Plastics and polymers
Tool developments (bobbin tool, dual rotation tool, skew stir tool, com stir tool, re-stir tool, retractable tool)	Pin offset (parameter of dissimilar joint)	Pipe joint	Reactive materials
FSW tool for repair application of casting defects	Workpiece material positioning (parameter of dissimilar joint)	Tube to tube-sheet metal joint	Ultra-thin thickness to higher section thickness (0.5 mm to 500 mm)

highly controllable robots in order to improve process efficiency. Recent developments of FSSW are refill FSSW, stitch FSSW, swing FSSW, and rotating anvil FSSW [45]. These advanced FSW processes are known for their capability to assist in avoiding key-hole formation at the end of the process.

5.5.1.2 Friction Bit Joining

Friction bit joining can be considered as advanced rivet joining technique in which consumable bit is inserted to the workpiece for the purpose of joining. The process is similar to the FSSW, but the tool is designed in such a way that the pin of tool acts as a bit. Friction between the bit and workpiece leads heat generation and subsequently bonding. Friction bit joining can also be applied for key-hole removal in FSW and FSSW [38, 39].

5.5.1.3 Friction Stir Extrusion

Friction stir extrusion is a new development for the dissimilar materials in which pin less tool is applied on the softer material to extrude material to the groove as shown in Fig. 5.12.

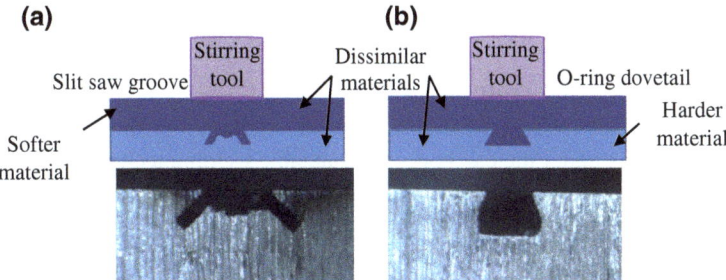

Fig. 5.12 Friction stir extrusion groove patterns: **a** slit-saw groove and **b** O-ring dovetail [46], with kind permission from Elsevier

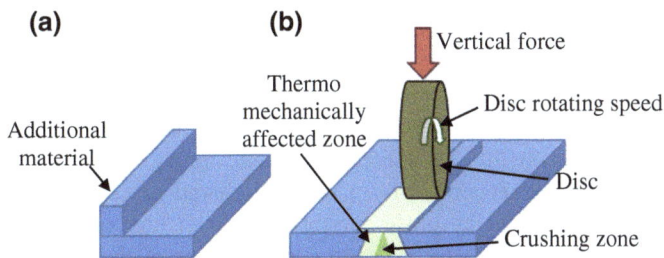

Fig. 5.13 Friction crush welding **a** workpiece preparation and **b** process principle [47]

Groove is required to be produced well in the advance on the harder material. This process is developed for the dissimilar combination of aluminum–steel and two different grooves such as slit-saw groove and O-ring dovetail. The quality joint is reported for slit-saw groove along with high strength and low amount of intermetallic compounds that to in a continuous layer form [46].

5.5.1.4 Friction Crush Welding

Friction crush welding is developed on the base of friction stir welding in which frictional heat and pressure is applied through disc as shown in Fig. 5.13b. Sheet metal edges are required to be prepared with extra material as shown in Fig. 5.13a. This extra material is crushed by means of rotating non-consumable disc with suitable pressure. Joining is established by plastic deformation of a material after the disc is traveled into a transverse direction. Concave shape is generally given to the disc surface in order to obtain defect free joint. Similar and dissimilar materials can be joined with this technique. Thickness of the workpiece is limited to thin sheets. Height of the additional material, disc design and its weight, vertical applied force, rotation and welding speeds of the tool are the important process parameters of friction crush welding [47].

5.5.2 Magnetic Pulse Welding

Magnetic pulse welding (MPW) is an environmental friendly process that uses the cleanest form of electromagnetic driving forces in order to join similar or dissimilar materials. MPW is under the category of sustainable welding technique, because of its characteristics such as no heat affected zone, no smoke, no radiation, high precision, no need of filler material, no distortion, no residual stresses and better repeatability. The mechanism for welding is similar to that of the explosive welding. The parameters such as collision angle and velocity are varying in the case of MPW while both of these parameters are kept constant for explosive welding. MPW improves quality of product and productivity through advantages such as solid state welding, less time operation, higher welding strength, process flexibility of combining the low cost and light weight dissimilar materials. Formation of brittle and hard intermetallic compounds is reduced to a great extent for dissimilar joints [48, 49].

MPW is a solid state welding process, in which electromagnetic forces are being applied between the coils and workpiece materials that are not in the contact with each other as shown in Fig. 5.14. Electromagnetic forces are applied with very short pulses, which are generated by a rapid discharge of capacitors. The high magnetic field is produced by pulse current along with high amplitude and frequency that subsequently creates an eddy current at one of the workpieces. This cause impact on another workpiece by high magnetic pressure created through repulsive Lorentz forces. Impact with sufficient collision velocity produces plastic deformation and results into the solid state bonding. Process parameters such as current, impact velocity, coil design, positions of tube, the gap between the coil and outer tube, impact angle and standoff distance are needed to be taken into the consideration to obtain successful weld. It is noted that, subsonic collision, sufficient impact velocity and pressure are the most important factors. Subsonic collision is required to achieve jetting condition. The high pressure regime is obtained

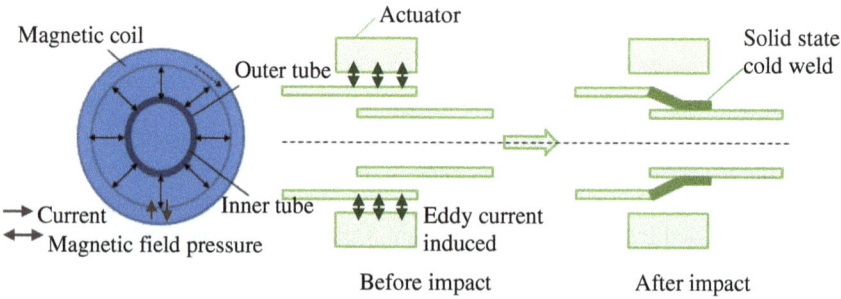

Fig. 5.14 Process principle of magnetic pulse welding [48], with kind permission from Springer

by optimum impact velocity. Optimum pressure allows proper plastic deformation of material and does not allow the melting and re-solidification of the material [48–50].

Applications

- MPW is most suitable process for dissimilar materials that includes dissimilar alloys of metals, dissimilar metals and metals to non-metals [50].
- To manufacture components of sheet metal products, drive shafts, and lightweight tubular structures [50]
- Applicable for all conductive materials [48–50].
- Sealing of tube with end plugs for metallic materials [50].
- Joining of materials applied to automobile and aerospace in order to reduce weight and improve strength [48].
- Applications in the nuclear industry like joining of refractory materials such as Niobium, Titanium, Molybdenum, and Zirconium [48].

5.5.3 Ultrasonic Welding

Ultrasonic welding is a solid state welding process in which frictional heat is being supplied by ultrasonic vibrations on workpieces to be joined. Bonding is established by shearing action and plastic deformation caused by ultrasonic vibrations without melting of material. Ultrasonic welding is having many advantages such as low energy requirement, fast automatic process, no filler wire needed, no shielding gas required, no oxide removal required, can able to join dissimilar materials and non-metallic materials [19, 51]. Therefore, ultrasonic welding is a sort of sustainable welding process. The basic set-up of ultrasonic welding is shown in Fig. 5.15. The major equipments of ultrasonic welding are transducer, booster, and sonotrode that generate ultrasonic vibrations. These vibrations are then transferred to the workpieces. Process parameters such as frequency, amplitude, pressure, and area are most important working conditions of ultrasonic welding [51]. Developments of ultrasonic welding are discussed in terms of ultrasonic seam welding and ultrasonic torsional welding along with its applications for plastics, polymeric composites, dissimilar materials and welding of workpieces having thickness variations [19].

5.5.3.1 Ultrasonic Seam Welding

Ultrasonic seam welding is also known as ultrasonic roll welding used to weld workpieces continuously with the help of the circular disk sonotrode and a transducer as shown in Fig. 5.16. The rotating disc is moved transitionally against the workpiece to produce vibrations. Long length sheets are easily joined with this

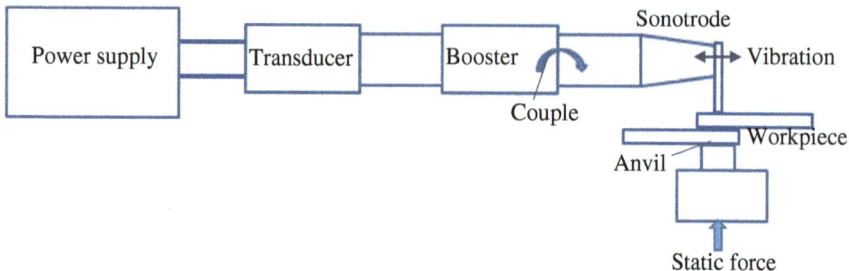

Fig. 5.15 Ultrasonic welding set-up of lateral drive [51]

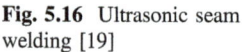

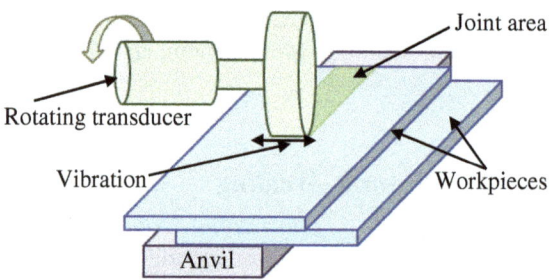

Fig. 5.16 Ultrasonic seam welding [19]

technique. Additionally, this process finds extensive applications in joining of dissimilar materials such as foils of copper and aluminum.

Ultrasonic seam welding is a flexible process in which circumferential weld of round parts can be processed using same sonotrodes on one welding device. Moreover, thermoplastic fibers of textile industry can also be welded with ultrasonic seam welding.

5.5.3.2 Ultrasonic Torsion Welding

Ultrasonic torsional weld is also called as ring weld in which torsional motion is given to the specially designed horns and tooling in order to produce circular vibrations from sonotrode and welding tip. This process is applied at components where circular weld pattern are required and it is most suitable for the spot welding configurations. Therefore, this technique can be applied to the spot welding of automobile body components. Torsional movement of sonotrode and tooling is created using an opposite rotation of transducers as shown in Fig. 5.17. Number of transducers required depends on the need of power. Use of four transducers produces up to 10 kW power at 20 kHz frequency [19].

Fig. 5.17 Ultrasonic torsional welding [19]

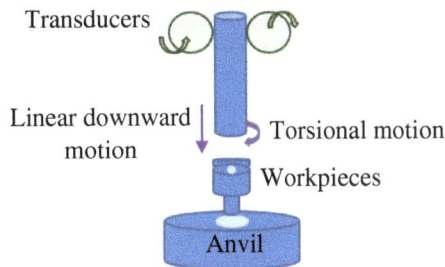

Applications

Plastics and Polymeric composites
Ultrasonic welding operated at high frequency–low amplitude vibrations leads to interfacial heating and fusion welding of plastic and polymeric components at suitable axial pressure. Thermoplastics such as woven, non-woven fabrics, and coated materials can be ultrasonically welded by different novel techniques such as plunge welding, sequential welding, continuous welding, and scan ultrasonic welding. Important factors of ultrasonic welding of plastics are viscoelastic dissipation of energy, intermolecular diffusion, heating at the interface between two workpieces and ability to transmit vibrations to this interface. Ultrasonic welding is successfully applied to thin sections while thick parts of thermoplastics are difficult to weld by ultrasonic welding [52, 53].

Dissimilar materials
The solid state nature of ultrasonic welding provides a range of capabilities for dissimilar combination joints. Ultrasonic welding is an excellent method for joining of dissimilar materials having high thermal and electrical conductivities such as copper and aluminum [54]. It is feasible for all dissimilar metallic combinations. Additionally, dissimilar joints with non-metals such as ceramic–metal, glass–metal, and plastic–metal can be achieved in some cases using metallized coating or transition layer [19, 51].

Workpiece thickness variations
Ultrasonic welding is able to join thin workpieces such as 0.025–0.250 mm. This allows joining of multiple foils with thick substrate in a single weld. Joining of multiple wires with substrate is obtained through ultrasonic welding (for example, weld of multiple copper wires to aluminum connector) [19, 51].

5.6 Micro-Nano Joining

Micro-joining and nano-joining are adopted to manufacture miniature components, devices and systems such as micro-gears, micro-pumps, micro-turbines, micro-motors, battery packs and cells, micro-sensors, micro-transducers, nano-electronics, nano-structures, nurostimulators, endoscope, and micro-electro mechanical systems (MEMS), etc. [55]. There is no exact definition of micro-nano joining techniques. But, joining of micro-components are usually having dimensions less than 100 μm and similarly for the joining of nano-components are having dimensions of the order of 100 nm. Micro and nano level systems are very complicated for joining and welding due to irregularities in microstructures, chemical compositions, and surface layers within the same component or system. It is very important to remove surface oxides and contaminants from the surface which is subjected for welding. Additionally, fixturing and handling of micro-nano component, joint quality inspection, testing methods, process repeatability, and process capabilities are some of the challenges in micro-nano joining [55, 56].

Different approaches of micro-nano joining are discussed for fusion micro-welding, solid state micro-welding, wafer nano-joining and nano-structured joining as below.

5.6.1 Fusion Micro-Welding

Beam welding processes are well-controlled processes that are applied for different micro-welding applications. Practical applications from laser welding are possible through small beam diameters of different types of lasers such as laser diodes (0.8–1.1 μm), solid state Nd:YAG laser (1.06 μm), fiber laser (1.04–1.5 μm), thin disk laser (1.03 μm) and CO_2 laser (10.6 μm). Furthermore, other process parameters of laser welding, such as wavelength, travel speed, power density, beam velocity and time are required to be selected in such a way that optimum energy is transferred for the purpose of micro-welding. A selection of these parameters depends on the material of workpiece to be welded [for example: (1) Relatively large diameter of beam is required for polymer than the glass due to linear and non-linear absorption phenomena), (2) Power and wavelength requirements for materials like silver and copper are different such as 900 W at 1.06 μm and 300 W at 1.06 μm respectively, due to conductivity differences)]. Rapid travel speed and low power are recommended for micro-welding applications. Laser welding is applied for joining of lithium ion battery of 0.2 mm aluminum alloy, micro-spot welding of hard-disk suspension of computer, welding of watch gear to shaft, thin foil welding, micro-wire welding and micro-thickness sheets. Furthermore, novel laser micro-joining processes such as laser droplet welding, laser spike welding, shadow welding, ultra-short pulse laser welding and single mode fiber laser welding are used to extend special micro applications [55].

Electron beam welding is also a capable process for the micro-joining application in which beam spot diameter is kept from 0.1 to 1 mm in order to reach 10^6 to 10^7 W/cm^2 power density. Set-up of scanning electron microscope (SEM) is found promising tool for micro-welding application. This set-up is able to do positioning, joining and analysis of joint in a single device. Electron beam welding is used to weld MEMS, micro-systems, micro-components with materials of plastics, aluminum, magnesium, silicon and copper at an accelerating voltage of 10 to 40 keV,

10 μm beam current and 1–3 μm beam diameter. Electron beam welding can be utilized for multi beam, filler addition, and superposition for micro-joining [55].

Soldering and brazing are famous techniques adopted for micro-joining of wires in electronics industries. The main advantage of these processes is that the melting of base material is not present. Also, micro-component welding wherein dissimilar materials are involved can be effectively joined at reduced formation of intermetallic compounds [55, 56].

Gas tungsten arc welding and plasma arc welding techniques are found in some applications of micro-joining especially for the micro-spot welding. But, these techniques are not recommended as controlling the joining parameters for micro applications is difficult [55].

5.6.2 Solid State Micro Bonding and Welding

Solid state processes such as ultrasonic welding, FSW, diffusion welding, forge welding, and cold welding can be employed for the micro-joining. Solid state bonding processes are associated with large plastic deformation of materials. In solid state welding, heating and forging or only forging are applied to the micro-system. Micro-forge welding and micro-cold welding includes three bonding mechanisms such as contaminant displacement/interatomic bonding, decomposition of the interfacial structure and dissociation of retained oxides. In order to reduce oxidation, most of the time the solid state diffusion process is carried out in vacuum (10^{-3}–10^{-5} mbar) or in the presence of an inert gas for micro-component application [55, 56].

Modifications in the set-up of solid state welding processes are mandatory due to many problems such as pressure requirement, heat dissipation, complex geometry, materials variety, and tool manufacturing. These challenges in micro-FSW can be avoided by different approaches such as use of insulating anvil, precise manufacturing of tool and fixture and use of high rotational speed and low welding speed [56]. In case of ultrasonic micro-welding of plastics, it is reported that the machine operates around at 35 kHz with tool width 150–300 μm, up to 800 μm in depth, 500 μm in height, and 8 mm length [57]. Diffusion bonding set-up requires a small heating furnace and less pressure. Diffusion bonding takes much longer bonding time. However, the bonding is achieved with minimum deformation [56].

5.6.3 Nano-Joining

As mentioned, nano-joining implies the joining at the order of 100 nm, i.e., joining at molecular level. Repair of carbon nanotube through nano-joining is possible. It is reported that, the application of electron beam welding in transmission electron microscopy (TEM) using high accelerating voltage (of 1.25 MV) and high specimen temperature (at 800 °C) can be used for nano-joining. Under the effect of electron irradiation and annealing at their contact region, the welding is obtained. Similarly, scanning electron microscopy (SEM) can also be utilized for nano-joining of micro-to-nano-scale (50 to 900 nm). It is reported that, spot sizes of 50–125 nm (approximate) are possible at SEM that can melt areas up to 100 nm from 50 μm. Joining of materials such as polysilicon, nickel, Alumel, Chromel, and Tophet C is possible through direct SEM or TEM. Indirect SEM and TEM are able to give resolution in sub-nanometer and can create spot diameter less than 1 nm. Nano-surface can be manipulated through depositing contamination selectively using SEM or TEM [56].

Ion beam nano-welding, resistance nano-welding, ultrasonic nano-welding, and laser nano-welding are techniques through which nano-joining can be obtained. Multi-walled carbon nanotubes are joined by ion irradiation. Resistance welding and ultrasonic welding are possible on welding of nano-wire with substrate using sophisticated set-up and process parameters. Laser pulses can be utilized to weld nano-particles. It is reported that, gold nano-particles are welded by irradiating laser pulses of 532 nm and 0.2 mJ for 10 min on a carbon-coated copper TEM grid in order to have ohmic nano-contact [56].

Nano structure joining can be done similar like powder metallurgy in which nano-particles are combined together at the first stage and next it is pressed at optimum pressure and heat which leads to the joining. Silicon wafer bonding is another example of micro and nano-joining that can be done by different micro-nano joining methods such as anodic bonding, bonding via an intermediate layer of silicides and bonding via solder [56]. The selection of method is dependent on specific application of micro-nano system.

5.7 Hybrid Welding Processes

Hybrid welding processes are combination of two or more welding techniques through which process capabilities are extended significantly. Hybrid welding processes are found effective where individual process is incapable. Double electrode arc welding discussed in Sect. 5.3.5 can be considered as one of the examples of hybrid welding processes wherein the deposition rate is increased at low heat input. Hybrid welding concepts of laser assisted hybrid welding, hybrid arc welding processes and external energy assisted FSW are discussed in this section.

5.7.1 Laser Assisted Hybrid Welding

The laser beam has focused heat source and high density that produces higher depth of weld with small heat affected zone (see Fig. 5.18a). Laser beam welding can be performed at higher welding speed. Besides these advantages, laser welding has

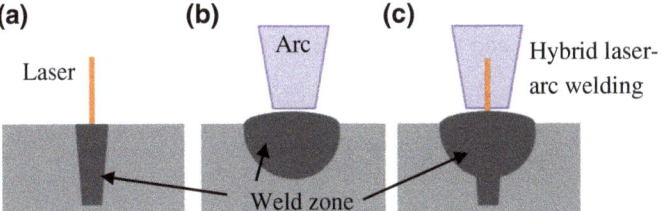

Fig. 5.18 a Weld bead of laser welding, b Weld bead of arc welding, and c Weld bead of hybrid laser-arc welding

some disadvantages such as higher power requirement and higher initial as well as service cost. These disadvantages can be minimized by hybrid laser-arc welding. Arc welding is operated at much low energy density and slow welding speed along with large heat source and low operating cost relative to laser welding that consequently results in shallow penetration and higher width (see Fig. 5.18b). Therefore, hybrid laser-arc welding serves advantages of both the process such as higher welding speed, improved weld quality, less deformation, an excellent gap bridging ability and better process efficiency (see Fig. 5.18c).

Additional process parameters such as distance between laser and arc, shielding arrangements, welding speeds and power level of arc and laser need to be taken into consideration along with individual basic process parameters.

Laser-arc welding concept is reported for laser-GTAW, laser-GMAW, and laser-plasma welding. Laser-GTAW hybrid welding can eliminate disadvantages of GTAW and laser welding, such as low penetration and absorptivity problems. It is reported that, higher process efficiency with deeper penetration is obtained through hybrid laser-GTAW technique. Similarly, laser-GMAW can solve distortion problems, improve weld bead profile, process performance, and improve weld strength. Laser-plasma hybrid technique improves process stability and weld bead profile [58].

5.7.2 Hybrid Arc Welding

Hybrid arc welding processes are those in which different arc welding processes are coupled in order to obtain benefits of each process. GTAW-GMAW and GMAW-Plasma arc welding are examples of hybrid welding techniques. GMAW-plasma welding can be used at different speeds at low current, low power density and high deposition rate that subsequently improves process capability and joint properties. GTAW-GMAW process is another hybrid process through which penetration level of GTAW is improved significantly with filler addition. Dual shielding can result into excellent properties of joint [19]. Hybridization in filler wire such as solid-flux cored, solid-metal cored improves process performances along with overall cost and time reduction [59].

5.7.3 Hybrid Friction Stir Welding

Hybrid approaches are applied to FSW technique in order to take benefits such as improve mechanical properties of joints, uniform heat generation and material flow, reduce load on the tools, and increase process parameters window. Electrically assisted FSW, laser assisted FSW, arc assisted FSW, ultrasonic energy assisted FSW, and cooling enhanced FSW are examples of hybrid FSW techniques [40, 60].

5.7.3.1 Electrically Assisted FSW

Electrically assisted FSW is a hybrid technique in which workpiece is subjected to resistance heating through electrical current. The Joule effect causes electro plastic heating that subsequently leads to the additional material softening to the workpiece. In this hybrid approach bulky set-up is not required like in arc assisted FSW and laser assisted FSW. This concept helps to reduce the forces generated on tool due to softening effect that in turn improve wear resistance and life of the tool. Increased welding speed of FSW is applicable as preheating of workpiece that softened the materials initially. Additionally, improved dissimilar joints can be obtained by temperature rise at single workpiece. However, electrically conductive materials are mandatory to provide resistance heating effect [60].

5.7.3.2 Laser Assisted FSW

Laser assisted FSW is another approach of hybrid FSW in which laser is applied as preheating source before the FSW tool in order to improve the process. The laser beam is a flexible and precise source of heating through which focused heat is supplied to the specific point. Preheating of dissimilar materials is done at specific material through flexible laser source. Properties of dissimilar joints can be significantly improved with this technique [40, 60].

5.7.3.3 Arc Assisted FSW

Arc assisted FSW is reported for GTAW assisted FSW and plasma assisted FSW for dissimilar combinations. The external torch of GTAW or plasma welding is attached in front of FSW tool for preheating purpose. In this technique, arc with shielding gas is supplied to a material which is harder than the other material [40, 60]. Shielding gas can prevent atmospheric contamination during preheating.

However, materials like copper affected by oxidation at higher preheating current and deteriorated joint properties of aluminum–copper joint [40]. This process can be applied to nonmetallic materials too.

5.7.3.4 Ultrasonic Energy Assisted FSW

Ultrasonic vibrations are applied similarly to the previously discussed process in order to preheat the workpiece. This process is advantageous in terms of solid state preheating where arc, laser, and shielding gases are not present. Therefore, this hybrid process can be considered as a sustainable hybrid FSW process. Ultrasonic energy assisted FSW is applied to dissimilar materials and similar materials [40, 60].

5.7.3.5 Cooling Enhanced FSW

Cooling enhanced FSW is a hybrid approach in which workpiece is welded under the effect of different cooling mediums such as water, liquid CO_2, and liquid nitrogen. The superior fine grain microstructure is obtained under this hybrid mode. Additionally, the formation of intermetallic compounds is significantly restricted due to cooling effect [40, 61]. However, forces acting on tool cannot be decreased as in thermally assisted FSW. Underwater FSW requires special purpose fixture where the workpiece is kept at under water. Cooling enhanced FSW is reported as an improved method for dissimilar welds by which the formation of intermetallic compounds has drastically reduced [62].

5.8 Summary

Recent developments of welding and joining techniques are discussed in this chapter. The special features of process background, parameters, capabilities, novel aspects, and variants are presented for recently developed welding and joining techniques. Developments in the area of fastening and bonding processes, arc welding processes, beam welding techniques, sustainable welding processes, micro-nano joining, and hybrid welding are covered. Advanced welding and joining techniques provides the tangible benefits such as favorable joint properties, synergistic mix of materials, significant reduction in the component cost, tremendous enhancement in productivity, excellent quality, ability to join dissimilar materials, repairability, attaining complex geometrical configuration, and the ability to deal with a variety of materials to manufacture new products.

References

1. Bodjona K, Lessard L (2016) Hybrid bonded-fastened joints and their application in composite structures: A general review. J Reinf Plast Compos 35(9):764–781
2. Moroni F, Pirondi A (2010) Technology of rivet: adhesive joints. In: Hybrid adhesive joints. Springer, pp 79–108
3. McGrath G, Jones I, Hilton P, Kellar E, Taylor A, Sallavanti P (2001) New advances in plastics joining for high speed production. SAE Technical Paper. doi:10.4271/2001-01-3398
4. Eshtayeh M, Hrairi M, Mohiuddin A (2016) Clinching process for joining dissimilar materials: state of the art. Int J Adv Manuf Technol 82(1–4):179–195
5. Lambiase F (2015) Clinch joining of heat-treatable aluminum AA6082-T6 alloy under warm conditions. J Mater Process Technol 225:421–432
6. Pirondi A, Moroni F (2010) Science of Clinch–adhesive joints. In: Hybrid adhesive joints. Springer, pp 109–147
7. Vidyarthy R, Dwivedi D (2016) Activating flux tungsten inert gas welding for enhanced weld penetration. J Manuf Process 22:211–228
8. Chai G, Zhu Y (2010) Spectra and thermal analysis of the arc in activating flux plasma arc welding. Guang pu xue yu guang pu fen xi = Guang pu 30(4):1141–1145
9. Kuo M, Sun Z, Pan D (2013) Laser welding with activating flux. Sci Technol Weld Joining 6(1):17–22
10. Huang H-Y (2010) Effects of activating flux on the welded joint characteristics in gas metal arc welding. Mater Des 31(5):2488–2495
11. Delany F, Lucas W, Thomas W, Howse D, Abson D, Mulligan S, Bird C (2005) Advanced joining processes for repair in nuclear power plants. In: Proceedings of 2005, pp 54–69
12. Ma L, Hu S, Hu B, Shen J, Wang Y (2014) Activating flux design for laser welding of ferritic stainless steel. Trans Tianjin Univ 20:429–434
13. Zhang H, Feng J, He P, Zhang B, Chen J, Wang L (2009) The arc characteristics and metal transfer behaviour of cold metal transfer and its use in joining aluminium to zinc-coated steel. Mater Sci Eng, A 499(1):111–113
14. Pickin CG, Williams S, Lunt M (2011) Characterisation of the cold metal transfer (CMT) process and its application for low dilution cladding. J Mater Process Technol 211(3):496–502
15. Xiurong Y (2006) CMT cold metal transfer process. Electric Welding Machine 6:005 (http://en.cnki.com.cn/Article_en/CJFDTOTAL-DHJI200606005.htm)
16. Almeida P, Williams S (2010) Innovative process model of Ti–6Al–4 V additive layer manufacturing using cold metal transfer (CMT). In: Proceedings of the twenty-first annual international solid freeform fabrication symposium, University of Texas at Austin, Austin, TX, USA, 2010
17. Zhang H, Feng J, He P (2008) Interfacial phenomena of cold metal transfer (CMT) welding of zinc coated steel and wrought aluminium. Mater Sci Technol 24(11):1346–1349
18. Kadoi K, Murakami A, Shinozaki K, Yamamoto M, Matsumura H (2016) Crack repair welding by CMT brazing using low melting point filler wire for long-term used steam turbine cases of Cr-Mo-V cast steels. Mater Sci Eng: A 666:11–18
19. Ahmed N (2005) New developments in advanced welding. Elsevier, pp 1–269
20. Praveen P, Yarlagadda P, Kang M-J (2005) Advancements in pulse gas metal arc welding. J Mater Process Technol 164:1113–1119
21. Palani P, Murugan N (2006) Selection of parameters of pulsed current gas metal arc welding. J Mater Process Technol 172(1):1–10
22. Lu Y, Chen S, Shi Y, Li X, Chen J, Kvidahl L, Zhang YM (2014) Double-electrode arc welding process: principle, variants, control and developments. J Manuf Process 16(1):93–108
23. Li K, Zhang Y (2008) Consumable double-electrode GMAW-Part 1: The process. Welding J-New York 87(1):11

24. Henon B (2011) Advances in automatic hot wire GTAW (TIG) welding. Arc Machines Inc
25. Nakamura T, Hiraoka K (2013) Ultranarrow GMAW process with newly developed wire melting control system. Sci Technol Weld Joining 6(6):355–362
26. Elmesalamy A, Li L, Francis J, Sezer H (2013) Understanding the process parameter interactions in multiple-pass ultra-narrow-gap laser welding of thick-section stainless steels. Int J Adv Manuf Technol 68(1–4):1–17
27. Feng J, Guo W, Francis J, Irvine N, Li L (2016) Narrow gap laser welding for potential nuclear pressure vessel manufacture. J Laser Appl 28(2):022421
28. Li R, Yue J, Sun R, Mi G, Wang C, Shao X (2016) A study of droplet transfer behavior in ultra-narrow gap laser arc hybrid welding. Int J Adv Manuf Technol:1–12, doi: 10.1007/s00170-016-8699-9
29. Klein R (2012) Laser welding of plastics. Wiley, pp 1–243
30. Jin Y, Y-l Li, Zhang H (2016) Microstructure and mechanical properties of pulsed laser welded Al/steel dissimilar joint. Trans Nonferrous Metals Soc China 26(4):994–1002
31. Wang P, Chen X, Pan Q, Madigan B, Long J (2016) Laser welding dissimilar materials of aluminum to steel: an overview. Int J Adv Manuf Technol:1–10. doi: 10.1007/s00170-016-8725-y
32. Mai T, Spowage A (2004) Characterisation of dissimilar joints in laser welding of steel–kovar, copper–steel and copper–aluminium. Mater Sci Eng, A 374(1):224–233
33. Borrego L, Pires J, Costa J, Ferreira J (2009) Mould steels repaired by laser welding. Eng Fail Anal 16(2):596–607
34. Vedani M, Previtali B, Vimercati G, Sanvito A, Somaschini G (2007) Problems in laser repair-welding a surface-treated tool steel. Surf Coat Technol 201(8):4518–4525
35. Díaz E, Tobar M, Yáñez A, García J, Taibo J (2010) Laser Powder Welding with a Co-based alloy for repairing steam circuit components in thermal power stations. Phys Proc 5:349–358
36. Peng W, Jiguo S, Shiqing Z, Gang W (2016) Control of wire transfer behaviors in hot wire laser welding. Int J Adv Manuf Technol 83(9–12):2091–2100
37. Węglowski MS, Błacha S, Phillips A (2016) Electron beam welding–Techniques and trends–Review. Vacuum 130:72–92
38. Mishra RS, De PS, Kumar N (2014) Fundamentals of the friction stir process. In: Friction stir welding and processing. Springer, pp 13–58
39. Mishra RS, Ma ZY (2005) Friction stir welding and processing. Mater Sci Eng: R: Rep 50(1–2):1–78. doi:10.1016/j.mser.2005.07.001
40. Mehta KP, Badheka VJ (2016) A review on dissimilar friction stir welding of copper to aluminum: process, properties, and variants. Mater Manuf Process 31(3):233–254. doi:10.1080/10426914.2015.1025971
41. Mehta KP, Badheka VJ (2015) Influence of tool design and process parameters on dissimilar friction stir welding of copper to AA6061-T651 joint. Int J Adv Manuf Technol 80:2073–2082. doi:10.1007/s00170-015-7176-1
42. Mehta KP, Badheka VJ (2014) Effects of tilt angle on properties of dissimilar friction stir welding copper to aluminum. Mater Manuf Process 31:255–263. doi:10.1080/10426914.2014.994754
43. Mehta KP, Badheka VJ (2016) Effects of tool pin design on formation of defects in dissimilar friction stir welding. Proc Technol 23:513–518
44. Mehta KP, Badheka V (2016) Experimental investigation of process parameters on defects generation in Copper to AA6061-T651 friction stir Welding. Int J Adv Mech Autom Eng (IJAMAE) 3(1):55–58. doi:10.15242/IJAMAE.E0316007
45. Gerlich AP, North TH (2010) Friction stir spot welding. Innov Mater Manuf Fabric Environ Safety:193
46. Evans WT, Gibson BT, Reynolds JT, Strauss AM, Cook GE (2015) Friction Stir Extrusion: A new process for joining dissimilar materials. Manuf Lett 5:25–28
47. Besler FA, Schindele P, Grant RJ, Stegmüller MJ (2016) Friction crush welding of aluminium, copper and steel sheetmetals with flanged edges. J Mater Process Technol 234:72–83

48. Garg A, Panda B, Shankhwar K (2016) Investigation of the joint length of weldment of environmental-friendly magnetic pulse welding process. Int J Adv Manuf Technol:1–12. doi: 10.1007/s00170-016-8634-0
49. Hahn M, Weddeling C, Lueg-Althoff J, Tekkaya AE (2016) Analytical approach for magnetic pulse welding of sheet connections. J Mater Process Technol 230:131–142
50. Shanthala K, Sreenivasa T (2016) Review on electromagnetic welding of dissimilar materials. Front Mech Eng:1–11, doi: 10.1007/s11465-016-0375-0
51. Matheny M (2014) EWI, Columbus, OH, USA Note: This chapter is a revised and updated version of Chapter 9 "Ultrasonic metal welding" by K. Graff, originally published in New Developments in Advanced Welding, ed. N. Ahmed, Woodhead Publishing Limited, 2005, ISBN: 978-1-85573-970-3. Power Ultrasonics: Applications of High-Intensity Ultrasound:259
52. YEH H (2013) Ultrasonic welding of medical plastics. Joining and assembly of medical materials and devices, p 296
53. Benatar A (2015) 12—Ultrasonic welding of plastics and polymeric composites. In: Power ultrasonics. Woodhead Publishing, Oxford, pp 295-312. doi:http://dx.doi.org/10.1016/B978-1-78242-028-6.00012-0
54. Zhang Y, Li Y, Luo Z, Yuan T, Bi J, Wang ZM, Wang ZP, Chao YJ (2016) Feasibility study of dissimilar joining of aluminum alloy 5052 to pure copper via thermo-compensated resistance spot welding. Mater Des 106:235–246
55. Zhou YN (2008) Microjoining and nanojoining. Elsevier, pp 1–786
56. Sithole K, Rao VV (2016) Recent developments in micro friction stir welding: a review. In: IOP conference series: materials science and engineering, vol 1. IOP Publishing, p 012036
57. Sackmann J, Burlage K, Gerhardy C, Memering B, Liao S, Schomburg W (2015) Review on ultrasonic fabrication of polymer micro devices. Ultrasonics 56:189–200
58. Lee C-M, Woo W-S, Kim D-H, Oh W-J, Oh N-S (2016) Laser-assisted hybrid processes: A review. Int J Precision Eng Manuf 17(2):257–267
59. Prajapati P, Badheka VJ, Mehta KP (2016) Hybridization of filler wire in multi pass gas metal arc welding of SA516 Gr70 carbon steel. Mater Manuf Processes. doi:10.1080/10426914.2016.1244847
60. Padhy G, Wu C, Gao S (2015) Auxiliary energy assisted friction stir welding–status review. Sci Technol Weld Join 20(8):631–649
61. Sabari SS, Malarvizhi S, Balasubramanian V (2016) Influences of tool traverse speed on tensile properties of air cooled and water cooled friction stir welded AA2519-T87 aluminium alloy joints. J Mater Process Technol 237:286–300
62. Mehta KP, Badheka VJ (2016) Hybrid approaches of assisted heating and cooling for friction stir welding of copper to aluminum joints. J Mater Process Technol, Elsevier 239:336–345. doi:10.1016/j.jmatprotec.2016.08.037

Chapter 6
Laser-Based Repair of Damaged Dies, Molds, and Gears

Sagar H. Nikam and Neelesh Kumar Jain

Abstract Dies, molds and gears are very expensive and difficult-to-manufacture components which develop some repairable damages during their use. These damages do not render them to be rejected due to various industrial and economic reasons but affect their proper functioning. Timely and economic repairing or re-manufacturing of these components can greatly extend their life without compromising their functional and quality aspects thus yielding rich financial and productivity benefits. This chapter describes various repair processes for different types of damages that occur in the dies, molds and gears with major focus on laser-based repair processes. It also defines various types of damages and discusses possible causes of their occurrence. This will help users in selecting the most appropriate repair process for the damaged industrial dies, mold, and gear depending upon availability of the resources and various constraints.

Keywords Dies · Gears · Laser · Layered manufacturing · Molds · Plasma · Repair · Welding

6.1 Introduction

Dies, molds and gears are complicated, critical and very expensive mechanical components generally made of difficult-to-manufacture materials. Specifically, dies and molds are used in manufacturing of high quality die cast, plastic molded, sheet metal and forged products. Finish and final characteristics of these products depend on accuracy and surface quality of the dies and molds used. Consequently, their manufacturing is always challenging and costly due to continuously increasing complexities in their design and material. Therefore, industries significantly invest in either manufacturing or procurement of highly accurate, precise, wear resistant, corrosion resistant, heavy duty and durable dies and molds. Gear is a rotating part

S.H. Nikam · N.K. Jain (✉)
Discipline of Mechanical Engineering, Indian Institute of Technology Indore, Indore, India
e-mail: nkjain@iiti.ac.in

© Springer International Publishing AG 2017
K. Gupta (ed.), *Advanced Manufacturing Technologies*, Materials Forming, Machining and Tribology, DOI 10.1007/978-3-319-56099-1_6

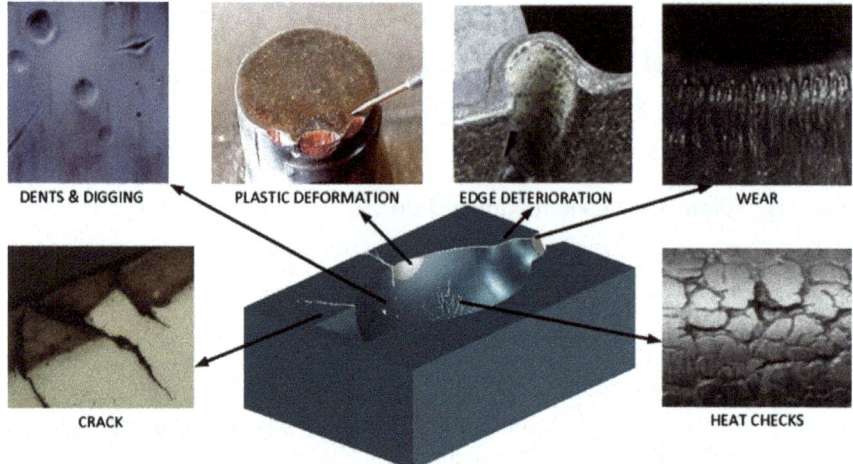

Fig. 6.1 Some typical damages that occur in dies and molds [2], with kind permission from Elsevier

having teeth on its periphery and meshes with another toothed part to transmit power and/or motion or to change direction of motion. Accuracy, quality, and condition of a gear affect its characteristics related to motion and power transmission, wear, vibration and noise, operating performance and service life.

Lifecycle of dies, molds, and gears are dependent on different aspects such as appropriate material selection, design and manufacturing quality, suitable heat treatment process, operating environment, and handling by the users [1]. These components develop different minor damages during their use which do not render them to be rejected outright but, if they are not detected and repaired timely then they may lead to premature failure of these components. Different types of damages in dies, molds and gears include various types of cracks, scratches, broken edges, dents, digging, thermal cracking, plastic deformation, surface irregularities, worn-out geometries, dimensional changes, surface damages, etc. Figure 6.1 illustrates some typical damages occurring in the dies and molds.

Important reasons causing the damages in dies, molds and gears are faulty design and manufacturing, defects in their materials, mishandling, mechanical strain, plastic deformation, local impacts, cyclic loading, high thermal stresses, stress corrosion, wear, fatigue, and major forces due to accidental conditions. Damages due to faulty design, manufacturing, and material of these components are generally observed during initial stage of their use. Damages such as cracks, scratches, broken edges, surface irregularities may occur during later stage of their usage and are caused due to elastic–plastic deformation, shape distortion, frictional wear and combined effect of distortion and wear. For example, main reason of damage to dies in the die-casting process is the surface cracking caused by the thermal fatigue phenomena. Other reasons may include tensile cracks induced due to manufacturing imperfections such as notches, grooves, pits, etc., and erosion caused by flow of the molten material [3–5]. Damages such as worn-out geometries, dimensional changes, and surface damages sometime may arise within their few operating cycles which can cause their premature failure while sometimes they may fail towards completion of their useful life after several thousand working cycles. Intensity and magnitude of damage depend on severity of the work environment and the required precision in their shape and size.

Damages in dies, molds and gears causes them to lose their accuracy, size and shape leading to manufactured products of poor accuracy and loose dimensional and geometrical tolerances. Specifically in gears, these factors cause downtime in their operation. This directly affects the economic aspects of industrial and manufacturing sector. Only two options are available to avoid this namely either stand-by components or quick and economic repair [6]. Dies, molds and gears are manufactured of very costly and difficult-to-manufacture materials and have very high precision and accuracy therefore their quick and economic repair or remanufacturing is the only viable solution [7, 8]. Timely repair of damaged dies, molds and gears increases their life economically and productively avoiding interruption in production and saving huge investments that would have been done for procurement of new dies, molds, and gears.

Typical repair process of the damaged dies, molds, and gears involves removal of the damaged volume or enlargement of cracks by milling or grinding followed by depositing the missing volume using suitable metallic deposition processes using an appropriate filler material [9]. Damaged components are heat treated before and after the deposition process to avoid solidification cracks and excessive residual stresses induced by it. Analysis of the damage mechanisms aids in identifying their causes which helps in selecting the most suitable repair process. Use of the finite element modeling and analysis can be of great help in predicting the impending damages.

Fusion welding process is generally used to repair these types of damages. Gears are generally subjected to impact and alternating loading which may be sometimes heavy also. It often leads to their failure due to fatigue, wear, and breakage. Therefore, it is necessary to study the repair processes for the large size heavy and costly gears. Advantages of laser cladding such as efficient and high strength deposition, good metallurgical bonding with the substrate and small heat affected zone (HAZ) make it to be the most attractive repair process for gears [10].

Since, repair of the damaged dies, molds and gears can yield substantial financial, production and competitive benefits therefore it has led to development of different types of repair processes for damaged dies, molds and gears. Traditional repair processes are arc-based and plasma-based deposition processes. Whereas, lasers, electron beam and micro-plasma transferred arc-based deposition processes have emerged recently as attractive, more competitive and sustainable repairing options due to controlled heating of the substrate and precise control over the geometry and rate of deposition. Modern machines are available for these processes having sophisticated manipulators which can be computer numerically controlled (CNC), micro-controller-based, pneumatically controlled, or a robotic arm.

This chapter briefly describes various types of repair processes with major focus on laser-based repair processes for the damaged dies, molds gears along with different types of damage modes and their causes.

6.2 Types of Damages and Their Causes

Figure 6.2 presents types of damages that occur in dies, molds and gears along with their possible causes.

6.2.1 Catastrophic Damages

Main causes of catastrophic damages to dies, molds, and gears are faulty design, extreme loading, defective material, improper storage, and materials handling. To avoid inappropriate design it is necessary to take additional care while designing corners shapes, cuts, and sudden change in the product design which increases pressure on contact surfaces and may lead to failure due to fatigue and surface wear [2, 11, 12]. Use of computer aided engineering (CAE)-based software can confirm preciseness in design and analysis of dies, molds, and gears. Simulation plays a crucial role in predicting the damage mechanism which helps in avoiding the early damages [9, 13]. Damages in dies, molds, and gears due to extreme loading result in

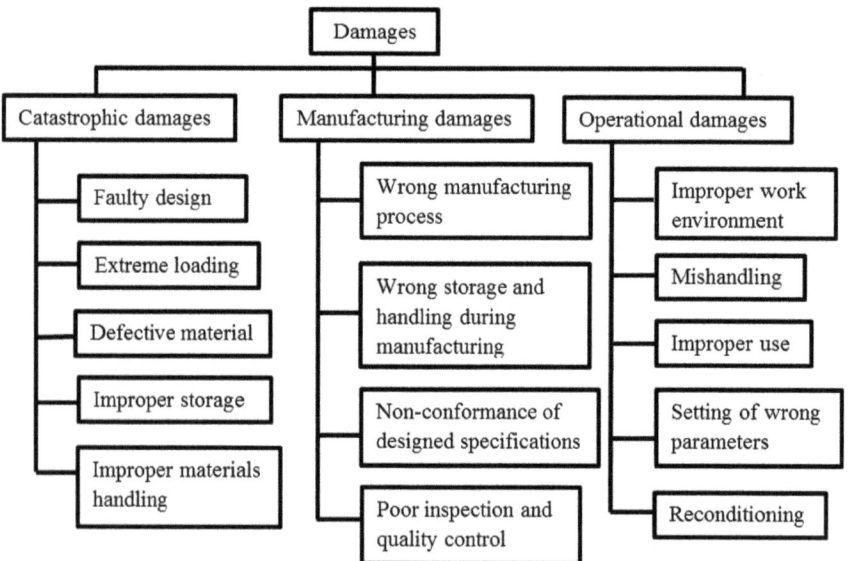

Fig. 6.2 Types of damages that occur in dies, molds, and gears and their possible causes

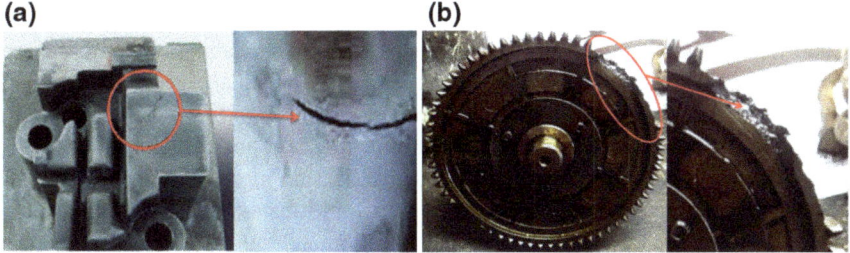

Fig. 6.3 Catastrophic damages due to extreme loading **a** in dies [15] (*courtesy* Looktech Co., Ltd); and **b** in gears [16] (*courtesy* tractorbynet.com)

cracking, mechanical fatigue, and plastic deformation. Dies and molds used in the hot forming processes are subjected to cracks, thermal distortion, elastic–plastic deformation, and corrosion because they are subjected to continuous heating and cooling cycles along with the mechanical loading. Whereas, dies and molds used in the cold forming process generally get damaged due to application of sudden loads and initiation of the cracks at the cross-sections having stress concentration as depicted in Fig. 6.3a [14]. Extreme loading in a gear causes deviation in its macro and micro-geometry and deteriorates its quality. Figure 6.3b depicts the catastrophic damages due to extreme loading in gears. Selection of defective raw material having undetected internal defects such as voids, porosity, flaws, etc., also leads to occurrence of catastrophic damages in dies, molds and gears. Improper storage and handling of the raw materials of dies, molds, and gears either before or after their manufacturing are also responsible of such type of damages.

6.2.2 Manufacturing Damages

Major causes of manufacturing damages are: use of inappropriate manufacturing process, wrong use of correct manufacturing process, non-conformance of their design specifications, poor inspection and quality control and inappropriate storage and handling of dies, molds and gears either during or after their manufacturing. Despite the availability of advanced design and analysis software and CNC machines; manufacturing of molds, dies and gears is still challenging in most of the cases. Use of wrong manufacturing, finishing and heat treatment processes can also induce damages in dies, molds and gears off-setting advantages of using materials having superior properties. Here, wrong process implies incapable or just capable manufacturing process which is unable of either processing the selected material or producing the required complicated geometry, size, dimensional and geometrical tolerances. Sometimes, too strict specifications of the dies, molds, and gears can make their manufacturing very difficult and even impossible. All these lead to non-conformance of the designed specifications of the dies, molds, and gears.

Wrong use of correct manufacturing process also induces some damages in dies, molds and gears, i.e., use of wrong or inadequate lubricating oil in a surface grinding may cause minor cracks on surfaces of dies, molds and gears [17]; a good heat treatment process but performed improperly can lead to lower fatigue strength caused by tensile residual stresses, thermal cracking and poor toughness.

Even after selecting the correct design, material and manufacturing process for dies, molds and gears, their damages and non-conformance of the designed specifications can go undetected due to poor inspection and quality control techniques. Therefore, carefully planned and accurately implemented inspection (which can be either online or offline), well defined quality assurance policies and selection of the most appropriate quality tools are very crucial for dies, mold and gears to detect defects and specification deviations induced during their manufacturing, finishing and property enhancing processes. This ensures conformance of their designed specifications.

Faulty transportation, storage and handling during or after manufacturing of dies, molds and gears can also damages them thus undoing the entire efforts put in achieving their best design, material, manufacturing, inspection and quality control. For example wrong alignment during usage of dies and molds or while operating gears amplifies the induced stress leading which can damage them [18]. Figure 6.4 illustrates some photographs of damages caused to a typical die (Fig. 6.4a) and gear (Fig. 6.4b) due to wrong storage during their manufacturing.

6.2.3 Operational Damages

Operational damages to dies, molds, and gears can be caused due to adverse work environment such as chemically reactive, strong alkaline, acidic or gaseous, highly humid, very high temperature and high impact loading which has not been anticipated during their design. Such work environment causes oxidation, corrosion, and

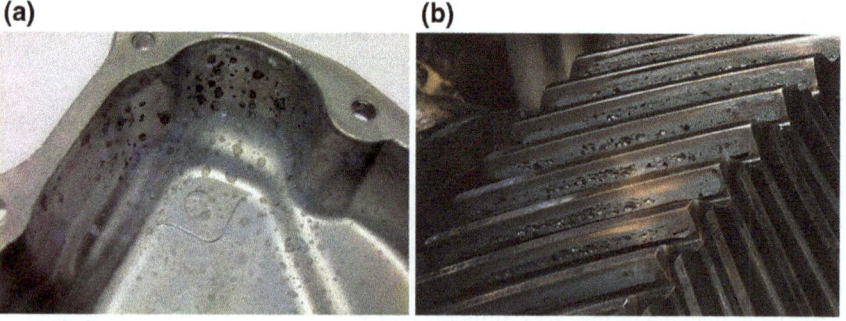

Fig. 6.4 Damages caused in **a** dies [19] (*courtesy* Empire die casting Co.); and **b** gears [20] (*courtesy* Novexa Inc.) due to wrong storage during their manufacturing

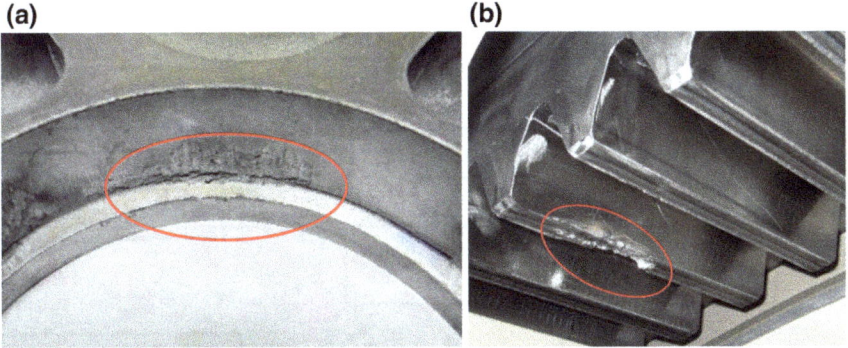

Fig. 6.5 Operational damage caused in **a** dies [19] (*courtesy* Empire die casting Co.); and **b** gears [20] (*courtesy* Novexa Inc.) due to their improper use

wear and tear of their surfaces. Mishandling may further accelerate their cracks and wear which may eventually lead to their premature failures. Often, poorly trained users or operators set the incorrect parameters during use of dies, molds and gears that lead to their faulty loading and hence causes damages as shown in Fig. 6.5. After using the dies and molds in hot and cold working operations, some molten material particles remain adhered that deteriorates the surface finish of dies and molds, and makes their reconditioning necessary. Prior estimation of operational damages can be made using the finite element modeling and analytical approaches. Such estimations have proven to be helpful in selecting optimum process parameters, proper raw material, shape, and geometry, and operating speed [21, 22].

6.3 Repair Process Sequence

Figure 6.6 depicts sequence of different activities in repair of the damages in the dies, molds, and gears. It begins with inspection of the damages occurred and determine their type so as to determine their causes as described above. Cleaning process is selected according to type and extent of the damages occurred. If the damages are caused due to improper storage, inappropriate storage and handling during manufacturing and improper work environment then chemical-based cleaning process is used in which the damaged surface or volume is cleaned by using acid, solvent, salt bath, vapor degreasing or oxidation. If the damage is caused due to extreme loading, defective material, use of wrong manufacturing process and setting of wrong process parameters than mechanical-based cleaning process is used in which the damaged surface or volume is cleaned by using polishing, abrasive blasting or machining process. The cleaned up cavity is filled with deposition material either in wire, powdered or hybrid form using deposition process. Selection of deposition process, deposition material, and its form depends on the

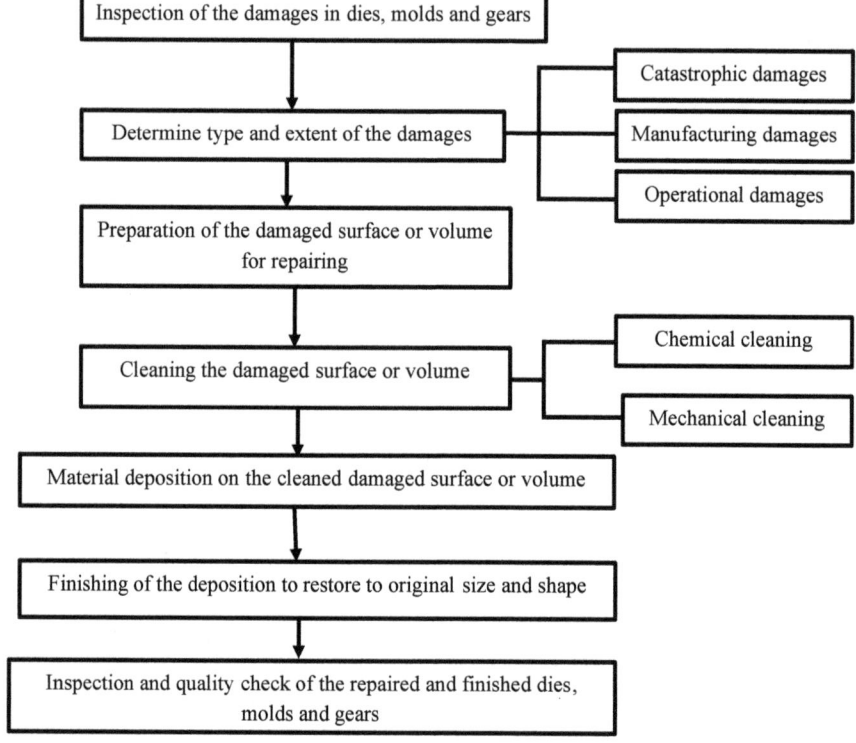

Fig. 6.6 Sequence of repair process of damages in the dies, molds, and gears

dimensions, shape, orientation, and location of the damaged surface or volume, its accessibility, ability of the deposition material to fill it, availability and cost of the deposition material, and total cost of repairing. Finishing of the deposition is done to restore the damaged die, mold or gear to its original shape and size which is ensured by inspection and quality check of the repaired component.

6.4 Repair Processes

Requirement of an accurate, fast, and economic repair process for the damaged dies, molds and gears has forced the industries to develop new superior processes. Figure 6.7 presents various repair processes and following paragraphs describes them briefly. Generally, repair process involves material deposition therefore metallurgical relationship between substrate and deposition material needs to be studied carefully due to their dependence on volumetric rate and geometry of deposition.

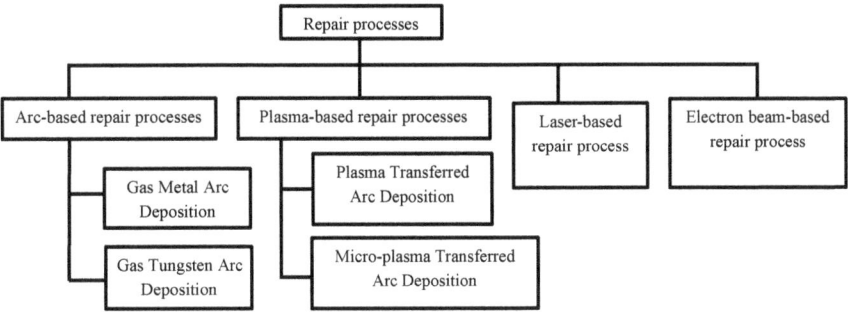

Fig. 6.7 Various processes that can be used to repair the damaged dies, molds, and gears

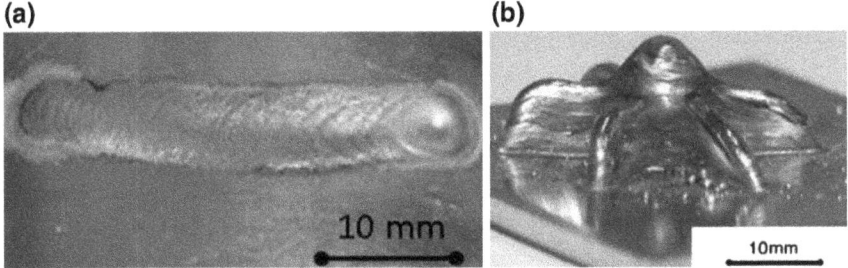

Fig. 6.8 a Surface crack repaired by GTAD process [2], with kind permission from Elsevier; and **b** material deposited using micro-GTAD process [26], with kind permission from Elsevier

6.4.1 Arc-Based Repair Processes

Most commonly used arc-based repair processes for the damaged dies and molds are gas metal arc deposition (GMAD) and gas tungsten arc deposition (GTAD). These processes are easy to operate, portable, yield higher deposition rate, incur lower initial investment, and give higher arc stability. This results in accurate and controlled metallic deposition and eliminates chances of tungsten inclusion due to absence of contact between the tungsten electrode and workpiece in the GTAD process. The process equipment are generally manually operated thus requiring highly skilled operators [23]. Dies and molds are made of heat sensitive materials therefore heating before the deposition and appropriate heat treatment after the deposition are necessary to avoid cracking and tensile residual stresses [24–27]. Figure 6.8a depicts the surface crack repaired by the GTAD process. Micro-version of GTAD process has also been developed and used for multi-layer deposition of an alloy of titanium and nickel in wire form to manufacture a 3-dimensional pin-shaped object depicted in Fig. 6.8b and having good quality of deposition [28].

6.4.2 Plasma-Based Repair Processes

Plasma transferred arc-based deposition (PTAD) processes use concentrated plasma arc for metallic deposition in the form of thin layers which leads to smaller heat affected zone (HAZ) and higher arc penetration depth. A pilot arc is formed between a non-consumable tungsten cathode and constricting nozzle by applying suitable potential difference between them whereas in the GTAD process, arc is formed between the tungsten electrode and the workpiece. Consequently, PTAD processes give better quality of deposition, lesser HAZ, minimizes tungsten inclusion in the deposition and better heat transfer efficiency as compared to the arc-based deposition processes [29]. Figure 6.9 presents a typical material deposition using PTAD process.

Both GTAD and PTAD processes use high value of current in the range of 100-300 A which results in higher dilution, thermal distortion, and adverse microstructure. These problems can be overcome by using micro-version of the PTA process known as micro-plasma transferred arc (μ-PTA) deposition process, which uses current in the range of 20-30 A. Power supply unit having digital controller and possessing very small increment in current values (up to 0.1 A or even smaller) makes it possible to produce a smaller, softer, bell-shaped and precisely controlled arc. Micro-PTA process offers many additional advantages such as improved steady arc direction, increased arc stability, less sensitivity towards changes in the arc length, higher energy efficiency and better material efficiency. Its equipment can be automated by using a CNC machine or robotic arm as manipulator. It is more suitable for the miniature or small amount of metallic depositions which makes it very useful in various repairing and remanufacturing applications.

6.4.3 Laser-Based Repair Process

Laser (an acronym of light amplification by stimulated emission of radiation) is a monochromatic coherent light which can be focused to a very small area.

Fig. 6.9 Surface repaired using plasma transferred arc deposition process [29], with kind permission from Elsevier

This results in highly localized heating of that area. Consequently, laser-based deposition process are preferred to the arc-based and plasma-based deposition processes due to their some unique advantages such very small area of laser beam concentration and consequently very small volume of the weld, HAZ and higher penetration depth, better deposition quality, and negligible undercut [30–33]. Though, deposition rate is smaller and the equipment is very costly but they have been successfully used for micro-scale repair of the damaged dies, molds, and gears using filler material in the powdered or wire form [9, 34, 35] and for very small sized and precise deposition of metallic material with least changes in the substrate material composition. Adaptability for automation, ease to mount the laser on a CNC machine or robotic arm, easier control, lesser post-processing requirements and smaller repair time are some additional advantages of using laser-based metallic deposition processes for repairing damaged dies, molds and gears.

6.4.3.1 Laser-Based Repair of Dies and Molds

Leunda et al. [36] have used laser cladding process to repair a die used in cold and hot forming of automotive components and was made of vanadium carbide tool steel. The die was developing the heat checks during the hot forming process which damaged its surface morphology as shown in Fig. 6.10. This problem was overcome by coating the die with the special purpose vanadium steel produced using crucible particle metallurgical (CPM) process. This material had better thermal properties as compared to the original material of the die which significantly reduced formation of heat checks in the die.

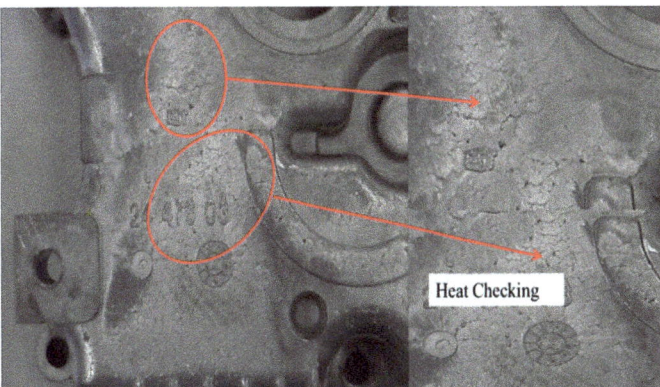

Fig. 6.10 Occurrence of heat checks in a die used for cold and hot forming of automotive components [37] (*courtesy* Badger Metal Tech. Inc.)

6.4.3.2 Laser-Based Repair of Gears

Mudge and Wald [38] have used laser cladding process to repair a coupler gear as depicted in Fig. 6.11a. This gear was coupled with an atomizer shaft whose rotation generated a thrust force generated on the flank surface face of the gear teeth causing their wear. Martensitic stainless steel of 420 grade (which contains minimum of 12% chromium) was deposited on the worn-out flank surfaces of the coupler gear teeth by the laser cladding process to enhance their wear and corrosion resistance. It was observed that the worn-out gear was successfully repaired as shown in Fig. 6.11b with enhanced wear resistance as compared to the original gear.

6.4.4 Electron Beam-Based Repair Process

Electron beam-based deposition process is used to repair the damaged gas nozzles used in the gas turbines, cracked combustion chambers, and broken shaft seals. Impingement of the electron beam causes rapid heating and melting of the substrate material. Its major advantages are lower heat input and smaller heat affected zone (HAZ). Advancements in the electron beam technology have made it possible to generate an electron beam having peak current density of the order of 104 A/cm^2 using comparatively low energy in the range of 10–35 keV with short pulse duration of 1 μs [40]. Major disadvantages of this repair process are: requirement of maintaining very high vacuum for its operation and applicability to repair simple geometries using the pre-defined deposition paths. Therefore, this process can possibly be used to repair minor external damages (i.e., cracks, dents, digging) in the dies, molds and gears.

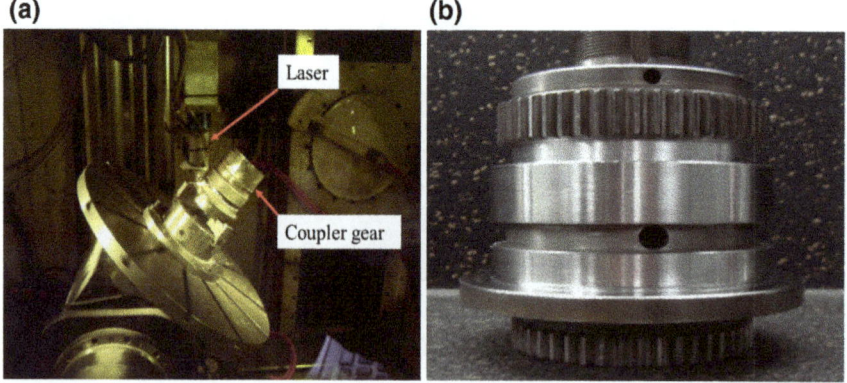

Fig. 6.11 a Repair of a worn-out coupler gear by the laser cladding process; b the repaired coupler [39] (*courtesy* RPM Innovations Inc.)

6.4.5 Comparative Study

Table 6.1 presents comparison of different processes which can be used to repair the damaged dies, molds, and gears in terms of deposition rate, HAZ, metallurgical properties, ability to repair the complex geometries, requirement of post-deposition treatment, flexibility in equipment usage and repair cost.

6.5 Details of Laser-Based Repair Process

6.5.1 Types of Lasers Used in Repair Process

Type of laser to be used to repair the damaged dies, molds and gears depends on the working principle, thermal efficiency, flexibility and ease to control. Most commonly and successfully used lasers for this purpose are carbon dioxide (CO_2) laser; Nd:YAG (neodymium: yttrium aluminum garnet) and Yb:YAG (ytterbium: yttrium aluminum garnet) laser due to their lower heat input, relatively very small resulting HAZ, ease of operation and control and low cost of operation and maintenance. Consequently, these lasers can be used for very small sized and precise deposition of the metallic materials with minimum changes to the substrate material and its composition. Initially, CO_2 laser was the only commercially available gas-based laser but it could not be used for commercial purposes in repair industry mainly due to very high investment, operating and maintenance costs, and requirement of highly skilled operators. Development of high power solid state lasers such as Nd: YAG and Yb:YAG lasers led to reduction in the operating and maintenance costs of

Table 6.1 Comparison of different repair processes which can be used to repair the damaged dies, molds, and gears

Criteria	Arc-based processes	Plasma-based processes	Laser-based processes	Electron beam-based processes
Deposition rate	●●●●	●●●	●●	●●●●●
Heat affected zone	●●●●	●●●	●●●	●●
Metallurgical properties	●	●●	●●●	●●●
Ability to repair the complex geometries	●●●●●	●●●	●●	●
Requirement of post-deposition treatment	●●	●●	●●●	●●●
Flexibility in equipment usage	●●●●	●●●●	●●	●
Repair cost	●	●●	●●●●	●●●

Lower to higher: ●, ●●, ●●●, ●●●●, ●●●●●

the laser-based deposition systems which increased their use to repair the critical and costly components specially made of those materials which are difficult to repair by other deposition processes.

6.5.2 Process Principle and Case Studies

6.5.2.1 Carbon Dioxide Laser

CO_2 laser is one of the early developed gas-based lasers which was explored for metallic deposition purposes. It is one of the highest power continuous-wave lasers currently available. Figure 6.12 depicts schematic of the CO_2 laser producing system. Discharge of CO_2 gas acts as the active laser medium which is air/water cooled. The discharge tube consists of a gaseous mixture containing 40–50% CO_2; 20–30% N_2; 5–10% H_2 and remainder as helium by volume. Vibrational motion of the nitrogen takes place as the helium atoms strike the walls of the discharge tube. Production of laser is achieved by the electron impact which excites vibrational motion of the nitrogen which is metastable and remains for longer duration. Collisional energy transfer between the nitrogen and the carbon dioxide molecule causes vibrational excitation of the carbon dioxide with sufficient efficiency for laser operation.

Figure 6.13 depicts schematic of CO_2 laser-based system to repair the damaged dies, molds, and gears. It consists of focusing head with two gold plated mirrors cooled with water, a straight lens, and a concave lens. The laser beam impinges on a concave lens mounted on the vertical axis of focusing head in such a way that it straightens the CO_2 laser beam which is then concentrated and allowed to pass through a hole in the nozzle through which deposition material is fed. Mostly the deposition material used for repair purpose with CO_2 laser is in powder form but in some cases deposition material in wire form can also be used [3]. The heat generated from CO_2 laser is used to melt the deposition material. This type of laser is mostly used to repair the damages caused due to improper storage, inappropriate storage and mishandling during manufacturing and improper work environment [3].

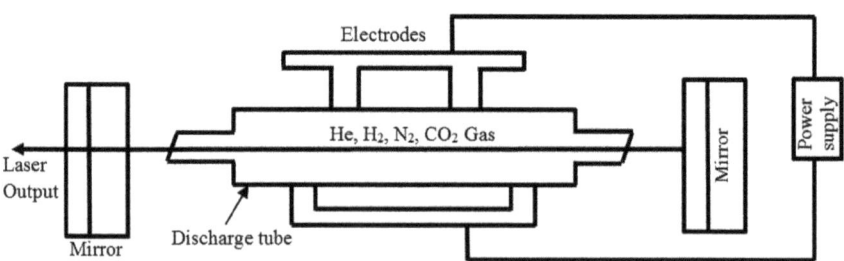

Fig. 6.12 Schematic of CO_2 laser producing system

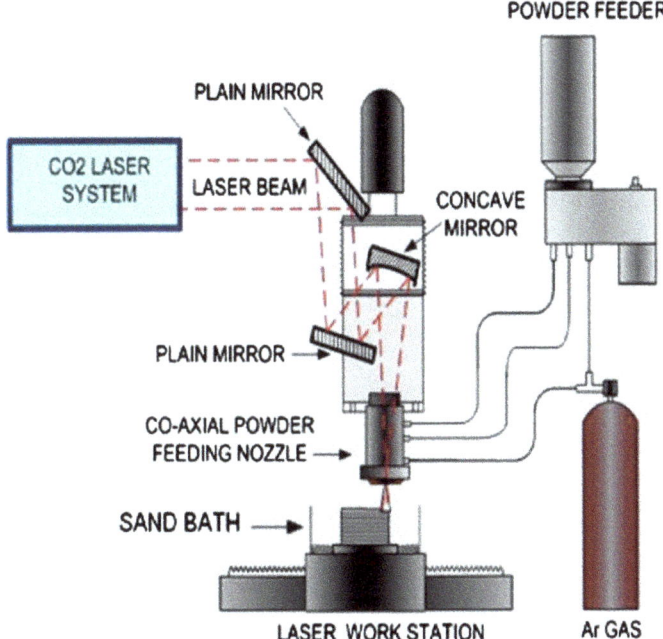

Fig. 6.13 Schematic of CO_2 laser-based repair system [41], with kind permission from Elsevier

It can also be used for mechanical cleaning of damages caused to surfaces due to corrosion, heat checks, etc.

Case Study

Grum and Slabe [3] studied the damages caused to the maraging steel die used to produce castings of aluminum and magnesium alloys. The damages were repaired by depositing Ni–Co–Mo alloy on the thermally cracked surface using CO_2 laser. After deposition, the die was heat treated and its microstructure, micro–hardness and residual stress analysis was carried out which revealed significant reduction in its thermal cracking. Srivastava et al. [42] have investigated occurrence of thermal fatigue cracks (as shown in Fig. 6.14) in a die made of tool steel and used for die-casting and hot forging applications. The die was subjected to the thermal fatigue caused due to sudden cooling. It initiated the fatigue cracks which propagated with increase in the fatigue load. This made the die unusable. They recommended that these cracks can be repaired by a laser-based metallic deposition process which can reduce tendency of thermal cracking thus increasing life of the die.

Fig. 6.14 Thermal fatigue cracks in a die made of tool steel due to sudden cooling [42], with kind permission from Elsevier

6.5.2.2 Nd:YAG Laser

Nd:YAG is a solid state laser developed by J.E. Geusic, H.M. Marcos and L.G. Van Vitert in 1964. Metallic deposition by Nd-YAG laser is an advanced repair process which is very flexible and has advantages over GTA- and PTA-based depositions. Figure 6.15 depicts the schematic of Nd:YAG laser producing system in which energy is generated at four levels. Pumping of Nd ions from the ground state to upper state, i.e., at 4th level is done by krypton arc lamp (also known as flash lamp) using light of wavelength 7200–8000 Å. From 4th level, the ions become non-radiative which drops their energy to 3rd level. Therefore, the laser emission occurs in the 3rd level which is upper laser emission level and 2nd level which is lower laser emission level. This laser has better thermal efficiency than CO_2 laser hence it can be used for depositing the material in powder and wire form. Figure 6.16a depicts schematic of Nd:YAG laser-based repair system with a powder feeder to supply deposition material in the powdered form. This type of laser can be used for small damages caused due to mishandling, defective material, reconditioning, poor inspection and quality control, setting of inappropriate parameters, and improper work environment. Figure 6.16b presents a typical material deposition obtained by Nd-YAG laser-based deposition system.

6 Laser-Based Repair of Damaged Dies, Molds, and Gears

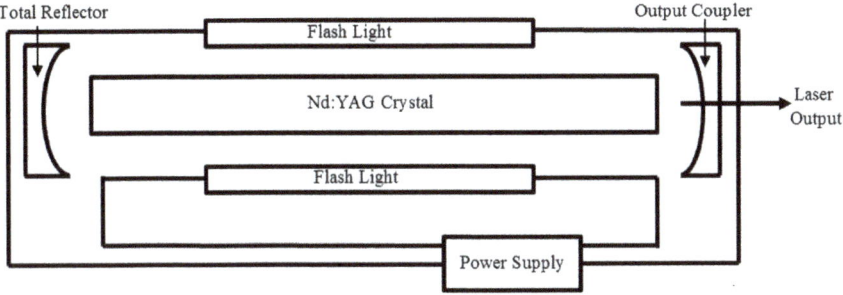

Fig. 6.15 Schematic of Nd:YAG laser producing system

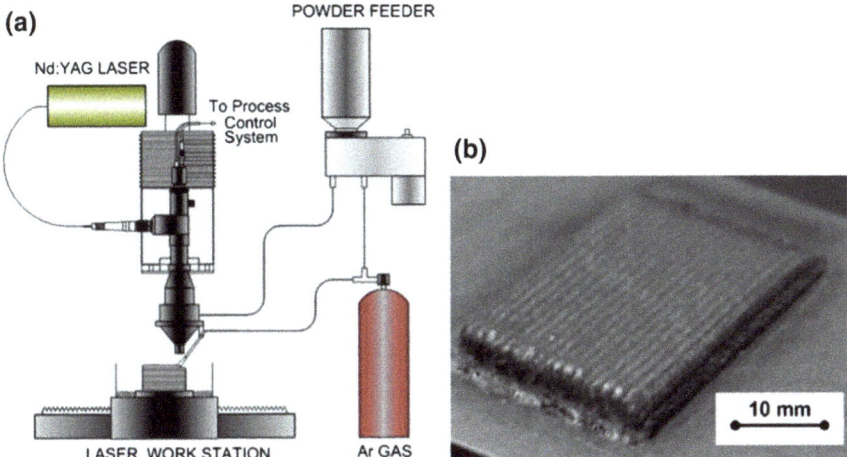

Fig. 6.16 a Schematic of Nd:YAG laser-based repair system; and **b** deposition obtained by this system [43], with kind permission from Elsevier

Case Study

Borrego et al. [31] used Nd:YAG laser to repair the damages such as wear and fatigue cracks (shown in Fig. 6.17a) caused due to high thermal–mechanical loads in the molds made of P20 and AISI H13 steel. The cracked volume was cleaned mechanically and the cavity was filled with similar deposition material in the wire form using Nd:YAG laser-based deposition system. Figure 6.17b illustrates the surface after deposition and Fig. 6.17c shows surface of the repaired mold after removing the excess deposition material by a finishing process. Fatigue strength of the repaired mold was analyzed to check improvement in its fatigue behavior which confirmed enhancement in it due to reduction in the crack formation.

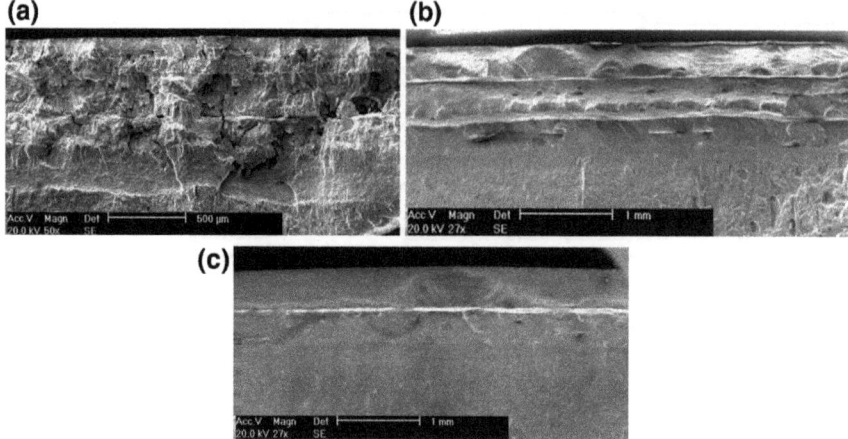

Fig. 6.17 Repair of a damaged mold using Nd:YAG laser **a** fatigue cracks on the mold surface; **b** mold surface after material deposition; **c** mold surface after removing the excess deposited material [31], with kind permission from Elsevier

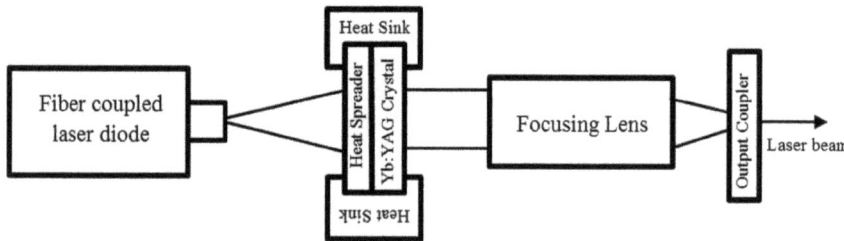

Fig. 6.18 Schematic of Yb:YAG laser producing system

6.5.2.3 Yb:YAG Laser

Yb:YAG crystal is one of the most promising laser active material as compared to the Nd:YAG crystal. It exhibits much larger absorption bandwidth which reduces requirement of the energy controlling system by storing upper level energy for longer lifetime. Figure 6.18 depicts the schematic of Yb:YAG laser producing system in which energy is generated at four levels similar to that in Nd:YAG laser producing system. The pumping of Yb ions to upper state (i.e., at 4th level) from the ground state is done by fiber coupled laser diode. From the 4th level, when the ions passes through focusing lens it becomes non-radiative which drops their energy to 3rd level. Therefore, the laser emission will occur in the 3rd level which is upper laser emission level and 2nd level which is lower laser emission level. This laser has higher thermal efficiency and high beam concentration as compared to Nd:YAG laser. Figure 6.19 depicts schematic of Yb:YAG laser-based deposition system. It can be used to deposit metallic material in wire and powder form

6 Laser-Based Repair of Damaged Dies, Molds, and Gears

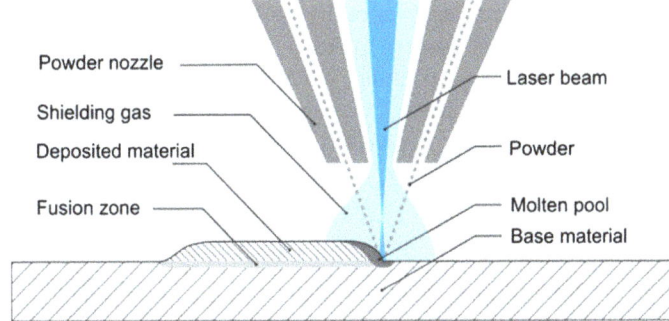

Fig. 6.19 Schematic of Yb:YAG laser-based deposition system [44], with kind permission from Elsevier

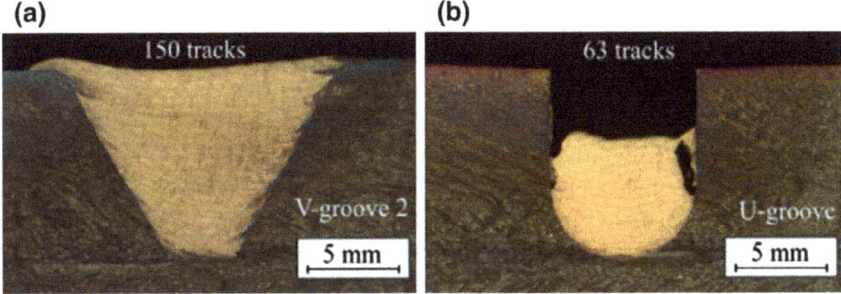

Fig. 6.20 Effect of groove shape on the repaired dies by Yb:YAG laser deposition **a** V-groove; **b** U-groove [45], with kind permission from Elsevier

especially for the super alloys. This laser is mainly preferred to repair deep damages caused due to accidental conditions, deep cracks, faulty design, reconditioning, and setting of inappropriate process parameters.

Case Study

Graf et al. [45] have used Yb:YAG laser to repair of cracks formed in dies made of stainless steel and titanium and experimentally studied the effect of groove shape on the fatigue strength of the repaired die. The cracks were enlarged by preparing V-shaped (Fig. 6.20a) and U-shaped (Fig. 6.20b) grooves in the die material by the milling process. These grooves were filled with the deposition material same as that of dies and molds. Their study revealed that U-shaped groove provides enough access to Yb:YAG laser for deposition and yields good fusion at the side walls as shown in Fig. 6.20b.

Impact and heavy load causes wear of the flank surfaces of a gear tooth. Shi and Bai [10] used laser cladding process to repair these worn-out flank surfaces of a gear made of MS 45 by depositing Ni45 powder as a cladding material. Morphology of the repaired gear tooth was analyzed for different clad thickness and at different process parameters. It revealed that wear resistance of the gear tooth flank surfaces improved significantly.

6.5.3 Form of Deposition Material

Deposition material can be used in either wire or powdered form to repair the damaged dies, molds, and gears. Wire form is chosen when the deposition is carried out continuously because deposition material in the wire form is directly fed to the molten pool of substrate material. For continuous and symmetrical deposition, a ratio between arc diameter and wire diameter is calculated. This ensures that the material is deposited smoothly and the deposited material gets strongly bonded with the substrate material. Any disturbance in wire feed rate and its positioning disturbs the molten pool which causes inaccuracy in shape and size of the deposition. Main advantages of using deposition material in the wire form are: (i) very high deposition efficiency which results better material utilization and lesser consumption of the deposition material; (ii) less hazardous due to less wastage of the deposition material. Whereas, main limitations are: (i) lower deposition rate; (ii) poor control over deposition geometry; and (iii) poor metallurgical bond with the substrate material. Maintenance of accurate position and feeding of wire is very difficult in the electron beam and laser-based deposition processes.

Powdered form of the deposition material is used when (i) electron beam or laser-based deposition process is to be used; and/or (ii) for those deposition materials which cannot be drawn in the wire form. Such materials include hard alloys (i.e., Cobalt-based, Ti-based, Ni-based), refractory materials (i.e., Ta, W, Mo), reinforcements for composite material (i.e., TiC, CrC), ceramics (i.e., silicon nitride, boron nitride), and functionally graded materials. Its main advantages are: (i) higher deposition rate; (ii) better control over deposition geometry; and (iii) better metallurgical bond with the substrate material. Its major limitations are: (i) lower deposition efficiency which results poor material utilization and more consumption of the deposition material; (ii) more hazardous due to more wastage of the deposition material. There are three different ways of supplying the deposition material in the powdered form as shown in Fig. 6.21. They are: (i) laser beam and powder delivery non-coaxial (Fig. 6.21a); (ii) laser beam coaxial with the continuous powder delivery (Fig. 6.21b); and (iii) laser beam coaxial with discontinuous powder delivery (Fig. 6.21c). Powder supply method shown in Fig. 6.21b gives higher powder utilization efficiency and better deposition than that depicted in

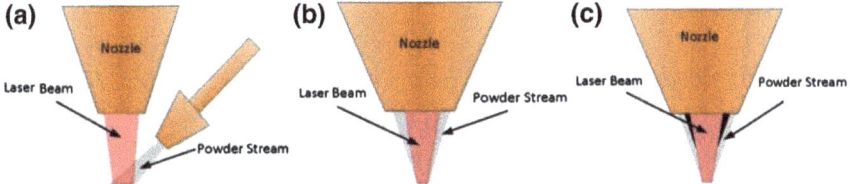

Fig. 6.21 Different methods of supplying the deposition material in the powered form in the laser-based deposition processes: **a** laser beam and powder delivery non-coaxial; **b** laser beam coaxial with the continuous powder delivery; and **c** laser beam coaxial with discontinuous powder delivery [2], with kind permission from Elsevier

Fig. 6.21a. Discontinuous powder delivery (Fig. 6.21c) is better than the continuous powder delivery (Fig. 6.21b) for manufacturing 3-dimensional shapes where inclined deposition is required which necessitates tilting of the laser beam [2].

6.6 Summary

Important causes for the damages to dies, molds and gears are; faulty design, defects in their materials, mishandling, mechanical strain, plastic deformation, local impacts, cyclic loading, high thermal stresses, stress corrosion, wear, fatigue and major forces due to accidental conditions. Traditional repair processes are arc-based and plasma-based deposition processes. Repair processes employing lasers, electron beam and micro-plasma transferred arc have emerged as recent attractive options because these processes have ability to provide techno-economical solutions for repair of the damaged dies, molds and gears. Lasers can be used for repair purposes because they yield more accurate material deposition with very thin deposition layer. The selection of particular laser from the available different type of lasers for repair purpose depends on the process principle, thermal efficiency of the individual laser, flexibility, and easiness to control. CO_2, Nd:YAG and Yb:YAG laser-based metallic deposition are most widely used to repair the damages in the dies, molds and gears because of their low heat affected zone, good metallurgical properties and bonding between substrate and deposited layer, ability to do repair of complex geometries of dies, molds and gears.

With technological advancements in automatic control of the process and power supply, the repair process of damages in the dies, molds and gears has become more competitive and environment friendly. Different types of filler materials, user friendly processes, various combination process parameters, and availability of post-deposition heat treatments have motivated the repair industries to develop various deposition materials. This chapter will facilitate the users in selecting the most appropriate repair process for the damaged industrial dies, molds and gears depending upon availability of the resources and various constraints.

References

1. He B (2011) Research on the Damage and material selection of Plastic mold. Proc Eng 23:46–52
2. Jhavar S, Paul CP, Jain NK (2013) Causes of damage and repairing options for dies and molds: a review. Eng Fail Anal 34:519–535
3. Cosenza C, Fratini L, Pasta A, Micari F (2004) Damage and fracture study of cold extrusion dies. Eng Fract Mech 71:1021–1033
4. The tool and die industry: contribution to us manufacturing and federal policy considerations. CRS Report for Congress. Congressional Research Service 7-5700
5. Timothy A (2003) With foreign rivals making the cut, toolmakers dwindle, Indian edn. The Wall Steel J
6. Pantazopoulos G, Zormalia S (2011) Analysis of the failure mechanism of a gripping tool steel component operated in an industrial tube draw bench. Eng Fail Anal 18:1595–1604
7. Grum J, Slabe JM (2003) A comparison of tool-repair methods using CO2 laser surfacing and arc surfacing. Appl Surf Sci 208–209:424–431
8. Pleterski M, Tuek J, Kosek L, Muhi M, Muhi T (2010) Laser repair welding of molds with various pulse shapes. METABK 49(1):41–44
9. Rao I (2012, May–June) Casting dies for a sustainable future. Eff Manuf (EM) Mag
10. Shi J, Bai SQ (2013) Research on gear repairing technology by laser cladding. Key Eng Mater 546:40–44
11. Garat V, Bernhart G, Hervy L (2004) Influence of design and process parameters on service life of nut hot forging die. J Mater ProcessTechnol 147:359–369
12. Moura GCR, Aguilar MTP, Pertence AEM, Cetlin PR (2007) The materials and the design of the die in a critical manufacturing step of an automotive shock absorber cap. Mater Des 28:962–968
13. Sun Y, Hanaki S, Uchida H, Sunada H, Tsujii N (2003) Repair effect of hot work tool steel by laser-melting process. J Mater Sci Technol 19:91–93
14. Pereira MP, Yan W, Rolfe BF (2012) Wear at the die radius in sheet metal stamping. Wear 274–275:355–367
15. http://looktech.en.ec21.com/Mold_Doctor-1605536_2399545.html
16. http://www.tractorbynet.com/forums/john-deere-lawn-garden/295609-metal-cam-gear-failure.html
17. Stavridis N, Rigos D, Papageorgiou D, Chicinas I, Medrea C (2011) Damage analysis of cutting die used for the production of car racks. Eng Fail Anal 18:783–788
18. Alaneme KK, Adewuyi BO, Ofoegbu FA (2009) Damage analysis of mould dies of an industrial punching machine. Eng Fail Anal 16:2043–2046
19. http://www.empiredie.com/empire-die-casting/resource-center/faq/q7-table.html
20. http://www.novexa.com/en/intervention/gears/defects-treated.html
21. Ebara R, Takeda K, Ishibashi Y, Ogura A, Kondo Y, Hamaya S (2009) Microfractography in Damage analysis of cold forging dies. Eng Fail Anal 16:1968–1976
22. Choi C, Groseclose A, Altan T (2012) Estimation of plastic deformation and abrasive wear in warm forging dies. J Mater Process Technol 212:1742–1752
23. Thompson S (1999) Handbook of mold, tool and die repair welding. William Andrew Publishing
24. Preciado WT, Bohorquez CEN (2006) Repair welding of polymer injection molds manufactured in AISI P20 and VP50IM steels. J Mater Process Technol 179:244–250
25. Branza T, Duchosal A, Fras G, Beaume FD, Lours P (2004) Experimental and numerical investigation of the weld repair of superplastic forming dies. J Mater Process Technol 155–156:1673–1680
26. Horii T, Kirihara S, Miyamoto Y (2008) Freeform fabrication of Ti–Al alloys by 3D micro-welding. Intermetallics 16(11–12):1245–1249
27. Horii T, Kirihara S, Miyamoto Y (2009) Freeform fabrication of superalloy objects by 3D micro welding. Mater Des 30:1093–1097

28. Mizuta N, Matsuura K, Kirihara S, Miyamoto Y (2008) Titanium aluminide coating on titanium surface using three-dimensional microwelder. Mater Sci Eng A: 199–204
29. Xu FJ, Lv YH, Xu BS, Liu YX, Shu FY, He P (2013) Effect of deposition strategy on the microstructure and mechanical properties of Inconel 625 superalloy fabricated by pulsed plasma arc deposition. Mater Des 45:446–455
30. Wang W, Pinkerton AJ, Wee LM, Li L (2007) Component repair using laser direct metal deposition. In: Proceedings of 35th international matador conference 14:345–350
31. Borrego LP, Pires JTB, Costa JM, Ferreira JM (2009) Mould steels repaired by laser welding. Eng Fail Anal 16:596–607
32. Borrego LP, Pires JTB, Costa JM, Ferreira JM (2007) Fatigue behavior of laser repairing welded joints. Eng Fail Anal 14:1586–1593
33. Zhong M, Liu W, Ning G, Yang L, Chen Y (2004) Laser direct manufacturing of tungsten nickel collimation component. J Mater Process Technol 147:167–173
34. Lim JS, Ng KL, Teh KM (2008) Development of laser cladding and its application to mould repair. SIMTech Tech Rep 9(3):142–147
35. Capello E, Colombo D, Previtali B (2005) Repairing of sintered tools using laser cladding by wire. J Mater Process Technol 164–165:990–1000
36. Leunda J, Soriano C, Sanz C, Navas VG (2011) Laser cladding of vanadium–carbide tool steels for die repair. Phys Procedia 12:345–352
37. http://www.badgermetal.com/
38. Mudge RP, Wald NR (2007) Laser engineered net shaping advances additive manufacturing and repair. Welding J:44–48
39. http://www.rpm-innovations.com/laser_cladding_technology
40. Zou JX, Grosdidier T, Zhang KM, Dong C (2009) Cross-sectional analysis of the graded microstructure in an AISI D2-steel treated with low energy high-current pulsed electron beam. Appl Surf Sci 255:4758–4764
41. Desale GR, Paul CP, Gandhi BK, Jain SC (2009) Erosion wear behavior of laser clad surfaces of low carbon austenitic steel. Wear 266(9–10):975–987
42. Srivastava A, Joshi V, Shivpuri R (2004) Computer modeling and prediction of thermal fatigue cracking in die-casting tooling. Wear 256:38–43
43. Paul CP, Alemohammad H, Toyserkani E, Khajepour A, Corbin S (2007) Cladding of WC–12 Co on low carbon steel using a pulsed Nd:YAG laser. Mater Sci Eng, A 464(1):170–176
44. Graf B, Ammer S, Gumenyuk A, Rethmeier M (2013) Design of experiments for laser metal deposition in maintenance, repair and overhaul applications. Proc CIRP 11:245–248
45. Graf B, Gumenyuk A, Rethmeier M (2012) Laser metal deposition as repair technology for stainless steel and titanium alloys. Phys Proc 39:376–381

Chapter 7
Friction Stir Welding—An Overview

Arun Kumar Shettigar and M. Manjaiah

Abstract Friction stir welding is generally recognized as a solid state welding process and developed to overcome the difficulties of joining of aluminium alloys. Later, this process has been adapted to join copper, steel, dissimilar metals, magnesium, composites, etc. This concept can be further used in friction stir processing of metals, production of micro composites and coating of coppers on steel. This chapter elucidates the concept of friction stir welding process, material flow pattern, evolution of microstructure at weld region, and effect of process parameters on mechanical properties.

Keywords Aluminium · Composites · Friction stir welding · Joining · Microstructure

7.1 Introduction

The joining process has extended its applications in various sectors in the vein of automobile, aerospace and locomotive and other industries due to manufacturing constrains. Joining process has gained popularity due to continuous development in the demand of light weight and high strength to weight ratio. Even though joining process has more advantages, but some category of aluminium alloys may not be welded by adopting a fusion welding process. A unique technology was developed by "The Welding Institute" (TWI) in 1991 at Cambridge UK [1]. In this technology the concept is mainly to weld the non weldable aluminium alloys under the frictional heat. The welding process is performed below the melting temperature of the

---

A.K. Shettigar (✉)
Mechatronics Engineering Department, Manipal Institute of Technology,
Manipal, Karnataka, India
e-mail: arunkumarshettigar@gmail.com

M. Manjaiah
Department of Mechanical Engineering Science, University of Johannesburg,
Johannesburg, South Africa

base material; hence, it is belong to solid state welding process. The heat required to perform the weld is obtained from the friction developed at the weld region so it is called as Friction Stir Welding process (FSW). It offers numerous advantages as follows:

(1) Defect related to solidification are eliminated,
(2) The liberation of gases during welding is eliminated,
(3) Filler materials are not required,
(4) Less distortion,
(5) Formation of inter-metallic phases is reduced,
(6) Less Power consumption, and
(7) Green technology.

This chapter sheds a light on the FSW process, describes its working principle, process mechanism; and discusses microstructure formation, important process parameters and other important aspects of this process.

7.2 Equipment and Working Principle

The Friction stir welding requires a stable frictional heat generating element, constant movement of the work piece (welding speed), and homogeneous deformation to consolidate the material. A non-consumable tool consisting of shoulder and pin and made of harder material is used. The rotating tool is plunged into the butting surfaces until the shoulder makes contact with the work piece surface and moves along the joint line. As the welding tool is plunged into the butt joint, the pin helps to produce frictional heat followed by plastic deformation of the metal, and the shoulder helps to produce heat as well as compaction of the material below it. The joint is formed due to the frictional heat as primary heat source and deformation heat under the influence of applied normal force as secondary heat source [2]. Figure 7.1 shows the schematic representation of FSW machine. The basic requirements are automatic welding feed control of x-, y- and z-axis. The z-axis must have tilt facility to provide tool tilt angle. It must have long range of weld feed and spindle speed ranges. The fixture assembly is used to hold the component together rigidly.

The three primary functions of the tool are:

(1) Produce sufficient heat in the workpiece by friction between the rotating tool (pin and shoulder) and the butted workpiece.
(2) Impart proper movement of the plasticized material to form the joint.
(3) Provide reservoir of hot metal beneath the tool shoulder.

The localized heat produced by friction softens the material around the pin and combines with the tool rotation and translation, which leads to the movement of the plasticized material from the leading side to the trailing side of the pin. Thus, softer

7 Friction Stir Welding—An Overview

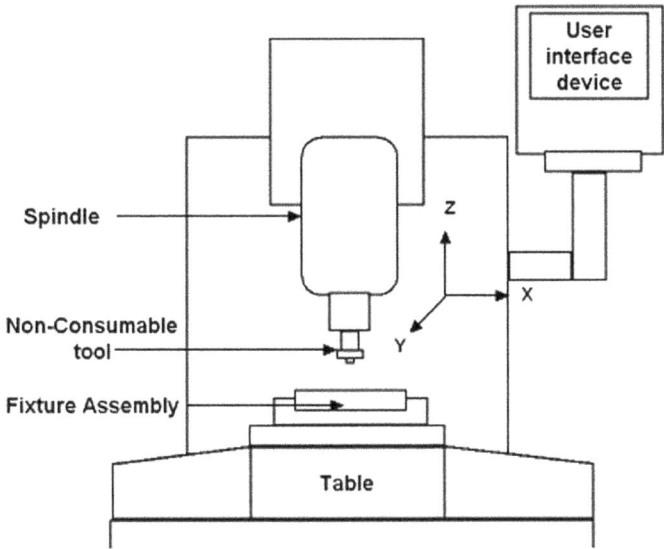

Fig. 7.1 Friction stir welding machine

plasticized material will fill the gap as the tool moves in forward direction. The tool shoulder will restrict the escaping of soft plasticized material flow up to a level equivalent to the shoulder position, i.e. approximately to the top surface of the workpiece. As a result, the thrust force is exerted by the revolving tool against the work piece along the butt line, thereby making the softened and deformed material underneath the tool shoulder compact. Due to this, solid state joint is produced without melting.

Figure 7.2 is the schematic representation of FSW process and its terminologies. The advancing side (AS) refers to the side where tool rotation velocity vectors and direction of welding tool movement are in same direction. In the case of Retreating side (RS), the tool rotation velocity vector, and direction of welding tool movement are in opposite direction. The front side of the tool while moving is called as leading side of the tool and back side of the tool is called trailing side of the tool. The process parameters are tool geometry, axial force, tool rotation speed, traverse speed and tool tilt angle. The angle between the tool axis and normal to the surface of the work piece is called tool tilt angle.

The schematic diagram as shown in Fig. 7.3i–vi depicts the step by step procedure followed in FSW process. The initial stage is plunging of the tool as shown in Fig. 7.3i. During this stage the rotating tool at constant speed approaches the work piece. The plunging depth depends on thickness of the work piece and thrust force. Figure 7.3ii indicates the dwell time after the plunging operation, where velocity vector in each axis is zero. The next stage is welding, where the rotating tool is moved along the direction of the weld and other two axis velocity vectors remain zero as shown in Fig. 7.3iii. Material flow is mainly dependent on two effects. First is the extrusion process where the plasticized material is propelled by

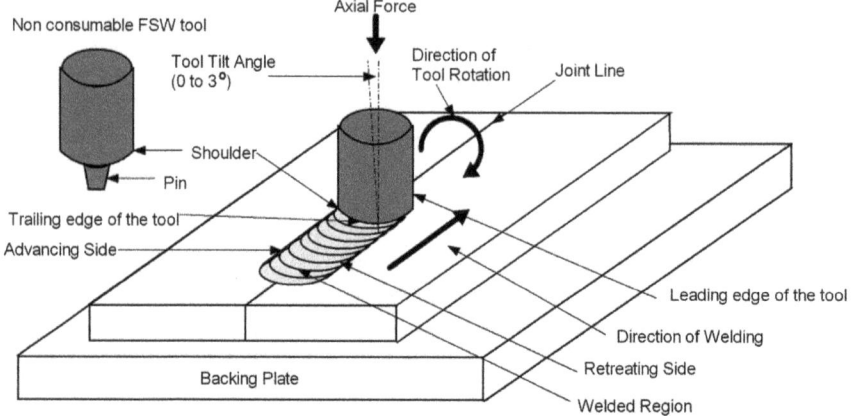

Fig. 7.2 Friction stir welding terminology

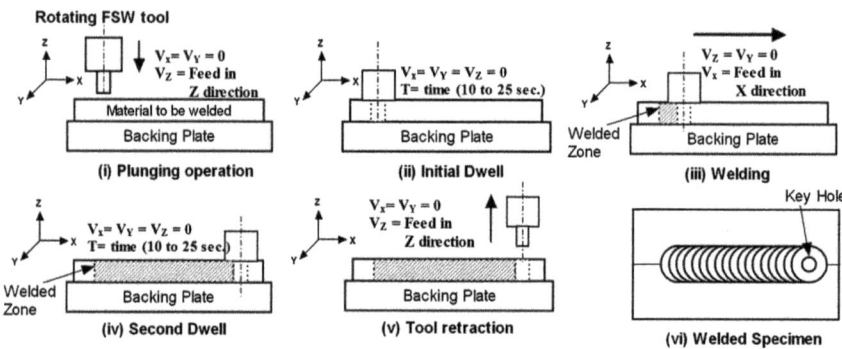

Fig. 7.3 Schematic representation of various steps in friction stir welding process

combined effect of axial force and movement of the tool pin. Second is the driving force, which is produced due to the rotation of the pin. The details of material flow are being discussed in succeeding sections. Figure 7.3iv indicates the second dwell period after the rotating tool reached the final stage. The velocity vector in each axis is zero. The final stage of the welding process is shown in Fig. 7.3v where the tool is retracted from the weld region. During this stage, the z-axis velocity vector is constant and the other two axis velocity vectors remain zero. The top view of the weld region has been shown in Fig. 7.3vi. A key hole is produced at the end of the weld. It is pin occupied region at the end of the weld. It can be eliminated by providing extra length to the component.

So many mechanisms have been proposed so far to describe the material flow during FSW. It involves complex material movement and plastic deformation [3]. Material flow pattern is predominantly dependent on the welding parameters, tool geometry and type of joint design. As soon as the rotating tool plunges into the base material, a cavity is formed in the base material. The shape of the cavity is

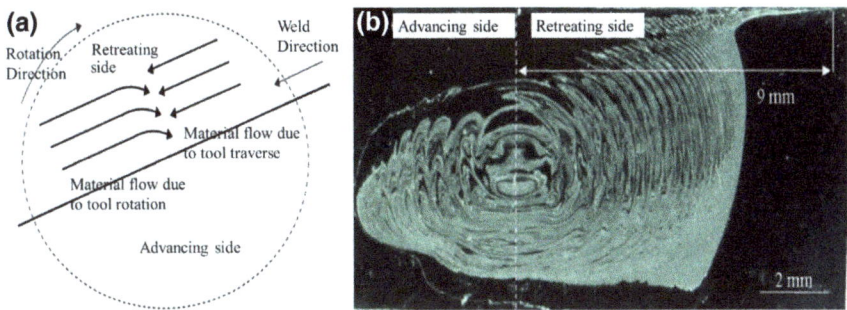

Fig. 7.4 Schematic of metal flow pattern **a** top of the nugget region **b** cross section of nugget region [7] with kind permission from Elsevier

dependent on the pin profile. There is a plasticized material formed around the pin and beneath the shoulder. This plasticized material is enclosed by the surrounding cooler base material and backing plate at the bottom. This arrangement along with tool rotational direction and tool movement decide the material flow path. During the tool movement, the soft metal from the leading side is progressively plasticized and moves to the trailing side through the retreating side by two different modes, namely shoulder- and pin-driven flows.

Soft plasticized material flows through retreating side and the gets deposited behind the tool forming the joint. Figure 7.4 represents the material flow pattern. Colligan et al. [4] made an effort to explain the material flow pattern in AA6061 and AA7075. Small steel balls known as tracer were embedded into the previously made grooves which are parallel to the welding direction at different distances. Radiographic examination of the welded components revealed different material flow at different parts of the material. The thread profile in the pin assisted in forcing the stirred material in downward direction. Remaining material in the weld zone is extruded around retreating side of the pin and deposited. Seidel [5] stated that the flow pattern is not symmetric on advancing side and retreating side. The plasticized material flow, at the top surface, up to one third thickness of the weld zone is governed by the tool shoulder rather than the thread of the pin [6]. Near the top surface of the weld, a considerable quantity of plasticized material moves from retreating side to the advancing side due to the shape of the tool. Hence, movement of the material causes vertical mixing in the weld and complex circulation of the material along the longitudinal axis of the weld.

7.3 Mechanism of Friction Stir Weld Formation

In case of shoulder-driven FSW, the material flows from the retreating side and gets deposited on advancing side of the base material with forging action induced by the tool. Where as in the case of pin drive, material flows layer by layer around the pin,

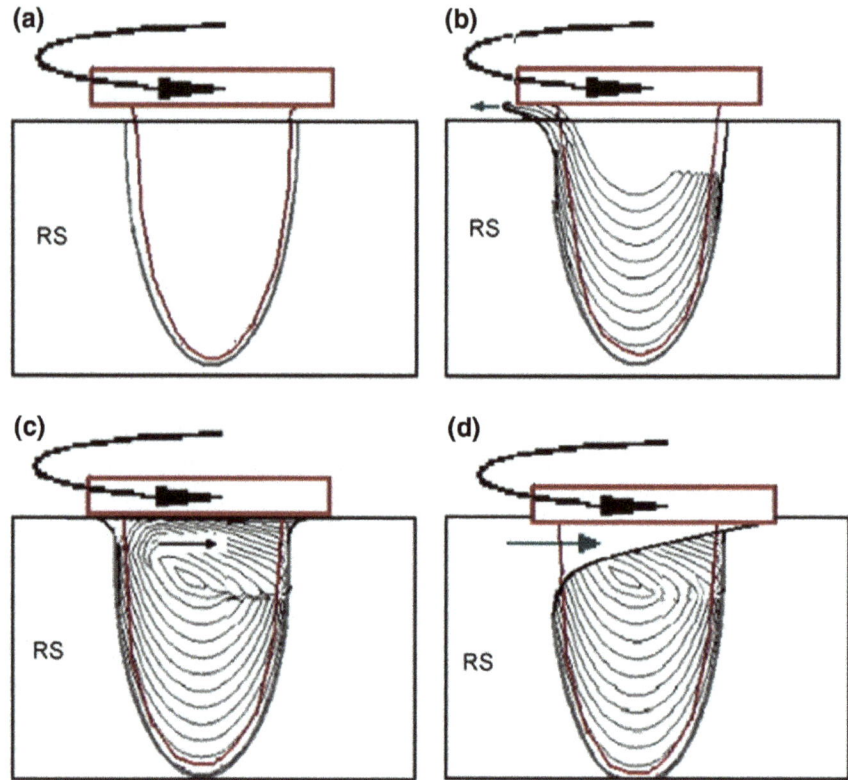

Fig. 7.5 a Creation of weld cavity during plunging, **b** Cross section of the layers in the pin-driven flow, **c** Merging of pin- and shoulder-driven material flow region and **d** Drawing of base material into weld nugget

due to the extrusion phenomenon and the layers continuously get stacked in the weld line. The shearing action of the tool shoulder and extrusion of the soft plasticized material around the pin cause layer by layer material transfer. Once this pin-driven material interacts with the shoulder-driven material on the retreating side, the plasticized material gets transferred from the retreating side to advancing side. There will not be any change in the structure details of layers in the pin-driven material. Further increase of shoulder interaction results in merging of pin-driven and shoulder-driven material. The joint will be formed if the process develops sufficient temperature and hydrostatic pressure to transfer the shoulder-driven and pin-driven material to fill the weld cavity [8]. The material flow in the FSW process is summarized schematically in Fig. 7.5.

7.3.1 Role of FSW in Material Flow

As reported by Arbegast et al. [9], resemblance was found between the resultant microstructure of hot worked aluminium extrusion and forging, with that of the features obtained through friction stir welding. Figure 7.6a, b show the different mechanical processing zones formed during friction stir welding. Hence, modelling of FSW process can be treated as metal working, which includes conventional metal working zones of preheat, initial deformation, extrusion, forging, and cool down. Preheating zone is ahead of the pin. The rise in the temperature at preheating zone is due to frictional heat developed by rotating tool and deformational heat. The heat rate expansion of preheat zone is dependent on the thermal properties of the work piece material and rate at which welding process is carried out. Initial plastic zone is formed as the tool moves in the forward direction. As a result, material is heated above the critical temperature and the amount of stress exceeds the critical flow stress of the material, resulting in material flow. The plasticized material in this zone is forced to move upward in the shoulder zone and downward in the extrusion zone.

Vertex swirl zone is formed beneath the pin tip due to rotation of the pin. A small amount of material experiences the vertex flow pattern in this zone. A finite amount of material moves around the pin from the leading side to the trailing side of the pin known as extrusion zone. The magnitude of temperature and stress is not enough to allow the material flow in this zone and critical isotherm defines the width of the extrusion zone. Next to the extrusion zone is the forging zone where plasticized material is forced into cavity formed by forward movement of the tool under hydrostatic pressure condition. The shoulder of the tool assists in moving the material in cavity and also provides downward forging force. The softer material at shoulder zone is forcefully dragged from the retreating side to the advancing side. At the final stage, the material gets cooled under natural or forced cooling condition post heat zone. The plasticized material at weld zone experiences three types of flows. Initially at near the tool, a slug of softened material is rotated around the tool

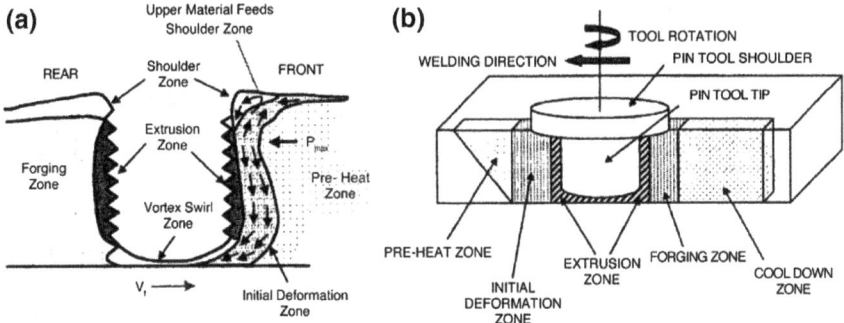

Fig. 7.6 **a** Zones of different mechanical processes, and **b** mechanical interactions determining these processes in FSW [10] with kind permission from Elsevier

under the influence of the rotating tool and also results in friction between the work piece and the tool. Secondly, during tool rotation, the threaded portion of the pin moves the softened plasticized material in the close proximity of the pin in the downward direction, which in turn, results in driving an equal amount of plasticized material which is farther away from the pin, in the upward direction. Finally, there is a relative motion between the tool and the work piece. The combined effect of this material flow results in formation of the joint. The rate at which the heat is transferred into the tool along with the parameters of welding and properties of material affects the width of the recirculating plasticized material flow region.

7.3.2 Evolution of Microstructure at Weld Zone

Significant attention has been paid to study the microstructural evolution of friction stir welded aluminium alloys. Aluminium alloys demonstrate verity of crystallographic textures, grain size and grain morphologies that depend on the material composition and heat treatment. The first attempt at classifying microstructures was made by Threadgill et al. [11]. Figures 7.7 and 7.8 show different zones of microstructure developed at weld region after welding. The system divides the weld zone into distinct regions as follows:

A. Nugget Zone or Stir Zone (NZ or SZ)
B. Thermo-Mechanically Affected Zone (TMAZ).
C. Heat Affected Zone (HAZ).
D. Unaffected Base Material (BM).

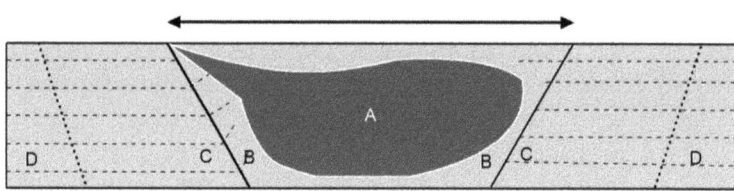

Fig. 7.7 Generalized butt joint profile proposed by TWI

Fig. 7.8 Welded region of FS W process and different regions

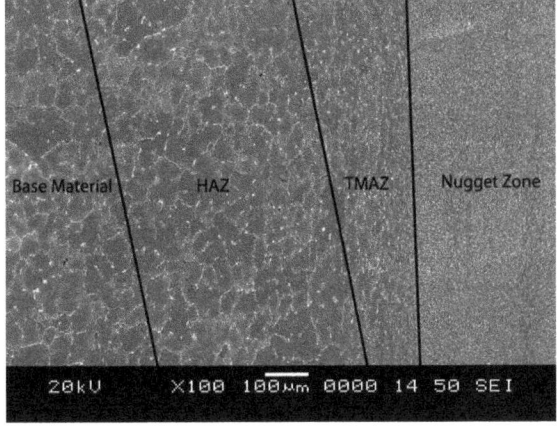

Fig. 7.9 Typical SEM images of the TMAZ, the HAZ, BM and Nugget region Grain distribution

Figures 7.8 and 7.9 show the typical Scanning Electron Microscopy (SEM) images of the TMAZ, the HAZ and the BM.

7.3.2.1 Nugget Zone

A nugget zone is characterized by fully recrystallized area, sometimes called as stir zone. It also refers to the zone previously occupied by the tool pin. The variation in the size of the nugget zone depends on welding process parameters, built temperature, tool geometry and thermal conductivity of the material. Mishra et al. [10] reported that the nugget shape can be sub categorized by basin-shape and elliptical shape. The wider nugget shape is formed on upper surface due to extremely higher heat generated at tool shoulder and material interface. The lower rotational speed produces basin-shape and higher rotational speed produces elliptical shape.

7.3.2.2 Thermo-Mechanically Affected Zone

TMAZ is a zone which is characterized by the plastic deformation of grains. It is also a zone which is in close proximity to the nugget zone and thus it is exposed to higher temperature. The cause of the plastic deformation in TMAZ is due to the shearing of grains induced by traverse of the tool and the tool rotation. The plastic deformation in the TMAZ varies with its proximity to the nugget zone as well as to its depth in the weld.

The higher degree of deformation in the grains is identified towards the tool shoulder and nearer to the weld. The grains are less deformed further from the centerline of the weld. It is hard to recognize the boundary between the TMAZ and HAZ. However, to define the outer boundary of the TMAZ, a method has been developed based on the angular distortion. Since, the TMAZ experiences significantly higher temperature, the strengthening precipitates in the vicinity of the nugget zone gets dissolved and the strengthening precipitates in areas near to HAZ coarsen, leading to significant deterioration of strength. The exact boundary between the coarsened particles and dissolved particles is highly dependent on welding parameters. Thus, the final distribution of precipitates is a function of the time-temperature history of the zone. TMAZ is a transition zone, between the parent metal and nugget zone, created by the FSW. The grains appear to be elongated in the upward flowing direction around the nugget zone. This transition zone between the parent metal and nugget zone on the advancing side is sharp while on the retreating side appears to be relatively spread over.

7.3.2.3 Heat Affected Zone

In this zone, there will not be any plastic deformation. But, thermal energy of the FSW process is experienced by this region and facilitates reforming of the grain at HAZ. Therefore, grains in this zone are slightly larger as compared to the base material. In precipitation strengthening alloys, thermal energy causes over ageing of precipitates which results in deterioration of mechanical properties. The heat input to the work piece at HAZ is a function of the welding parameters. So far, the researchers have highlighted that welding parameters considerably depend on the nature and purpose of the process. Therefore, there is considerable variation in the width and property of the HAZ.

7.3.2.4 Unaffected Base Material

This material lies far away from the weld premise that has not been deformed and it may have experienced a thermal cycle of negligible magnitude during the welding. This thermal cycle does not affect the microstructure or mechanical properties of the weld material.

On the advancing side elongated and extremely extruded grains have been observed in the TMAZ. But on the retrieving side, an unclear interface between TMAZ and nugget zone has been observed. However, the HAZ exhibits a grain structure which is similar to the BM, since HAZ is not subjected to any thermal cycle and not undergoing any plastic deformation.

7.4 Process Parameters of FSW

The weldability of the aluminium alloy and aluminium metal matrix composites (AMMC's) welded by friction stir welding is dependent on the process parameters, material and tool geometry. Rotational speed, welding speed and thrust force are the some of the important process parameters in FSW.

7.4.1 FSW Tool

FSW tool is an essential component to achieve success in the welding process. A non-consumable rotating tool usually comprises of a round shoulder and a pin of different shapes. The primary purpose of the tool is to heat the workpiece by friction and deformation, and moves the soft plasticized material around it to form the joint. Due to severe stresses and high temperature developed at the pin during welding, it is considered as a weakest component [12]. Improper selection of tool material and shapes lead to tool wear, which impacts not only the tool life but also the weld characteristics [13, 14].

Deb Roy et al. [12] have studied the durability of the tool for FSW of aluminium alloys. The durability of the tool is dependent on the type of tool and work piece, process parameter, joint thickness and shape of the tool. Chen et al. [13] have performed the joining of dissimilar metal, which consisted soft metal aluminium on one side and hard metal AISI 1080 steel on the other side. The observation had confirmed the presence of small particles of tool material at weld zone. The mechanical properties of AA6061 using threaded tool was studied by Zeng et al. [14]. The effect of the tool wear on the mechanical properties of the welded composites was quite evident from the tensile test of the welded components. Therefore, it is essential to elucidate on the features of tool material as well as tool shapes.

7.4.1.1 Tool Materials

Several studies are available on the selection of tool material. The basic requirements of tool material properties are good dimensional stability, greater strength, good creep resistance, good fracture toughness, greater thermal fatigue strength, higher compressive strength at elevated temperature, lower thermal coefficient of expansion and it should not react with workpiece material. Before selecting the tool material it is important that, tool should be as simple a shape as to manufacture, to reduce cost of production and generate sufficient stirring effect [15]. Initially the friction stir welding was developed to join the soft metal like aluminium alloys, which are easily stirred with tool steel. Steel materials are easily available and it can be machined to desired shape and size at low cost, and also material characteristics can be established. But in the case of welding of composites, the presence of hard particles stimulates more tool wear [16]. The rotational speed is directly proportional to the tool wear.

From the work carried out so for on selection of the tool material and self-optimized tool wear, it appears that there is definitely an improvement in tool as far as its wear and stirring effect are concerned. Shindo et al. [17] have conducted study on the welding of Al359–20% SiC using threaded profile pin made of steel. A self-optimized shape (without threads) without any further additional tool wear of the pin produced excellent, homogeneous weld. Prado et al. [16] have studied the tool wear of threaded screw tool made of steel for friction stir welding of AA6061-20% Al_2O_3 composite. Self-optimized tool shape had been obtained after some experiment with the FSW tool without threads. Such optimized shape produces good quality of weld without any tool wear. In most of the cases, the tool wear occurs due to improper selection of process parameters. High rotational speed and low welding speed leads to tool wear rapidly. Fernandez and Murr [18] have reported the process parameters for self-optimized tool shape. The study suggested that, when rotational speed decreases and welding speed increases, the tool wear initially goes down due to filling of work material in the threads. After some distance is travelled by the tool (>3 m), there will not be any wear due to self-optimized shape.

7.4.1.2 Tool Geometry

FSW tool comprises of shoulder and pin profiles. The heat is generated due to the friction between the work piece and the pin during plunging stage. Some heat is also generated due to the deformation of the material. As soon as the tool shoulder makes contact with work material, the area of contact between the shoulder and the work piece increases. Hence, frictional contact between work piece and shoulder increases, which leads to generate larger amount of heat. Majority of the heat generated in thin sheet is produced by the shoulder. However, in thick sheets the pin produces majority of the heat. From heating point of view, the relative ratio of shoulder to pin is important. In the study carried out by Padmanban et al. [15], it is revealed that the shoulder also confines the plasticized softer material. During welding process, the plasticized soft material is extruded from advancing side to the retreating side of the tool. These materials are trapped by the shoulder and deposited along the joint to produce smooth surface finish. The second purpose of the tool is to stir and move the material. Mishra [10] has reported that the homogeneity of microstructure and mechanical properties are governed by the tool design and as well as welding speed and rotational speed of the tool. Lorrain et al. [19] reported that the most dominant factors are pin to shoulder diameter ratio, shoulder surface angle, pin shape and pin size. The various tool geometry parameters which affect the weld quality are discussed below.

Tool Shoulder Diameter
Significant attention has been devoted to study the effect of tool shoulder diameter of the FSW tool. The tool shoulder generates most of the heat [1, 20], and its control on the plasticized material largely establishes the material flow field. Mehta et al. [21] studied the influence of tool shoulder diameter on thermal cycle, power requirements, peak temperature and torque during FSW of AA7074-T6. The study revealed that the temperature increases with increasing diameter of the shoulder. For defect free weld, the tool shoulder must prevent the escape of the plasticized material, and the total torque, even the traverse force should not be extreme. The amount of heat generated increases with the increase in tool shoulder diameter due to larger contact area, thus resulting in a wider TMAZ and HAZ regions. As a consequence, it has been perceived that as the diameter decreases the amount of frictional heat generated reduces. The weld quality deteriorates due to lesser friction

7 Friction Stir Welding—An Overview

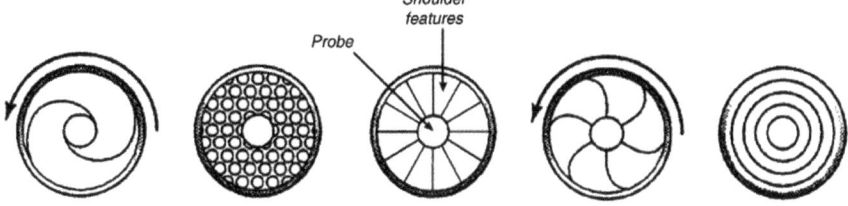

Fig. 7.10 Tool shoulder geometries, viewed from underneath the shoulder [24]

leading to lack of weld merging. Therefore, only a tool with an optimal shoulder diameter results in the highest strength.

Tool Shoulder Surface

Literature survey reveals that, the tool shoulder surface is also an essential aspect of the tool design. Several features were adopted in the surface to increase material deformation, to ensure proper mixing of material, to obtain smoother surface finish and to act as reservoir for the forging action. Commonly used features are shown in Fig. 7.10. The concave shaped tool produces good quality weld. It is simple in design and easy to manufacture. A small angle (6–10°) is made in the tool from the edge of the shoulder to the pin base. A small quantity of material is forced into this cavity during final stage of plunging of the tool and the material thus stored acts as a reservoir for the forging action of the tool shoulder. During forward traverse movement of the tool, new soft material is forced into this cavity and displaces the stored material into the flow of pin. Pedro Vilaca et al. [22], have investigated friction stir welding of AA5083-H1 alloy using concave shaped shoulder. The concave shape of shoulder produced smooth surface finish. Venkateswaralu et al. [23], have further studied the surface concavity effect on mechanical properties. The study involved, varying the surface concavity by addition of flat surface at the shoulder periphery. The result showed higher tensile strength obtained for 2 mm flat shoulder surface followed by concavity of 7°, which was perceived to be more appropriate for achieving adequate tensile strength.

Tool Pin Geometry

There have been many tool pin geometries reported in the literature to obtain improved mechanical properties. The profile of the pin designed must be in such a way that it retains the maximum plasticized material in weld cavity. The most commonly designed and used pin profiles are cylindrical or tapered with or without threads [24]. Complex pin profiles such as square profile, triangular profile, Flared-Triflute and Skew-stir have also been used. In threaded profile, the pin produces higher heat and vertical flow of the material in the direction of thread. Generally clockwise rotation is used for left hand threaded tool and anticlockwise rotation of the tool is used for right hand tool. The threaded profile pin with high pitch acts like a drill rather than stirrer, which spill the material in the form of chips. The improved mechanical property is obtained for a threaded profile pin with 0.8 and

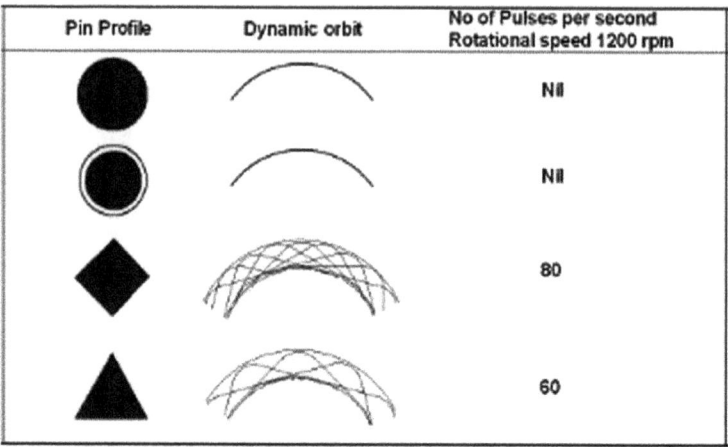

Fig. 7.11 Effect of pin profile on dynamic orbit and pulsating action

1.1 pitches. Elangovan and Balasubramanian [25], used different tool geometries for processing of AA2219. The shape used for the study and its dynamic orbit is shown in Fig. 7.11. The square and triangle shaped tool pin are associated with eccentricity which allow the incompressible material movement around the pin profile. Because of the eccentricity of the pin, the dynamic orbit is related to the eccentricity of the rotating plasticized material. The path for the flow of plasticized material is identified by the ratio of static volume and dynamic volume. It in turn leads to the pulsating stirring action in the plasticized flowing material due to flat surface.

FSW Pin Diameter

The pin diameter is responsible for deciding the volume of material being stirred. Suppose, the tool diameter is smaller, then the volume of material stirred also is less and vice versa. The combined effect of smaller pin diameter, lower welding speed and higher rotational speed causes higher heat input to a smaller volume of material which results in turbulent material flow and coarse grain structure. On the other hand, large pin diameter with lower welding and rotational speed causes lower heat input to maximum volume of material. This effect will lead to inadequate material flow and insufficient plasticization. Larger pin diameter with respect to shoulder diameter resulted in insufficient heat generation due to wider contact area and produced defects in advancing side of TMAZ region [26]. The shape of the tool pin influences the flow of plasticized material and affects weld properties. Figure 7.12 shows the shapes of some of the commonly used tool pins. A triangular or 'trifluted' tool pin increases the material flow as compared to a cylindrical pin [24, 26, 27].

The effects of tool profile on microstructure and tensile strength of Al–TiB$_2$ MMCs were studied by Vijay et al. [28], using eighteen various FSW tools by varying the tool pin profile, ratio of shoulder diameter to pin diameter (D/d). It is concluded that the joint made of taper profile pin tool has shown coarse grains when compared to the joints made with straight pin profiled tools. The tensile strength of

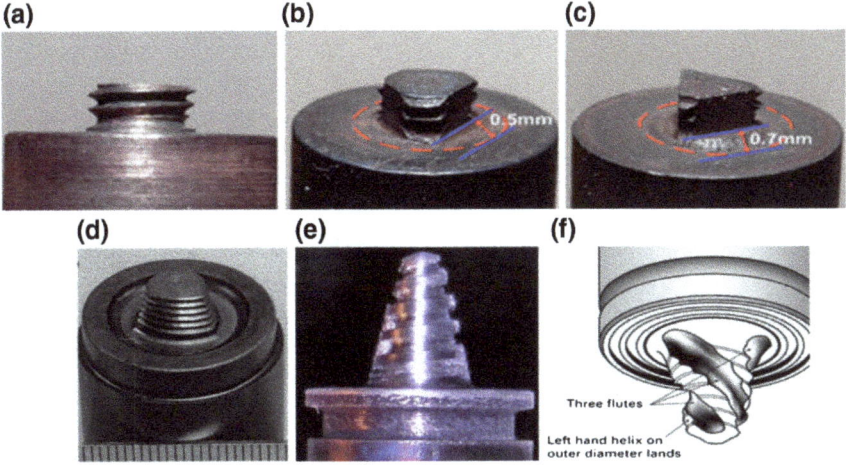

Fig. 7.12 Shapes of some of the commonly used tool pins. **a** Cylindrical threaded **b** Three flat threaded **c** Triangular **d** Trivex [26] with kind permission from Elsevier **e** Threaded conical **f** Triflute [24]

the friction stir welded specimen is also affected by the tool pin profile. The joints produced with square tool exhibited better tensile strength when compared to joints produced with other tools due to the pulsating stirring action of square flat faces. Since the tapered square pin tool sweeps less material when compared to that of straight square pin tool, these joint exhibits reduced tensile properties. There is not much pulsating action in the case of octagonal profiled tool pin because it almost resembles a straight cylindrical profiled tool pin at high rotational speed. Figure 7.13 shows the various tool pin profile used for study.

The improvement in the mechanical properties of the joint produced by square pin tool is due to the difference in dynamic orbit created by the eccentricity of the rotating tool of the FSW process. The relationship between the static volume and dynamic volume decides the path for the flow of plasticized material from the leading edge to the trailing edge of the rotating tool properties. Table 7.1 shows the effect of tool pin profiles on the % elongation of the FS welded aluminium matrix composites (AMCs). It is evident from the table that the effect of tool pin profile on the % elongation is insignificant except for the octagon pin tool. The reason is the presence of ceramic particles in the parent metal which considerably reduces the % elongation of the material. The effect of tool profile on the joint efficiency is also similar to that of tensile strength. The joint efficiency is high when the AMC is welded using square pin tool and low when it is welded with tapered square pin tool.

Features such as threads and flutes on the pin are believed to increase heat generation rate due to larger interfacial area, improve material flow and affect the axial and transverse forces. Elangovan et al. [25], have studied five tool profiles, namely, straight cylindrical, threaded cylindrical, tapered cylindrical, square and triangular for the welding of AA6061 aluminium alloy and found that the square pin profiled tools produced defect free welds for all the axial forces used.

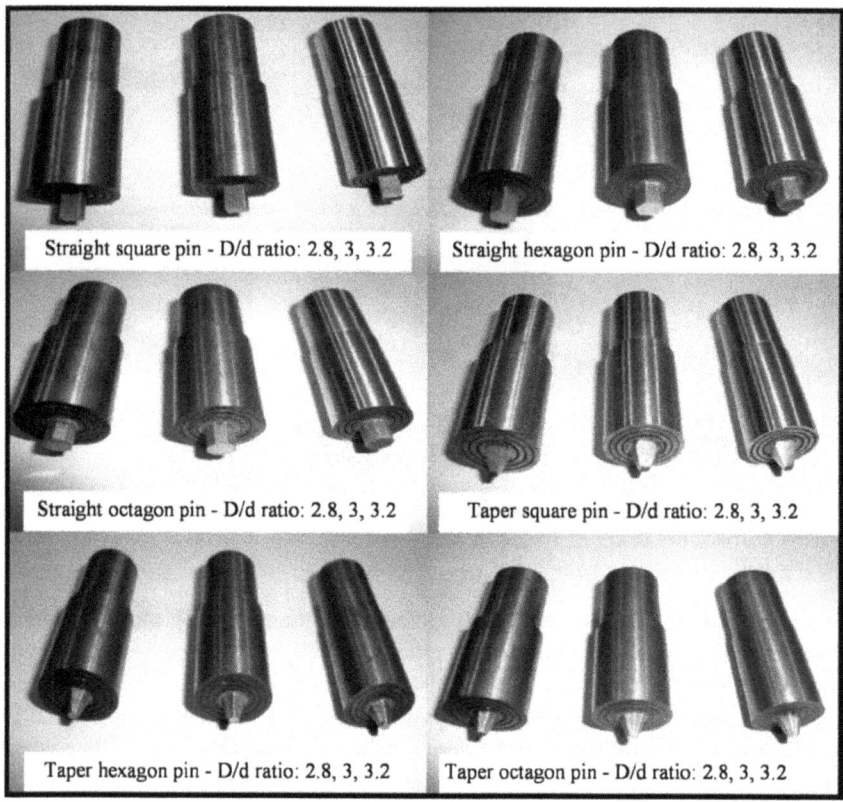

Fig. 7.13 Various FSW tools [28] with kind permission from Elsevier

Table 7.1 Mechanical properties of friction stir welded Al-TiB$_2$ composite [28] with kind permission from Elsevier

Profile of the tool pin	Average tensile strength (MPa)	Average (%) elongation	Joint efficiency (%)
Tapered square	223.33	5.32	78.92
Tapered	247.89	6.67	87.59
Tapered	245.27	6.22	86.67
Square	281.51	6.37	99.47
Hexagon	262.29	5.83	92.68
Octagon	240.00	3.39	84.81

Tool Rotational Speed

The study of the effect of tool rotation speed on weld quality showed that proper mixing and adequate heat generation is dependent on the rotational speed. At lower rotational speed, heat generated is less and also mixing of material is not proper, irrespective of the welding speed. Rajkumar et al. [29], have studied the effect of rotational speed on ultimate tensile strength (UTS). The result revealed that, heat liberated due to friction is predominantly dependent on rotational speed. If heat generated is less then heat supplied to the base material is also less, and it affects the softening of the material and consequently the material flow. As a result lower value of UTS is found in FSW of composites. At high rotational speed, heat generated is more. Therefore heat supplied also is more. Hence, softer plasticized material is available at nugget zone (NZ) which results in turbulence in material flow and coarse grains formed at NZ. The rotational speed increases, the UTS also increases up to a certain value and further increase in rotational speed decreases the UTS.

Welding Speed

The welding speed only decides the quantity of heat supplied to the base material to be joined. If the heat generation is less, then heat supplied will be relatively less and vice versa. Characteristics of friction stir welded joints are influenced by material flow and temperature distribution across the weld which are dictated by pin/shoulder geometry and welding parameters. The UTS, percentage elongation (EL) and joint efficiency decrease with increase in welding speed. Further, the minimum hardness region shifts from heat affected zone to weld nugget zone on increasing the welding speed and decreasing the tool rotational speed. Stirring becomes insufficient at higher welding speeds. The material then does not travel enough to the Retreating Side (RS) from the Advancing Side (AS) of the tool. The increase in the welding speed leads to lower heat generation with faster cooling of the plasticized material thereby reducing the softened area. The lower traverse speeds reduce grain size of specimens with SiC particles, in the NZ because of higher level of segregation of SiC particles. On the other hand, higher traverse speed causes accumulation of SiC particles and reduces micro hardness values. The ratio of rotational to welding speed (ω/v) in friction stir welding plays vital role. By increasing (ω/v), a slight decrease in the effective tensile properties is observed. This was due to increased heat input and softening of the material in these regions. Furthermore, increasing (ω/v) ratio results in the formation of a larger weld nugget because of an increase in heat input and an easier material flow. Therefore, the probability of formation of "incomplete root penetration" defect is reduced when (ω/v) ratio increases. Peng Dong et al. [30] have reported that, as welding speed increases the tensile properties also increases. This is due to lower heat input at higher welding speed which weakens the growth and transformation of precipitates. This leads to narrowed softened region as well as shifting of the weakest location to TMAZ just adjacent to NZ with reduced strength loss during tensile testing.

Axial Force

Another parameter which has to be considered during FSW is axial force. Only limited data is available to understand the effect of axial force on tensile strength. The material flow pattern mainly depends on the axial force. Inadequate plasticized material was obtained at top surface of the weld due to lower axial force. The bonding occurs when a pair of surfaces is brought in the vicinity of inter atomic forces. Axial force propels the plasticized material in the weld zone to complete the extrusion process. Axial force is also responsible for the plunge depth of the pin. When the axial force is relatively low, there is a tunnel defect found at the bottom of the weld zone. If axial force is excessive, the material flow in the form of flash, on either side of the weld region, that is, on both AS and RS. This leads to thinning of the joint thereby reducing the strength of the weld. Elangovan et al. [25] have expressed their thought on the effect of axial force increasing the tensile strength of AA6061 alloy joint fabricated by FSW. Coarse grains with clustered strengthening precipitates were obtained at axial force of 6kN. When axial force is 8 kN, coarse grain with fine and uniformly distributed strengthening precipitates were obtained. The tensile strength increases with increasing axial force up to a certain limit and then it decreases. At lower axial force, improper material flow forms defect in the weld zone. Hence, UTS was low. At higher axial force, the non-uniform distribution of Si particles and thinning effect led to a decrease in the UTS.

Tool Tilt Angle
The forward tool tilt angle has crucial significance on material flow, frictional force and heat input under the same process parameters condition. Lower tilt angle results in improper flow of plasticized material leads to producing bell-shaped nugget at the lower part of weld zone with defects. Increase in tool tilt angle will increase the heat generation. Additional heat input can improve the flow of plasticized material near the end of the pin and extrude in downward direction. When tool tilt angle more than the recommended, the plasticized material zone is widened and sheared drastically. This results in producing onion rings in weld zone.

7.5 Mechanical Properties

7.5.1 Hardness

The discussion on mechanical properties given here is pertaining to the FSW of metal matrix composites (MMCs). Amirizad et al. [31] have reported that the hardness of weld zone (Nugget and TMAZ) is more than the base material [20, 32] for A356/15% SiC composite. Fragmentation of SiC particles resulted in increase in the number of SiC particles in the nugget zone with a decrease in its size. The homogeneous distribution of this fragmented particles lead to the increase in hardness. Huseyin et al. [33] have carried out hardness test of heat-treated (T4) AA2124 alloy reinforced with 25% SiC particle composite. They concluded that the hardness value obtained was low at HAZ due to the annealing effect in the region. However, hardness at TMAZ was slightly higher on both sides of the weld region which was attributed to the second phase particle dissolution. Ceschini et al. [34] have investigated the effect FSW on evolution of microstructure and mechanical properties in extruded AA7005 reinforced with 10(Vol.%) Al2O3 composite plates.

They reported that the maximum hardness was noticed at TMAZ as compared to NZ, HAZ and base material. Feng et al. 2008 [35] have reported that the hardness of the friction stir welded AA2009/15% SiC heat-treated (T6) composite is dependent on distribution of strengthening precipitates, grain and dislocation density. The coarse precipitates undergo partial dissolution during the course of FSW followed by re-precipitation during natural ageing. This contributes to increase in the hardness. Subsequently, recrystallization of grains and fragmentation of hard particles also contributed to the increase in hardness at NZ. Chen et al. [36] have reported that hardness at NZ decreased after heat treatment of friction stir welded AA6063 reinforced with (0–10.5%) B4C composite. Nami et al. [37] have explored the effect of number of passes of the tool during FSW of Al-Mg2Si composite. With the increase in number of passes of FSW tool, hardness at NZ was decreased while the width of the weld zone was increased. This is due to increase in heat at the weld zone and lower cooling rate which resulted in grain growth. Ni et al. [38] have reported on the W shape hardness profile obtained for AA2009 reinforced with 17 (vol.) % SiC composite. Effect of tool shoulder diameter on hardness increases with increase in the shoulder diameter and tool rotational speed.

7.5.2 Tensile Properties

Feng et al. [35] have investigated the mechanical properties of friction stir welded AA2009–15(Vol.%) SiC composite. Enhancements in the tensile and yield strength are obtained due to precipitation strengthening which is resulted from the result of FSW thermal cycle. Fine grain and hard small particles also contribute to in increasing the strength of the composite. After heat treatment at T4 condition, both longitudinal and traverse direction tensile strength are less than the base material.

Vijay et al. [28] have investigated the effect of tool pin profile on the joint strength of Al reinforced with 10 (wt.) % TiB_2 composite. Strengths obtained with different profiles of the pin are listed in Table 7.1. Joint fabricated with square tool has showed higher tensile strength than the joint fabricated with other tool profiles and least strength had been noticed for tapered square tool. Tensile strength of the joint increases with increase in the rotational speed up to certain limit and reaches to maximum of 100% of the base material. With further increase in the rotational speed, voids and tunnel defects were observed. As a result, tensile strength is decreased. UTS increased with decrease in the welding speed. The UTS of FS welded composite increases with increase of axial force up to a certain limit of 9 kN and with further increase in the axial force, a decrease in the UTS was observed. Ni et al. [38] reported that the strength and ductility decreased due to softening at HAZ because of heat generation during FSW.

7.6 FSW Defects

Defects in any form adversely affect the functionality of a component. Defects related to melting and solidification, like porosity, crack, deleterious phase, intermetallic phases, etc. are completely eliminated due to the solid state process. However, defects like pin hole tunnel defect, worm holes, piping defects, less penetration depth and kissing bond are found in the weld zone due to improper selection of process parameters. Under optimum process condition, both material flow and energy balance were obtained. In hot working process, sufficient heat was generated to reduce the resistance of the material to deform. Meantime, heat will also influence the changes in the microstructure like coarse grain, recrystallization, grain reorientation and dissolution of strengthening precipitate. Chemical composition of the material plays a vital role in material resistance to deform and changes its microstructure. Some materials undergo changes in metallurgical characteristic at very low temperature, and some material does not undergo any changes until higher temperatures are attained. Too cold processing condition refers to non-bonding and void formation. Hot processing condition refers to excessive material flow and it leads to material expulsion like flash and unwanted degradation of mechanical properties.

Hot processing conditions are achieved in two ways. (1) The temperature generated is higher than the solidus temperature of the material. (2) Temperature generated approaches the solidus but the heat loss from the direct deformation zone is sufficiently slowed down so as to result in unwanted thermal softening of the work material. This results in degradation of the mechanical properties of the welded materials. Due to excessive heat generated by the FSW process, the surface contains blisters or surface galling. Further increase in the heat causes thermal softening of the material beyond the boundary of tool shoulder. Subsequently the tool shoulder, instead of confining the material, starts ejecting the material in the form of surface flash. Hence, insufficient material available at the weld zone leads to the shortage of material to form joint and the defect formed at the surface of the weld region. In some cases, due to excessive thermal softening during FSW which is carried out under thrust force instead of tool position control system, the material directly under the shoulder will no longer be able to withstand the axial thrust force acting on it. This leads to thinning of the material whereby material thickness reduces. Due to fixed length of the pin and thinning effect of the material, the pin may rub against the backing plate, which ruptures the material at the root of the weld and excessive material will spill out from the weld zone in the form of flash. This is shown in Fig. 7.14 as "Excessive Flash". Excessive flow of soft material from shoulder-driven zone to pin-driven zone leads to formation of nugget collapse. The tunnel or worm defects occur due to insufficient or extremely high rotational speed accompanied with too low a thrust force which leads to inappropriate mixing of material. The defects also occur due to improper design of the tool. Screw type of tool is able to join only few components as compared to conical type tool. The high rotational speed and welding speed led to the formation of tunnel defect at the

7 Friction Stir Welding—An Overview

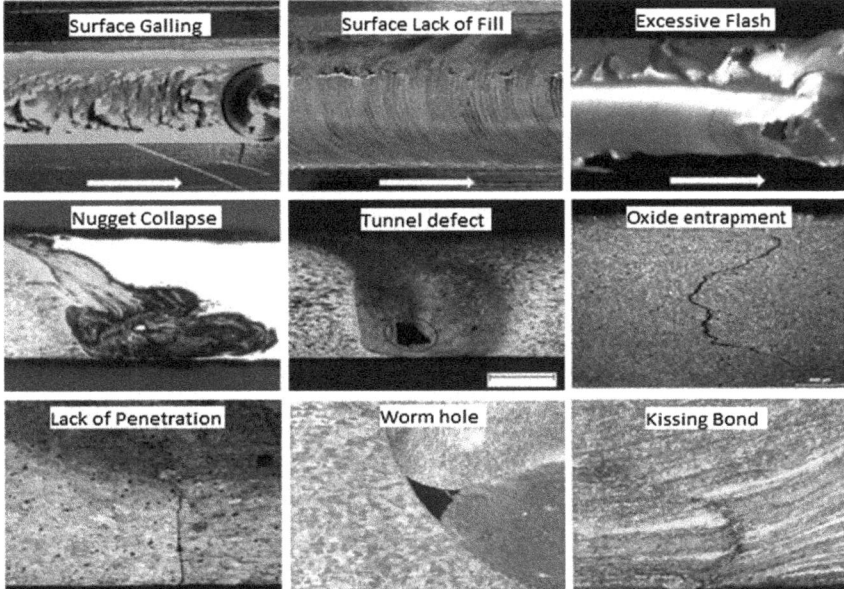

Fig. 7.14 Type of defects in Friction Stir Welding

intersection of NZ and TMAZ. The development of tunnel defect in weld zone and revealed that the tunnel defect is formed at higher as well as lower welding speed with higher rotational speed.

The formation of kissing bond in FSW was due to the oxide layer break up. The breakup of oxide layer was due to inadequate stretch of the contact surfaces around the welding pin. During the process of FSW, the oxide layer on the butt surface is fragmented into particles and the particles are distributed along the cross section of zigzag line. The kissing bond was formed due to insufficient frictional force and heat generation. It is very difficult to find out kissing bond using non-destructive process. The presence of kissing bond promotes cracks. Lack of penetration depth occurs due to insufficient depth of penetration or too short a tool length. Shorter tool length may be due to wearing out of the tool. The wormhole initiation near the bottom of the weld is due to increase in the welding speed with rotational speed being held constant. At low processing temperatures, tools made of brittle materials will lead to the formation of warm holes. The size of the wormhole increases with the welding speed because of inadequate material flow towards the bottom of the weld. There are indications that the welding speed to rotational speed ratio is an important factor in the formation of the wormhole defect. The various defects are presented in Fig. 7.14. They concluded that the square tool pin profile has higher eccentricity and produced least defect content in the weld. The eccentricity is described as the ratio of the dynamic volume swept by the tool to the static volume of the tool. The flat faces led to a pulsating action which in turn led to more effective stirring.

7.7 Applications

FSW plays an imperative role in various sectors. Initially shipbuilding and marine industries have opted and adopted the process for commercial applications. Later on, FSW was incorporated in aerospace industry for fabricating prototype and production parts. FSW extended its application in military and civilian aircraft to weld skins to spars, ribs, and stringers. The Eclipse 500 aircraft, in which ∼60% of the rivets are replaced by friction stir welding. Thus, FSW offers numerous significant advantages like reduced manufacturing costs and weight savings as compared to the fabrication process and machining from solid. The concept of FSW, i.e. longitudinal butt weld is utilized in space vehicles to weld the Al alloy-based fuel tanks. The foresaid process could be expanded to increase the size of commercially available sheets by welding them before the forming process. Even in the railway sector such as for bullet trains, goods wagons, rolling stock of railways, railway tankers, underground carriages and trolley cars FSW concept is adopted.

7.8 Summary

FSW has gained extensive application in various sector due to higher joint strength, simplicity, and benefits of solid state welding process, and achieved remarkable success. Different verities of tool geometry and shape have been used to join various materials by this modern welding process. However, desired long life of tool with reasonable prices are quiet unobtainable. Material flow pattern and heat generated during process depend on the process parameters. Hence, process parameters plays dynamic role in obtaining better quality of joint. The widespread use and acceptance of FSW process compels more research and development in this area to facilitate the welding and joining industry.

References

1. Thomas WM, Nicholas ED, Needham JC, Murch MG, Templesmith P, Dawes CJ (1995) Friction stir butt welding, Int Patent App PCT/GB92/02203 and GB Patent App. 9125978.8, Dec. 1991. U.S. Patent No. 5,460,317
2. Buffa G, Donati L, Fratini L, Tomesani L (2006) Solid state bonding in extrusion and FSW: process mechanics and analogies. J Mater Process Technol 177:344–347
3. Li Ying, Murr LE, Mcclure JC (1999) Solid-state flow visualization in the friction-stir welding of 2024 Al To 6061 Al. Scripta Mater 40(9):1041–1046
4. Colligan K (1999) material flow behavior during friction stir welding of aluminum. Weld J 75 (7):229s–237s
5. Seidel TU, Reynolds AP (2001) Visualization of the material flow in AA2195 Friction-stir welds using a marker insert technique. Metall Mater Sci A 32(11):2879–2884

6. Guerra M, Schmidt C, McClure JC, Murr LE, Nunes AC (2003) Flow patterns during friction stir welding. Mater Charact 49:95–101
7. Hamilton C, Dymek S, Blicharski M (2008) A model of material flow during friction stir welding. Mater Charact 59(9):1206–1214
8. Kumar K, Kailas Satish V, Srivatsan TS (2008) Influence of tool geometry in friction stir welding. Mater Manuf Process 23(2):188–194
9. Arbegast W (2003) Modeling friction stir joining as a metal working process. Hot deformation of aluminum alloys III, TMS, San Diego, CA, pp 313–327
10. Mishra RS, Ma ZY (2005) Friction stir welding and processing Mater Sci Eng R 50:1–78
11. Threadgill P (1997) Friction stir welding in aluminium alloys—preliminary microstructural assessment, TWI Bulletin, 30–33
12. DebRoy T, De A, Bhadeshia HKDH, Manvatkar VD, Arora A (2012) Tool durability maps for friction stir welding of an aluminum alloy. Proc R Soc Lond A 468(5):3552–3570
13. Chen CM, Kovacevic R (2004) Joining of Al 6061 alloy to AISI 1018 steel by combined effects of fusion and solid state welding. Int J Mach Tool Manuf 44:1205–1214
14. Zeng WM, Wu HL, Zhang J (2006) Effect of tool wear on microstructure, mechanical l alloy stir welded 6061 A. Acta Metall Sin 19(1):9–19
15. Padmanaban G, Balasubramanian V (2009) Selection of FSW tool pin profile, shoulder diameter and material for joining AZ31B magnesium alloy—An experimental approach. Mater Des 30(7):2647–2656
16. Prado RA, Murr LE, Soto KF, McClure JC (2003) Self-optimization in tool wear for friction-stir welding of Al 6061_/20% Al2O3 MMC. Mater Sci Eng A 349:156–165
17. Shindo DJ, Rivera AR, Murr LE (2006) Shape optimization for tool wear in the friction-stir welding of cast Al359-20% SiC MMC. J Mater Sci 37:4999–5005
18. Fernandez GJ, Murr LE (2004) Characterization of tool wear and weld optimization in the friction-stir welding of cast aluminum 359 + 20% SiC metal-matrix composite. Mater Charact 52:65–75
19. Lorrain O, Favier V, Zahrouni H, Lawrjaniec D (2010) Understanding the material flow path of friction stir welding process using unthreaded tools. J Mater Process Tech 210:603–609
20. Shettigar AK, Salian G, Herbert M, Rao S (2013) Microstructural characterization and hardness evaluation of friction stir welded composite AA6061-4 . 5Cu-5SiC (Wt .%), 63(4): 429–434
21. Mehta M, Arora A, Debroy T (2011) Tool geometry for friction stir welding optimum shoulder diameter. Metall Mater Trans A 42(A): 2716–2722
22. Vilaça P, Pepe N, Quintino L (2006) Metallurgical and Corrosion Features of Friction Stir Welding of AA5083-H111. Tech Uni Lisbon, Portugal, pp 1–13
23. Venkateswarlu D, Mandal Nr (2013) Tool design effects for FSW of AA7039. Weld J 92:41s–47s
24. Thomas WM, Staines DG, Norris IM, de Frias R (2003) friction stir welding tools and developments. Weld World 47(11–12):10–17
25. Elangovan K, Balasubramanian V (2008) Influences of tool pin profile and welding speed on the formation of friction stir processing zone in AA2219 aluminium alloy. J Mater Process Techno 200(1–3):163–175
26. Rajakumar S, Balasubramanian V (2012) Correlation between weld nugget grain size, weld nugget hardness and tensile strength of friction stir welded commercial grade aluminium alloy joints. Mater Des 34:242–251
27. Thomas WM, Johnson KI, Wiesner CS (2003) Friction stir welding–recent developments in tool and process technologies. Adv Eng Mater 5(7):485–490
28. Vijay SJ, Murugan N (2010) Influence of tool pin profile on the metallurgical and mechanical properties of friction stir welded Al-10 wt.% TiB_2 metal matrix composite. Mate Des 31:3585–3589
29. Rajakumar S, Muralidharan C, Balasubramanian V (2011) Influence of friction stir welding process and tool parameters on strength properties of AA7075-T6 aluminum alloy joints. Mater Des 32:535–549

30. Dong P, Li H, Sun D, Gong W, Liu J (2013) Effect of welding speed on the microstructure and hardness in frictio stir welding joints of 6005A-T6 aluminum alloy. Mater Des 45:524–531
31. Amirizad M, Kokabi AH, AbbasiGharacheh M, Sarrafi R, Shalchi B, Azizieh M (2006) Evaluation of microstructure and mechanical properties in friction stir welded A356 + 15% SiCp cast composite. Mater Lett 60(4):565–568
32. Kumar A, Nayak CV, Herbert MA, Rao SS (2014) Microstructure and hardness of friction stir welded aluminium—copper matrix-based composite reinforced with 10 wt- % SiCp. 18:84–89
33. Uzun H (2007) Friction stir welding of SiC particulate reinforced AA2124 aluminium alloy matrix composite. Mater Des 28:1440–1446
34. Ceschini L, Boromei I, Minak G, Morri A, Tarterini F (2007) Effect of friction stir welding on microstructure, tensile and fatigue properties of the AA7005/10 vol.%Al2O3p composite. Compo Sci Technol 67:605–615
35. Feng AH, Xiao BL, Ma ZY (2008) Effect of microstructural evolution on mechanical properties of friction stir welded AA2009/SiCp composite. Compos Sci Technol 68:2141–2148
36. Chen X, Silva M, Gougeon P, St-georges L (2009) Microstructure and mechanical properties of friction stir welded AA6063—B_4C metal matrix composites. Mater Sci Engg A 518:174–184
37. Nami H, Adgi H, Sharifitabar M, Shamabadi H (2011) Microstructure and mechanical properties of friction stir welded Al/Mg_2Si metal matrix cast composite. Mater Des 32:976–983
38. Ni DR, Chen DL, Wang D, Xiao BL, Ma ZY (2013) Influence of microstructural evolution on tensile properties of friction stir welded joint of rolled SiCp/AA2009-T351 sheet. Mater Des 51:199–205

Chapter 8
Ultrasonic Spot Welding—Low Energy Manufacturing for Lightweight Fuel Efficient Transport Applications

Farid Haddadi

Abstract High power ultrasonic spot welding is an alternative manufacturing process which recently has been developed for joining automotive bodies. This technique is a very low energy process and forms effective welds in less than a second. The present chapter covers different aspects of High Power Ultrasonic Spot Welding HPUSW technology towards lightweighting for automotive applications. The chapter will include but not limited to similar and dissimilar bonding, mechanism of bonding, interfacial reaction, grain structure, texture development and process model of high power ultrasonic spot welding.

Keywords Automotive · Ultrasonic spot welding · Grain structure · Texture

8.1 Introduction

High price of gasoline early this decade and new strategy of new fuel economy in addition to emission regulations for 2017–2025 to minimize fuel consumption to 54.5 mpg and reduction in CO/CO_2 emissions to 163 per mile in 2025, all strongly stress the necessity of lightweight vehicles than ever before [1]. Therefore, within the next coming years, lightweighting scheme in automotive industry will be significantly accelerated through the use of advanced light materials, design optimization and up-to-date manufacturing technologies at greater performances, comfort level and safety features [2]. To meet the requirements of mass reduction, as systematic approach to efficient design is needed that takes advantage from physical properties of specific lightweight alloys and materials. The process of lightweighting encompasses low density materials such as magnesium, aluminium, and carbon fibre reinforced polymers (CFRPs), etc. to obtain 10–70% weight

---

F. Haddadi (✉)
BIW Manufacturing, Process Engineer Specialist–Faraday Future Inc.,
Los Angeles, CA, USA
e-mail: farid.haddadi@gmail.com

© Springer International Publishing AG 2017
K. Gupta (ed.), *Advanced Manufacturing Technologies*, Materials Forming,
Machining and Tribology, DOI 10.1007/978-3-319-56099-1_8

saving when these light materials are integrated and replace predominantly used conventional steel alloys [2].

It was projected that advanced high strength lightweight alloys (including high strength low alloyed steels) will contribute to 30–70% of commercial automotive body structures by 2030 [3]. Non-metallic materials (e.g. plastics and composites) roadmap for automotive applications by American Chemistry Council has focused underpinned research and development in order to sustain the pace of innovation needed by automotive industry in pursuit of weight saving opportunities required to meet 2025 CAFÉ regulations proposal [4]. In the United States, the target has been determined by Department of Energy (DOE) Vehicle Technologies to approaching 50% mass saving in Body-in-White applications by end of this decade. Towards this target, development and validation of advanced materials and manufacturing technologies must be carried out [1].

Advanced materials such as aluminium alloys, advanced high strength steels, and carbon fibre reinforced—already in use of nowadays vehicles—are a few of several available opportunities that could be integrated in lightweighting themes in case they meet cost and manufacturability requirements. These advanced materials/alloys are extensively used in aviation industries where manufacturing cost is not the first requirement. However, replacement of conventional heavy components by lightweight materials in automotive sector depends on three main factors; (i) selection of material candidates and (ii) nature of intended application and (iii) production cost [1]. So far, Carbon Fibre Reinforced Polymers (CFRPs) offer the highest mechanical properties and most weight saving potentials but at a higher cost of 1.5–5 times of conventional steels. Advanced high strength steels, aluminium, titanium and magnesium alloys are under consideration for lightweighting applications [5, 6].

Above the technical requirements, the ability to integrate these materials into an automotive body structure at the lowest possible cost is essential for the sensitive transport sector. Therefore, and with all things considered, it is now accepted that no one material or class of materials will be satisfactory for achieving the lightweighting and functionality targets of future vehicles; innovative multi-material body structures featuring the proper use of "the right material for the right component(s)" is the best way forward. Initial steps in this direction have been taken by several OEMs, and can be already seen, though on a limited scale, in multiple production vehicles. Almost every car today incorporates more than four different grades of steels, and many of them incorporate several aluminium components. In Germany, the first true champion of all-aluminium space frame now incorporates several critical ultra-high strength steel components in the body structure. In spite of these early steps, the design and implementation of hybrid-material structures is highly challenging, leading to the confinement to mostly steel structural components with aluminium closures.

The process of joining dissimilar materials presents major challenges, due to the large differences in physical and thermal properties between the two materials [7]. In particular, the use of fusion bonding in most cases results in the formation of brittle intermetallic compounds at the bondline can limit the mechanical properties and lead to low fracture energy in joints [8]. Approaches for joining dissimilar

Table 8.1 A brief comparison of various dissimilar spot joining techniques for aluminium to steel

	Advantage	Disadvantage	Energy/time (s)
Self-pierce riveting	Good mechanical properties Cold process	Consumable cost	Low/ <1
Resistance spot welding	Fast process Cheap process	Intermatallics formation High energy cost	<50 kJ/ <0.2
Friction stir spot welding	Efficient energy process Cheap process	Too slow Key hole	2–4 kJ/2–5
Ultrasonic spot welding	Very low energy process Cheap process	Surface damage	<2 kJ/ <1.5

materials, such as mechanical fastening [9], friction stir spot welding [10] and ultrasonic spot welding [6, 11, 12] are being investigated in an attempt to overcome these difficulties. This is primarily, because fusion boding has very limited potential for welding due to the rapid formation of intermetallic phases in the liquid phase [8]. Self-piercing rivets have high consumable costs and increase the weight of construction [9]. Friction stir spot welding technique is an efficient potential solution that could replace the above welding techniques and avoid many of the problems that are caused when the material melts in the weld region [13]. However, an undesirable keyhole is also produced by the tool probe which can lead to a reduction in weld strength. Due to the requirement for a clamping system, this can restrict access to the joint [14]. Moreover, it is often difficult to form a full metallurgical bond which limits the lap shear strength and other mechanical properties of the joints. The process was also seen to be too long [13]. Table 8.1 shows the advantages and disadvantages of the main spot joining methods for body-in-white.

An alternative solution is to take advantage of ultrasonic metal welding, which has been used in manufacturing since the 1950s for thin foil welding application [15]. Recently, high power ultrasonic welding facilities have become more applicable and can be used to weld similar aluminium welds in automotive body structures up to 3 mm in thickness [16, 17]. This welding solution performs joint in less than 0.5 s. [18–21]. This method causes a friction weld by inducing small high frequency linear movement (10 µm) across the weld interface [22]. However, in high power ultrasonic welding there is currently little systematic knowledge of the influence of the welding parameters on weld performance and the joining mechanism is still poorly understood.

8.2 High Power Ultrasonic Spot Welding

The high power ultrasonic spot welding is fast, energy efficient, non-contaminating, environmentally friendly and requires no consumables. With this welding technique it is possible to weld thin materials to thick materials. A wide range of similar and dissimilar material combinations can also be joined by this technique [23].

8.2.1 Applications

Ultrasonic metal welding has a broad range of applications such as electronics, medical, aerospace, automotive and consumer goods industries. Examples include heavy duty electrical connections for miniature air switch (MAS)-type applications in automotive industry to meet ultrahigh current requirements. This could potentially replace multiple stranded aluminium wires on copper eyelet terminals. Formation a eutectic reaction at Al-34%Cu facilitated bonding between copper and aluminium at the interface. In electronics industry, ultrasonic welding has played a significant role for construction of sensor terminals, braided wire connections, and Brass bar terminals for automotive and trucking applications. Further, field coils, transformers, capacitors and electric motors are the other examples that ultrasonic welding was performed for the purpose of assembly and joining. Microbonding for microelectronic interconnection application has been seen as the most extensive uses of low power ultrasonic welding [24].

Ultrasonic spot welding is also used for battery and fuel cell manufacturing where the technique is applied in thin gauge copper, nickel and aluminium tab welding. This low energy manufacturing is a great opportunity for bonding multilayer foil-to-tab joints in prismatic lithium-ion battery consolidation in electric and hybrid vehicle market. However, this technique has not been far used in Body-in-White applications although; it has advantages to conventional welding methods [24].

8.2.2 Principles

During an ultrasonic spot welding, a trail of high frequency (e.g. 15–40 kHz) linear vibration is introduced to the welding materials through sonotrode tips that are linked to transducers. The clamping pressure increases the friction and thus more heat is generated at the interface of the materials being welded. Figure 8.1 schematically illustrates the ultrasonic spot welding process [7].

Generally, in ultrasonic welding setup, the materials need to be fixed in the required orientation and clamped in an overlapping manner over an anvil that supports the parts during welding. Thus, the requirement for rear face access can limit accessibility, particularly when welding large sheets [25]. In ultrasonic spot welding the teeth on the sonotrode tip penetrate the surface of the materials and the resulting irregularity of the surface can accelerate the corrosion rates [26]. High noise levels can also be produced by secondary vibration modes caused by the high frequency oscillation [26]. Ultrasonic spot welding is not only used for joining metals, but can also be used for welding polymers such as polypropylene. Ultrasonic plastic welding involves heat generation at the interface of the parts due to mechanical vibrations normal to the sheet surfaces, using a higher frequency (27-180 kHz) than for ultrasonic metal welding. Anelastic mechanical vibration

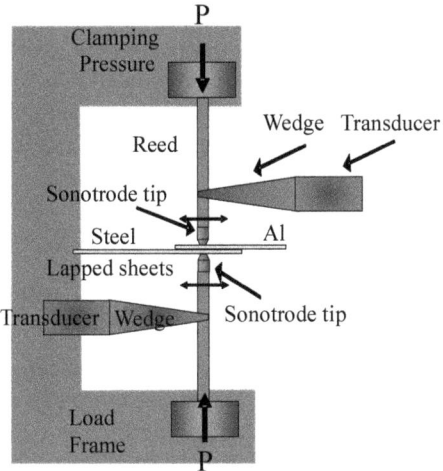

Fig. 8.1 Schematic illustration of the ultrasonic spot welding process [7] with kind permission from Elsevier

causes an increase in temperature at the faying surfaces. The materials at the interface soften as a result of heat generation and intermixing consequently occurs. Thus, a high quality weld can be achieved at a lower temperature compared to in conventional plastic welding methods [27].

8.2.3 Mechanism of Bonding

In ultrasonic spot welding, high temperatures and pressure are generated by friction due to the high frequency vibration and clamping system at the work-piece interface [15]. Contact, first occurs at the asperities on the sheet surfaces. In the initial stages of welding energy is dissipated at sliding frictional condition. In welding soft metals such as copper and aluminium, galling and thereafter, microbonding rapidly takes place when the oxide film scatters at these early contact points [16]. Therefore, more energy is dissipated due to increasing high degree of plastic deformation and microwelds spread in density and size, along the weld interface with increasing welding time [18]. In dissimilar materials welding, it is seen that, delay in weld interface deformation in the harder metal could delay formation of weld [28]. The process is also potentially less complex as there is likely to be less displacement of the weld line than in similar welding, such as with aluminium, where the interface is deformed locally, once microbonds and galling are created through exposing new surface [29].

In case of dissimilar metal to CFRP composites, Balle et al. 2011 [30, 31] demonstrated that this technique results in ~2.3 kN lap shear strength. It was shown that this fairly good mechanical property is caused by adhesion between aluminium and the polymer matrix which is promoted by mechanical interlocking when the carbon fibres intercalated with the aluminium at the interface. Interlocking was seen to be very limited in this practice as a result of low formability of the polymer matrix which suggests possibilities for even higher joint strengths if the matrix flow-ability is promoted. It is noteworthy that cracks were detected by indentation welding tool that may cause fibre damage and therefore, crack initiation in the polymeric composites, which would also facilitate galvanic corrosion on aluminium side [31].

8.2.4 Heat Generation and Temperature Evolution

In order to understand the heat generation at the work piece during welding, it is necessary to take a closer look at the distribution of the energy dissipation along the deformation zone and its development during the weld cycle. At the earliest stage of welding, deformation and microbonds form along the interface after the oxide layer is disrupted locally at contacting asperities across the weld interface. The heat generation and its subsequent temperature may therefore, not be uniformly distributed along the weld interfaces, as local plastic deformation can be caused heterogeneously. As the microbonds spread to form a continuous bond and the full power is transferred to the work piece, the heat generation becomes more uniformly distributed, as shown in Fig. 8.2 [32].

The heat flux at the weld interface has been measured in nickel to copper ultrasonic spot welding by Cheng and Li [33]. As shown in Fig. 8.3, the heat flux profile matches the thermal history. At the initial stage of the welding, a relatively

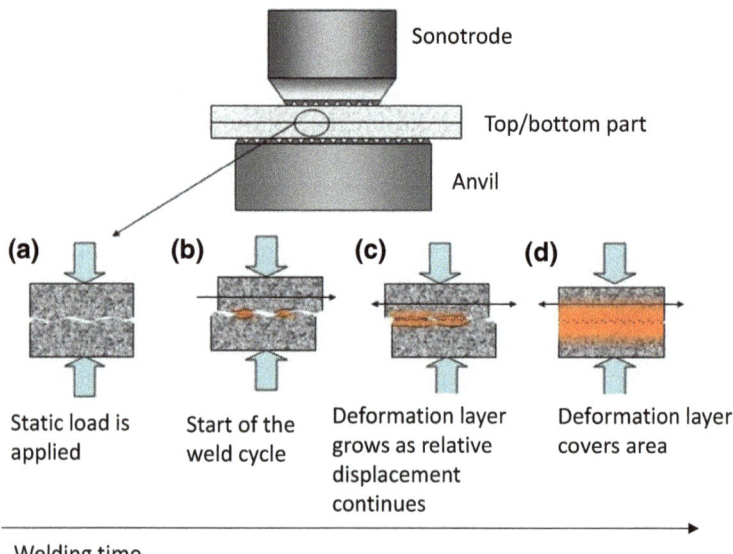

Fig. 8.2 Schematic presentation of heat distribution of ultrasonic spot welding [32]

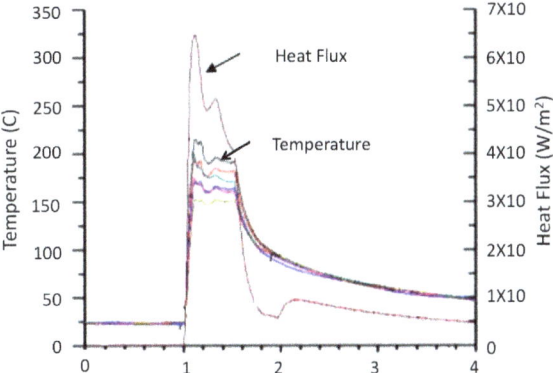

Fig. 8.3 Heat flux variation and temperature during nickel to copper high ultrasonic welding process [33] With kind permission from IOP Publishing

large amount of heat generation is caused by friction between the two sheets, leading to the mechanical cleaning of the interface where bonding begins to take place (Fig. 8.2a). Formation of microbonds then takes place due to an increase in pure metal to metal contact area at the interface. The microbonding (Fig. 8.2c, d) limits the further heat generation by friction between the work pieces. Thereafter, heat is generated as a result of plastic deformation at the faying interface, while the bonding area is gradually caused. The heat flux is reduced to a steady state level (plateau) suggesting less relative movement between the two work pieces when continuous weld is produced (Fig. 8.2d). However, the microbonds can be broken by the subsequent vibration cycles to generate small peaks in heat generation. The process repeats until a full welding area is developed [33].

A good example of the effect of weld energy on the thermal history in high power ultrasonic spot welding has been reported by Bakavos and Prangbell (2010). For similar aluminium ultrasonic spot welding produced with a 2.5 kW power machine. Weld temperatures were measured close to the weld edge as well as at the centre line. Figure 8.4 shows the temperature measurements as a function of energy delivery (welding time) under a 1.4 kN axial pressure [18].

The peak temperatures increased progressively with increasing welding time (energy delivery) from ~ 75 °C, after 0.1 s, to 220 °C after 0.18 s and approached the level of 400 °C for times over 0.43 s. It has also been claimed that the temperature at the centre of weld region will be at least 50 °C greater than the one recorded close to the edge for the joint produced within 0.3 s [18]. These temperatures measured are higher than have been reported previously (e.g. equal to 40% of melting point) in low power ultrasonic welding [23].

8.2.5 Power and Vibration Evolution

Vibration investigation and online monitoring of signals during an ultrasonic welding process are of primary tasks for obtaining robust and reliable joint

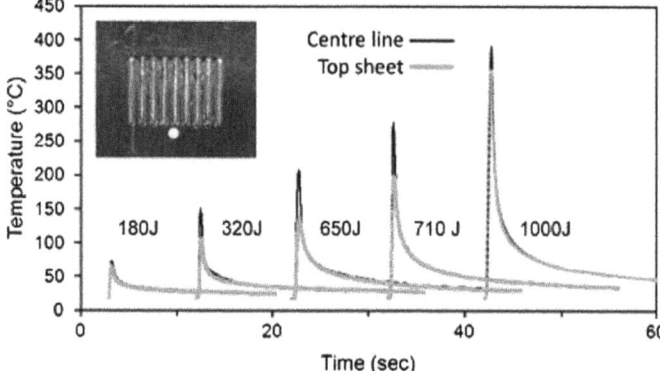

Fig. 8.4 Weld thermal cycles for increasing welding times, measured as close as possible to the join line. *Curves* are displaced in time for clarity and the thermocouple position is indicated in the insert [34] with kind permission from Elsevier

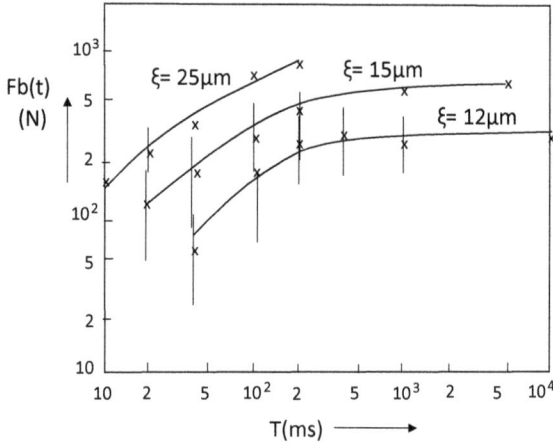

Fig. 8.5 Effect of vibration amplitude on aluminium weld strength with an increase in the welding time [15]

performances. The power is related to the amplitude of vibration in an ultrasonic welding system. By increasing the power more electric current enters the converter and, the piezoelectric part expands when more of it is excited by greater electrical energy producing higher amplitude of vibration [25].

So far only a few works have been published for vibration evolution during ultrasonic welding [15, 22, 35]. In a very old report it was shown the effect of different amplitudes of vibration on the weld strength for wide range of materials; such as aluminium, steel, etc. [15]. For example, in aluminium welding (Fig. 8.5), it was shown that an increase in amplitude of vibration leads to a higher shear strength at a constant welding time. In addition, increasing the amplitude of vibration provides the same weld strength within a significantly shorter welding time, as shown in Fig. 8.5. Welding aluminium along with a higher strength also requires higher power (up to 25%), to achieve the same amplitude of vibration [15]. It was also observed that the weld strength with the amplitude of vibration did not follow a linear behaviour reaching an optimum level with increasing welding time. This was because of the saturation of microbonds at the interface with increasing welding time [32].

However, the characteristic of vibration is not only dependent on the signal delivery by the electronics control system. In a recent work it was shown that the interface reaction (e.g. formation of liquid at the interface) is effective on vibration delivery to the work piece [22]. In this work the effect of melting zinc on vibration behaviour was studied when aluminium was welded to both hot dipped DX56-Z and galvannealed DX53-Z steels. Figure 8.6 compares the amplitude behaviour in aluminium 6111-T4, to different zinc-coated (DX53-ZF, DX56-Z) and un-coated steels (DC04) steel ultrasonic joints produced in a 0.26 s welding time, under a 1.4 kN clamping force [22]. In the initial stages of welding a spike of amplitude can be observed in all the samples. On the other hand, with welding time, the average plateau amplitude in the Al-DC04 joint could be seen to increase in order, from the hot dipped DX56-Z, to the galvannealed DX53-ZF hard zinc steels, to a less

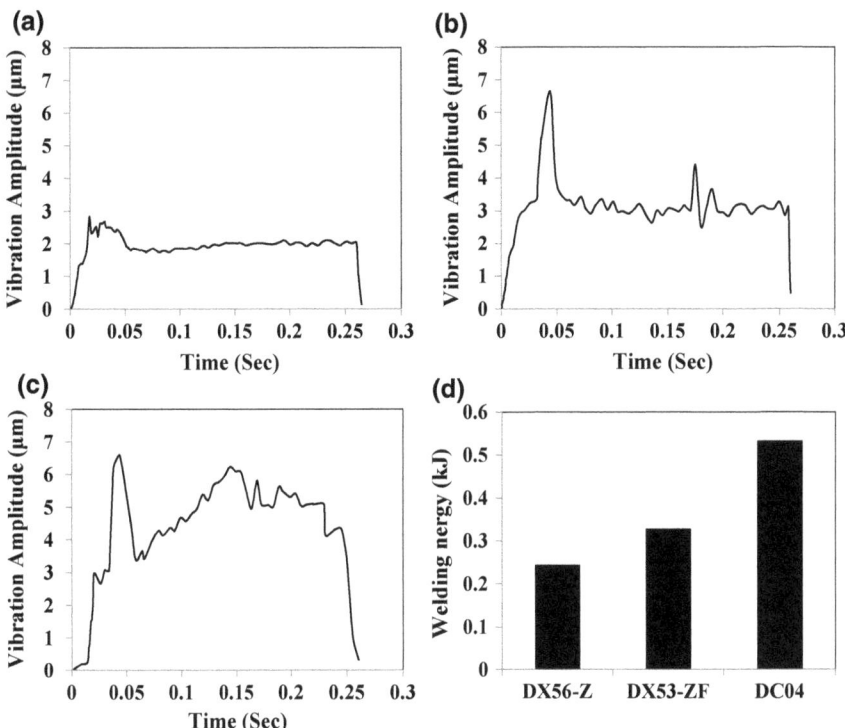

Fig. 8.6 Trace of vibration amplitude for aluminium when welded to **a** hot dipped soft zinc coat (DX56-Z), **b** galvannealed hard zinc coat (DX53-ZF), **c** non-coat (DC04) steel within a 0.26 s welding time and **d** the net welding energy for each weld [22] with kind permission from Elsevier

uniform but higher value with the DC04 bare steel sheet. This can be concluded that the power delivery is functioned by the interface reaction when zinc melts at the interface. This was also confirmed by investigation of the net energy delivery for each weld set up, while a constant welding time was set (Fig. 8.6d). Although, the machine is set to deliver a consistent power, a lower vibration amplitude was detected when there coupling was poorer by lower friction rates at the interface because of a softer surface and, or, melting of the zinc coating. This resulted in a lower total welding energy. Therefore, the power delivery is increased in order of resistance at the interface [22].

8.2.6 Similar Ultrasonic Spot Welding

In order to examine the performance of ultrasonic spot welding, it is necessary to have a closer look to microstructure, textures, defects and mechanical properties of joined materials.

In an outstanding work by Bakavos and Prangnell, several optical images of full weld cross sections of similar aluminium 6111-T4 ultrasonic spot joints demonstrate the sequence of welding development during ultrasonic spot welding under a constant clamping force of 0.9 kN [18]. Jahn et al. have also published similar images as a function of welding tip geometry and welding time [16]. As shown in Fig. 8.7, increasing the welding time (energy) metal softening, due to the rising temperature, can result in progressively more plastic deformation. Further, extensive plastic deformation can be caused throughout the sheet thickness as the welding tip penetrates into the work piece [18].

As shown in Fig. 8.7a, it was observed that the weld formation in the initial stages non-uniformly develops at specific locations under the teeth of the sonotrode, for short welding time (low energy) and there is only localized plastic deformation at the interface between the two sheets [18]. Bond formation at the periphery of weld region can be caused by a heterogeneous contact force between opposite teeth, which is maximized near the edge of the tip [36]. With increasing welding time, the bonding area develops outwards to the tool edge, as well as inwards, while maintaining a macroscopically flat interface, until it spreads across the whole interface as shown in Fig. 8.7b, c. This takes place as a result of saturation of microbonds at the interface as the temperature rises [37]. However, at higher welding energies, there is a noticeable evolution in behaviour, when plastic deformation is observed to be more severe and develops throughout the weld nugget, as shown in Fig. 8.7d, e. The weld line then becomes displaced, giving rise to a convoluted wave-like appearance and shear bands appear between the teeth, where minimum resistance of the material occurs due to a local reduction in the thickness and the concentration of stress (Fig. 8.7d). At extremely high weld energies (1075 J) the pattern changes to shearing between teeth, which are displaced to form a male-female registry (Fig. 8.7e). Thus, a uniform macroscopic

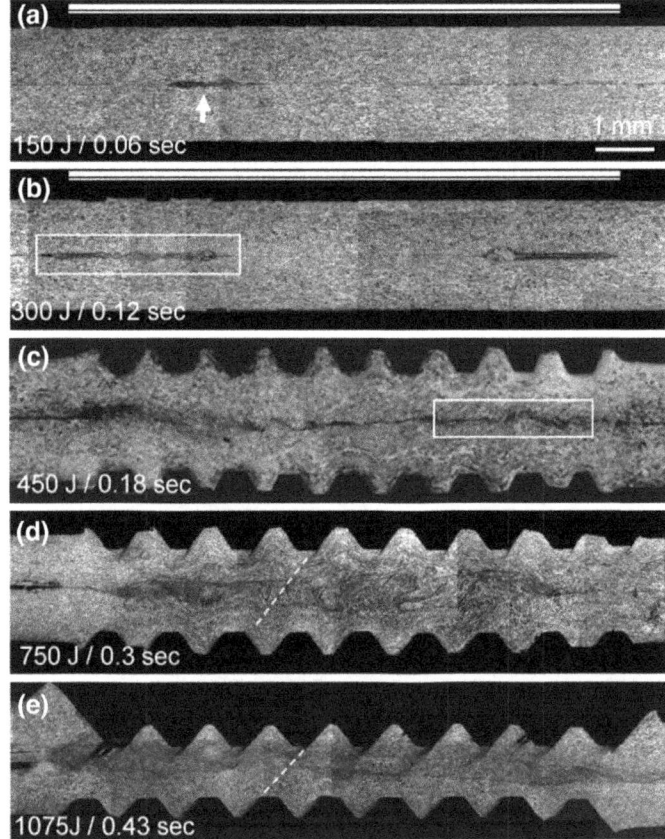

Fig. 8.7 Aluminium 6111-T4 weld cross-sections performed under a clamping force of 0.9 kN, when weld time (or energy) increased: **a** 0.06 s/150 J, **b** 0.12 s/300 J, **c** 0.1 s/450 J, **d** 0.3 s/750 J, and **e** 0.43 s/1075 J [18] with kind permission from Elsevier

displacement of the welding tips out of alignment can be observed across the joint line [18].

To date little research has been reported on the deformation mircostructures seen in ultrasonic spot welding. Electron backscatter diffraction (EBSD) measurements provide statistical data on the changes that occur with respect to the grain, subgrain boundaries and local texture in welds [5, 16, 34, 38]. Using EBSD, Bakavos and Prangnell have reported interesting grain structure features and missorientation variations at the weld interface in aluminium 6111-T4 ultrasonic spot welds [17].

Bakavos and Prangnell have reported the deformation in aluminium welds performed with a dual reed high power ultrasonic welder fitted with a knurled tips. Figure 8.8 shows an EBSD map from a slice throughout the centre of a weld produced under the optimum conditions of 0.3 s (750 J); regions of interest are revealed at higher magnification. In this work the weld region was described by three zones [18, 34];

I. Close to the interface, where severe deformation occurs, showing in an ultrafine grain structure.
II. A shear zone at mid thickness in the aluminium sheet.
III. And a forging zone when sonotrode tips penetrate into the sheet surfaces.

In zone (I) close to the displaced interface region, an ultrafine structure was observed, consisting of a mixture of small (~2 μm) grains and subgrains (Fig. 8.8b–d). There were also elongated grains within some of the shear bands in

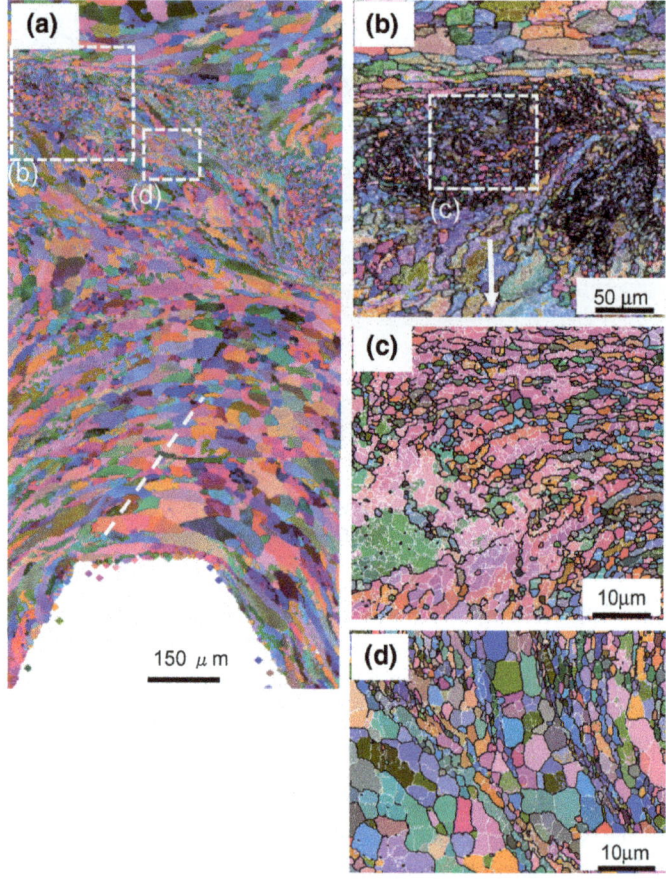

Fig. 8.8 Euler EBSD grain structure maps from a weld produced within 0.3 s (750 J, 40 MPa), demonstrating: **a** A typical portion across the weld centre. **b** a weld interface 'swirl' and the microstricture (grain structure) are shown at a higher magnification. **c** and **d** shear bands near the weld interface, [18] with kind permission from Elsevier

zone (II). However, there was a lower density of subgrains because less deformation occurred than at the weld interface. The occurrence of a forged zone on the top of the aluminium sheet is correlated with the geometry of knurls on the sonotrode welding tips [18, 34].

EBSD maps of the grain structure in the weld zone showed that plastic deformation takes place throughout the entire zone between the sonotrode welding tips. The complex dynamic strain cycle provides high plastic deformation in specific areas close to the weld interface, accompanied by shear bands. The majority of the interface microstructure appeared to predominantly contain low angle boundaries suggesting a recovered deformation structure. Formation of ultrafine grain structures within the local strain of shear bands was attributed to a dynamic recrystallization process [18]. In addition grain structure evolution, it is important to investigate the texture changes under actions of cyclic deformation at high temperature. In a similar work Haddadi and Tsivoulas performed investigation on texture evolution of Al-6111-T4 automotive sheets welded using the welding facilities and parameters [5].

EBSD ODF texture sections were obtained from the as received material, in addition to from the weld interface with increasing welding times up to 0.40 s (Fig. 8.9a–e). ODF of the forge region and area adjacent to the teeth were also investigated as a means of comparison to the microstructure of other regions. It can be seen that the texture is changed with welding time and hence deformation, while a transition takes place from mainly Cube texture to a strong typical β-fibre deformation components. The Cube texture demonstrated the highest volume fraction in the as received material, which is characteristic of recrystallization texture in deformed aluminium at high temperatures [5]. The claim that this condition is not heavily deformed is supported by the high fraction of random orientations which lie outside the misorientation spheres of the main texture components in aluminium. A rather uniform distribution of smaller fractions of other texture components was also present. At the initial stages of welding (0.10 s) deformation was to be low and the new texture were mainly scattered with no strong preference. With increasing welding time to 0.40 s which is considered as higher deformation level, the S, Brass and Copper components were observed alongside high fractions of Goss and P, and a substantially lower Cube content. Further, a drop in the fraction of random texture components is observed. The grains' orientation in the forge region of the 0.30 s weld indicatives of heavy deformation although, lower fraction of S were seen compared to the interface of 0.30 and 0.40 s joints [5].

Defects in ultrasonic spot welding mostly result from lack of bonding and material interactions with the compressive and cyclic shear loads generated by the sonotrode welding tip. Here, defects are classified into planar interface voids resulting from an un-bonded interface area and defects which are caused by material flow [16, 18, 39]. 3D X-ray tomography was used by Bakavos and Prangnell to investigate the interfacial voids, or unbounded areas, found at the jointline in aluminium 6111-T4 ultrasonic spot welding [18]. However, this technique is limited to resolution of ∼6 µm and thus cannot reveal fine defects seen at a microscopic level.

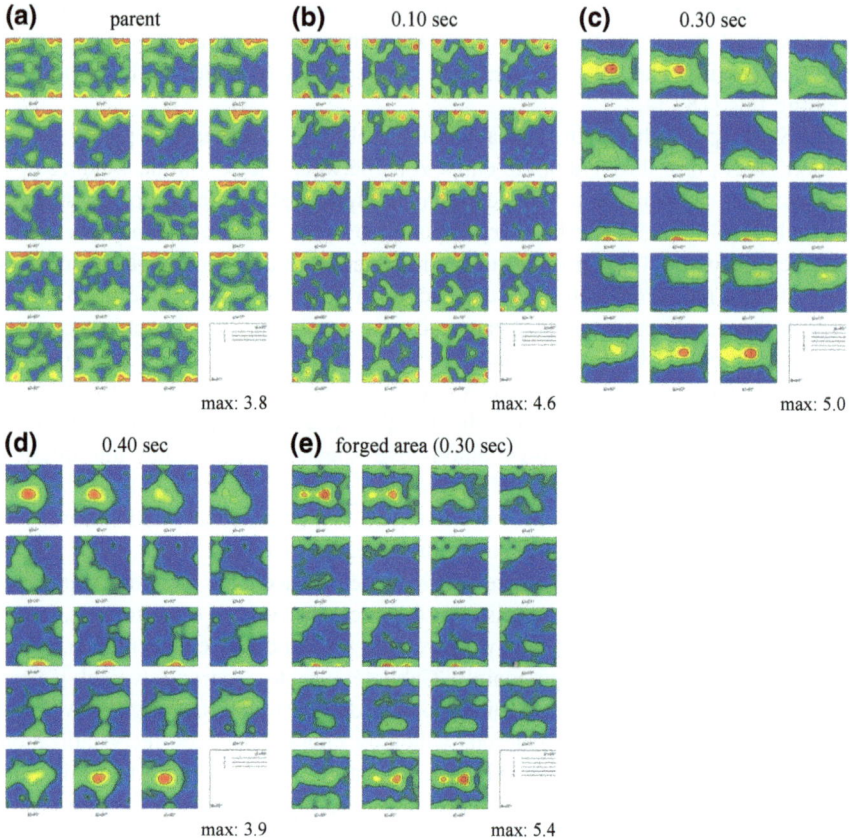

Fig. 8.9 ODF sections demonstrating changes in grain structure orientation with an increase in welding time, in addition to comparing with the unprocessed materials and the forge regions at the teeth near the top surface [5] with kind permission from Elsevier

For low weld energies (times), a high density of defects scattered randomly across the weld zone was observed (Fig. 8.10a). It was presumed that in a short welding time, the bondline is not affected by the local higher pressure of sonotrode tip teeth and unbounded areas can exist at the weld interface [18]. As the welding time and energy increased, the density of micorbonds increased rapidly and saturation of microbonds occurs at the weld interface [15, 36]. Therefore, less unbounded area was detected in the sample performed in 0.15 s (350 J), but two types of larger wavy planar where observed [18]. The formation of these voids can be interpreted as a result of folding at the sheet surface under compressive displacement of a bonded region [16, 39]. Bigger defects were found in areas of lower pressure, being concentrated between the teeth imprints, as can be seen in Fig. 8.10b. As the weld energy became larger, only remnants of the large defects could be detected by tomography at the weld edge (Fig. 8.10c) [18].

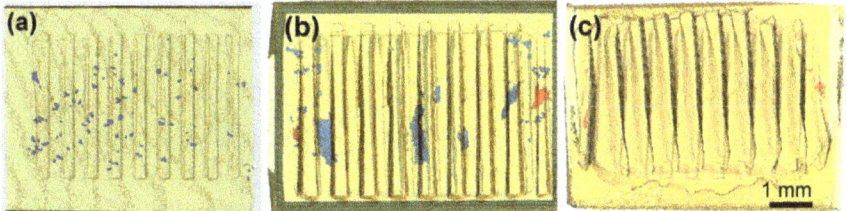

Fig. 8.10 3D X-ray tomography showing defects in the aluminium 6111-T4 ultrasonic spot welding produced. **a** 0.1 s, **b** 0.15 s and **c** 0.3 s [18] with kind permission from Elsevier

8.2.7 Dissimilar Ultrasonic Spot Welding

The substitution of light materials for conventional heavy steel alloys in automotive applications is not straightforward. Light alloys such as aluminium are significantly lower in strength and stiffness than steel and are more difficult to weld. In multi-material designs effective joining methods will be required to maximize the weight saving. However, interfacial reactions between weld components, poor joinability and high levels of distortion are significant barriers to integral aluminium to steel structures [40]. Problems associated with their metallurgical incompatibility include, the formation of brittle phases, differences in melting point, and large residual owing to their mismatch in expansion coefficients [41]. The success in a dissimilar metals welding process can also be dependent on other factors such as cost of operation, operation time, process reliability and quality that must be addressed, with the aim of introducing a new industrial vehicle welding technology [42].

In an interesting report, Zhang et al. demonstrated the interfacial reaction between aluminium 6111-T4 and TiAl6V4 alloys using ultrasonic spot welding [11]. In this study the weld interface was investigated using a high resolution transmission electron microscope (TEM). A series of bight field TEM microstructure images were obtained with increasing tilting angle as shown in Fig. 8.11.

No intermetallic layer was detected confirming that if any layer forms it must be extremely (<1 nm) fine [11]. Further investigation on obtaining diffraction pattern suggested bonding materials with no Al_3Ti at the interface. This is not very usual as this intermetallic compound was seen to form in welding aluminium to titanium alloys; however, similar "clean" interface was previously observed in metal–metal, metal–ceramic and metal–glass cases [11]. These clean interfaces—no reaction layers—are typically caused by low energy welding processes or a very high energy barrier for nucleating the intermetallic compounds.

The intermetallic growth at the interface was seen to be more significant which effectively influenced mechanical properties of joints. Examples include aluminium to steel and aluminium to magnesium welding with low activation energies for

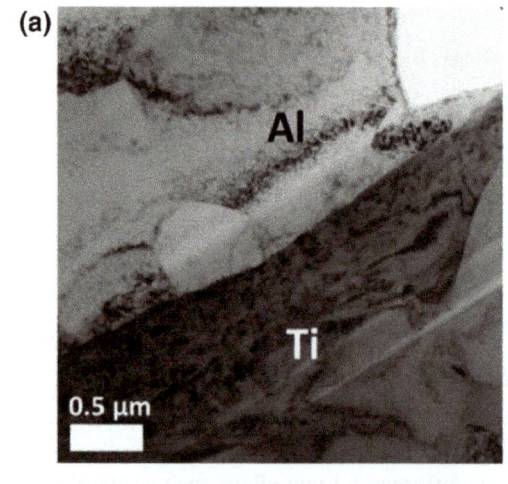

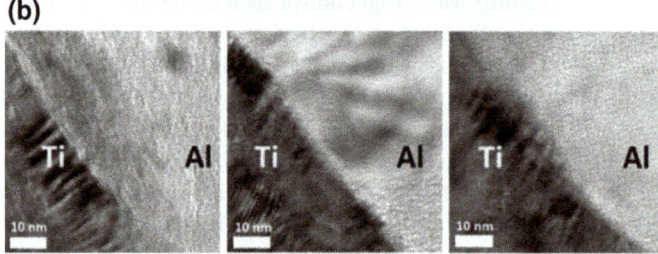

Fig. 8.11 TEM bright field microstructure images of Al–Ti interface using ultrasonic spot welding. **a** A low manufacturing image and **b** high magnification images at different tilting angle [11] with kind permission from Elsevier

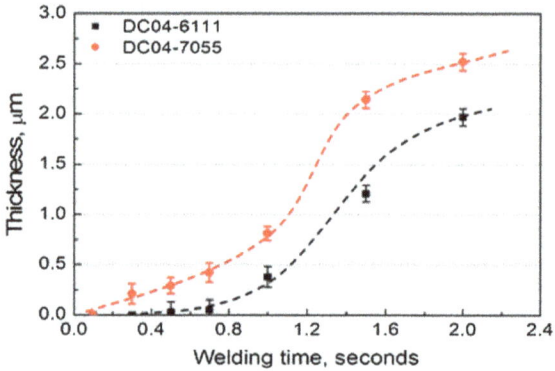

Fig. 8.12 The average thickness of the intermetallic layer with increasing time observed at the interface for DC04-Al6111 and DC04-7055 joints [44]

nucleating intermetallic compounds [19, 43]. In most of the cases failure was performed through the interfacial reaction as a result of intermetallic embrittlement.

Xu et al. have reported the Fe–Al intermetallic compounds growth at the interface of automotive Al-6111-T4 and DC04 steel sheets. A comparison was then made with

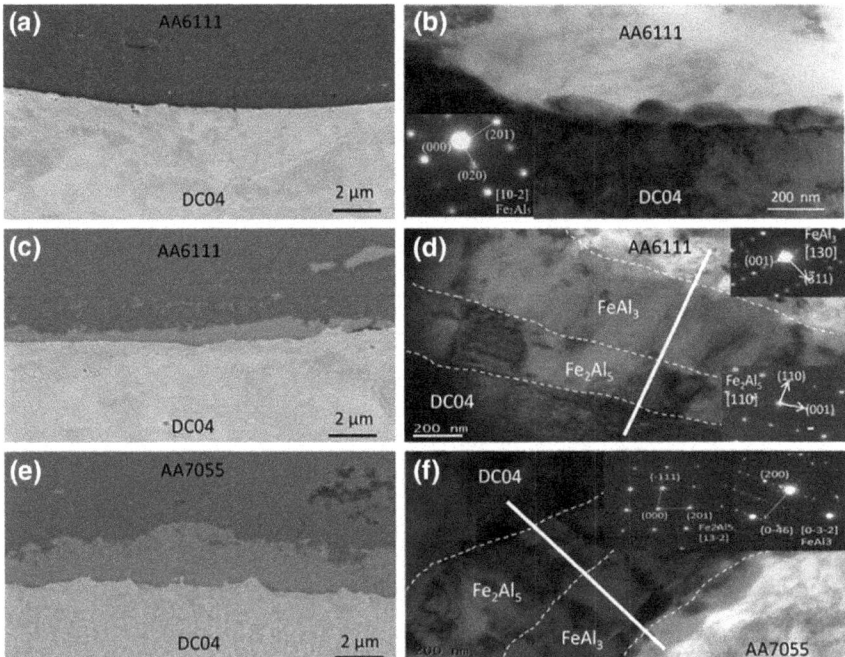

Fig. 8.13 *Left* SEM and *Right* TEM—microstructure images of the weld centre seen in the AA6111-DC04 steel weld; **a** and **b** within 0.3 s, **c** and **d** 1.5 s welding time compared to equivalent of AA5077-DC04 weld. The phases are identified with in the TEM images [44]

Al-7055 and DC04 combination to understand the effect of alloying elements on kinetic growth of intermetallic phases [44]. Figure 8.12 shows the average of thickness of Fe–Al intermetallic reaction as a function of welding time for both weld sets. The interfacial growth trend was observed to be similar for both case but at higher level for DC04-Al7055 combination. It was concluded that a limited concentration of zinc can dissolve into the intermetallic layer but the reason for higher growth rate requires further assessment. Further, Fig. 8.13b shows that in the initial stages of welding (<0.7 s) individual intermetallic islands are formed in a shape of single Fe_2Al_5 grains which latter join together causing a continuous layer as evident by Fig. 8.13c–f. Fe_2Al_5 compound forms first due to its lower Gibbs free energy compared $FeAl_3$. However, $FeAl_3$ phase appeared to have higher growth rate [44].

The use of zinc coating was found to inhibit formation of brittle intermetallic compounds when there is no direct contact between weld components. However, formation of liquid at the interface may result in precipitation of a series of unstable/meta-stable phases and structures (e.g. dendritic microstructures) during the solidification process [45]. Further, it is evident that, there is more limited indentation of the sonotrode tips into the aluminium sheet surface when there is a zinc coat at the interface. However, compared with the extensive plastic deformation seen throughout the weld area in Al–Al welds, with Al-steel welds it has

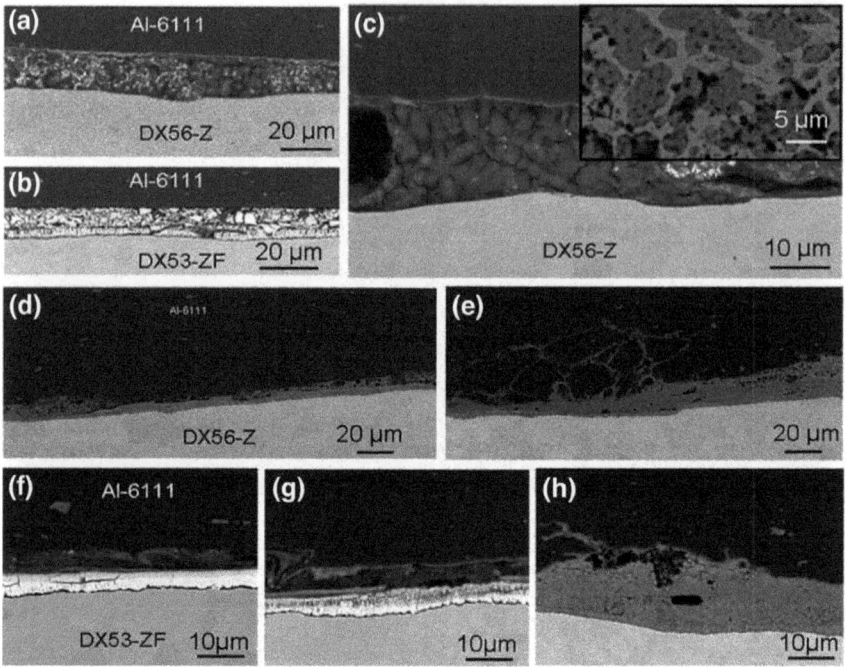

Fig. 8.14 Various weld interface structures detected along the weld edge interface containing **a** particles and melting hot dipped zinc coat (1.0 s welding time), **b** particles of the galvannealed zinc coat (1.0 s welding time), and **c** 100% melted, sheared/squeezed, and resolidified zinc-rich liquid from the hot dipped zinc steel weld at the weld edge (2.5 s welding time). **d, e** Examples of the squeezed zinc-rich liquid from the DX56-Z steel at the weld centre (2.5 s welding time). **f–h** The effect of welding time on the progressive melting of the hard galvannealed coating from the DX53-ZF steel, at the weld centre, an increase in welding times of 0.25 s, 1.5 s, and 2.5 s, respectively, [45]

been found that deformation due to ultrasonic vibration, as opposed to that caused by the forging force, is localized to the weld interface region and takes place almost entirely in the softer aluminium material.

In ultrasonic spot welding of aluminium to uncoated steel, it was seen that welding takes place by local breakdown of oxide layer at asperities on the contacting surfaces, which initiates diffusion of aluminium and steel causing formation of intermetallic layer products, mainly comprising Fe_2Al_5 and $FeAl_3$ phases [43]. However, when welding aluminium to zinc-coated steel, discussed here, four typical interface reaction can be seen as a function of position and type of coating in addition to welding time. This can be classified to: (i) Unbonded areas with granulated zinc particles of the original coat (Fig. 8.14a, b) (ii) Areas significantly brazed by solidified zinc-rich liquid (Fig. 8.14c) (iii) Regions with the partial dispersal of the zinc coat takes place by formation of zinc-rich liquid film due to eutectic reaction when zinc diffuses into the aluminium sheet by grain boundary

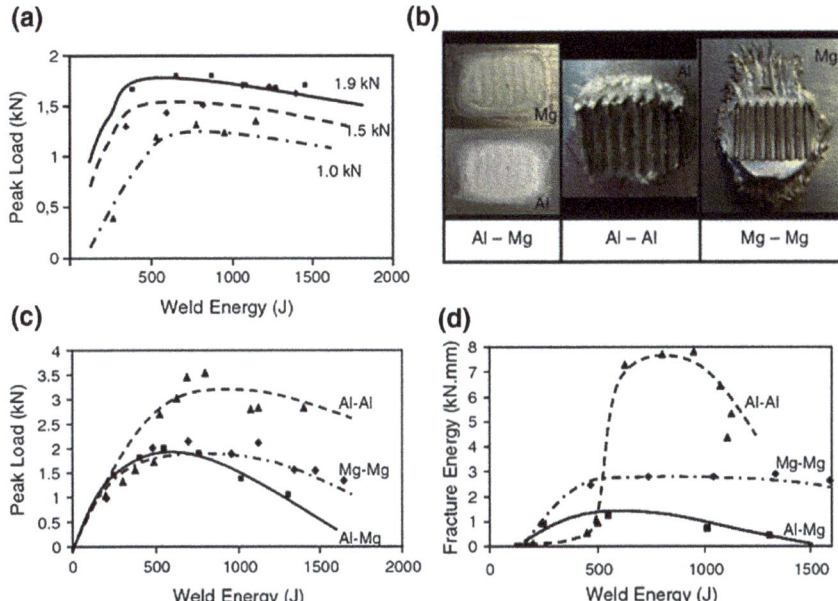

Fig. 8.15 **a** Mechanical performances as a function welding energy demonstrating the effect of a clampince forces for dissimilar Al–Mg joints, **c** and **d** comparison of similar Al–Al, Mg–Mg and dissimilar Al–Mg welds in terms of peak shear strength and facture energy for joints produced under a 1.9 kN clamping forces. **b** Examples of failure mode at optimized condition for different materials combinations [19]

penetration (Fig. 8.14f, g) (iv) Regions where a full dispersal of zinc-rich liquid occurs with extensive penetration along flow features and grain boundaries (Fig. 8.14d–h) [45].

Panteli et al. reported the effect of interface reaction on mechanical properties of dissimilar aluminium to magnesium ultrasonic spot welding compared to similar aluminium and magnesium joints [19]. In this work the effect of clamping force, welding energies and surface preparation on mechanical properties of the joints were studied. Figure 8.15a shows that laps shear strength is increased with increasing clamping forces for aluminium to magnesium joints. This is because of faster rate of surface abrasion when oxide first breaks down and greater metal to metal contact is achieved. In each set of experiment the strength increased with higher welding energy (time) before declining due to increasing the thickness of brittle Al–Mg intermetallic layer at the interface when the welding energy increased (>500 J) [19]. As shown in Fig. 8.15b failure is caused at the weld interface as a result of brittleness at the interface. However, greater mechanical performance (e.g. shear strength and fracture energy) was seen in similar aluminium and magnesium (Fig. 8.15c, d) welding when no intermetallic layer is caused at the interface leading to nugget pull-out failure mechanism.

8.2.8 Finite Element Process Modelling

Modelling work has so far been an important part of welding process optimization. Development of a reliable process model saves cost from production point of view when industrial parameter optimization becomes easier. Only a few attempts have been carried out to predict materials reaction processed under high power ultrasonic welding. Siddiq and Ghassemieh modelled thin foil aluminium ultrasonic consolidation using a rolling sonotrode [46]. In this work an experimentally friction coefficient, plastic deformation, thermal and ultrasonic softening was taken to account to model the materials behaviour; however, strain-rate dependence in yield stress was neglected although, high strain rate was exposed to the materials. The complexity of model was seen to limit the understanding of materials adjacent when subjected to deformation within a short period of welding time [47]. Another thorough finite element (FE) model by Elangovan et al. predicted temperature and stress distribution in aluminium joints. The model accuracy was found to be limited as materials thermal properties was independent of temperature. Further, the constant heat input was estimated form constant coefficient of friction and single value of yield strength [47]. FE model developed by Kim et al. [48] using Johnson-Cook materials deformation model, demonstrates similar limitation. It was explicitly mentioned that coupled investigation required extremely long run-time [48]. In some other reports FE model was developed for ultrasonic consolidation with only interface friction as the main source of friction plastic deformation [49, 50]. Their models predict every cycle during the process; however, no comparison was brought between simulated and experimental temperature investigation.

Jedrasiak et al. predicted the thermal histories within high power ultrasonic spot welding for different Al–Al, Al-Steel and Al–Mg welding components. The model was conducted with interfacial friction in addition to plastic deformation that occurred through the entire weld cross sections. In their report it was noteworthy that the plasticity was concentrated toward the interface as the gradient of temperature through the sheet thickness results in a high level softening close by the weld interface [47].

Figure 8.16 shows simulation model and experimental thermal history of the interface at both the weld centre and edge recorded by k-type thermocouple. It can be seen that the modelled temperature rises for the weld centre thermocouple reaching the peak value which is perfectly fit with experimental data. The predicted temperature is declining from the maximum level when the power is off and temperature is reduced accordingly although there is a bit scatter in this case. In the case of edge thermal history, the thermocouple data is consistently above the predicted temperature up to 50 °C. They explicitly demonstrated a greater uncertainty in prediction thermal history at the weld edge which can be because of lack of consistency in the rising part of the graphs [47].

Furthermore, as a comparison, Jedrasiak et al. showed the distribution of temperature at the weld plane of symmetry for three weld combination of Al–Al, Al-Steel and Al–Mg welding components (Fig. 8.17). Thermal maps were plotted

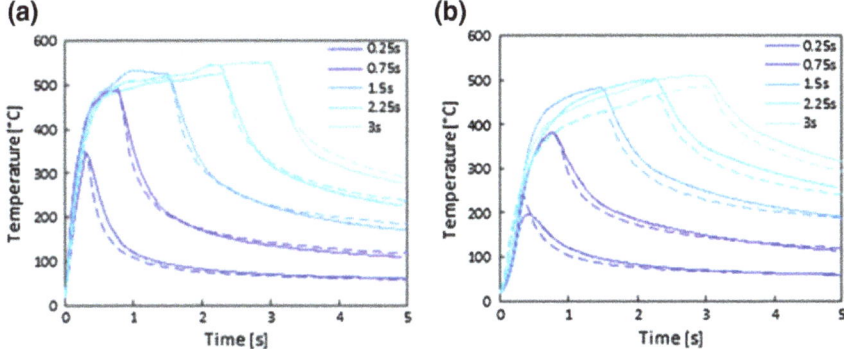

Fig. 8.16 Model simulation (*dash line*) and experimental for Al 6111 to DC04 steel welds with increasing welding time at the interface of **a** weld centre and **b** weld edge [47]

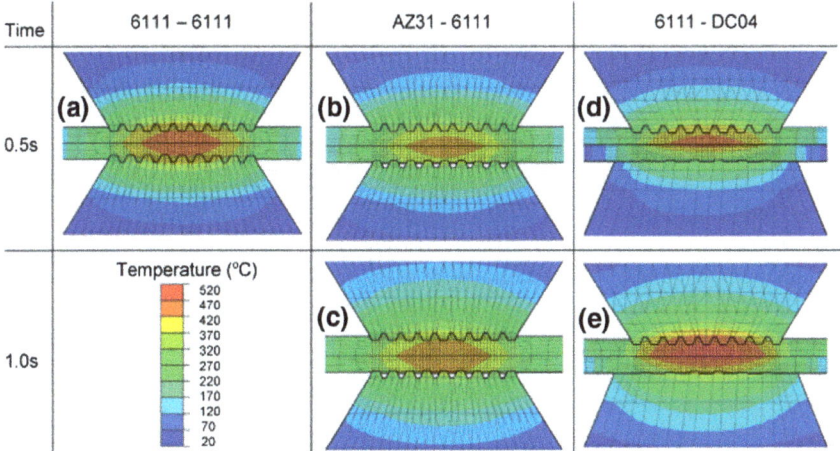

Fig. 8.17 Temperature distribution prediction at the weld plane symmetry for **a** Al–Al, **b, c** Mg–Al and **d, e** Al-Steel [47]

within the same scale of maximum welding time up to 1.0 s. It is noteworthy that the temperature fields are symmetrical in the case of Al–Al welds whereas, asymmetric temperature fields were seen in Mg–Al and Al-Steel as a result of differences in thermal properties. The minimum welding time was observed in the case of Mg–Al that is consistent with the lower efficiency of energy transfer to the workpiece [47].

8.3 Summary

In this chapter different aspects of ultrasonic spot welding, as a low energy manufacturing technique, were reviewed. Various factors influencing the welding process have been brought to this work as summarized here;

1. Ultrasonic spot welding technique is carried out in less than half a second and uses only 10% energy of conventional welding technologies such resistant spot welding.
2. The mechanism of bonding is involved with oxide film breakdown in the initial stages of welding and therefore, formation of microbonds that spreads in density and size with increasing welding time. Interfacial bonding takes place at longer welding time for dissimilar component as deformation of harder material occurs, it delays formation of microbonds.
3. Heat generation is caused by friction at the interface when welding components are in sliding condition in the initial stages of welding. With increasing welding time, plastic deformation is caused that results in heterogeneous heat generation.
4. Interfacial reaction in the main factor influencing the amplitude of vibration; with formation of liquid at the interface (e.g. Al-Zn) the vibration becomes in sliding condition and therefore, vibration amplitude is reduced compared to when there is no liquid due to interfacial reaction.
5. Various deformation regions were found across the thermomechanical affected zones including; severely deformed region at the interface, shear band at mid-cross section and forge zone where the sonotrode teeth penetrate to the sheets.
6. Formation of brittle intermetallic layer at the interface of dissimilar joints is the main factor of a low mechanical performance in ultrasonic spot welds. The use of coating such as zinc layers inhibits formation of intermetallic compounds.
7. Prediction of interface temperature has been carried out using Finite Element (FE) which is an important part of welding technologies as temperature had direct effect on microstructure and mechanical properties of the ultrasonic weld.

References

1. Das S, Graziano D, Upadhyayula VKK, Masanet E, Riddle M, Cresko J (2016) Vehicle lightweighting energy use impacts in U.S. light-duty vehicle fleet. Sustain Mater Technol 8:5–13. doi:10.1016/j.susmat.2016.04.001
2. Singh H, Kabeer B, Jansohn W, Davies J, Kan C-D, Kramer D, Marzougui D, Morgan RM, Quong S, Wood I (2012) Mass reduction for light-duty vehicles for model years 2017–2025. Department of Transportation, Washington, DC
3. Heuss R, Müller N, Sintern Wv, Starke A, Tschiesner A (2012) Lightweight, heavy impact. McKinsey & Company. www.autoassembly.mckinsey.com/html/resources/publications.asp
4. Council AC (2015) Implementing plastic and polymer composite lightweighting solutions to meet 2025 corporate average fuel economy standards. (PLASTICS DIVISION)

5. Haddadi F, Tsivoulas D (2016) Grain structure, texture and mechanical property evolution of automotive aluminium sheet during high power ultrasonic welding. Mater Charact 118:340–351. doi:10.1016/j.matchar.2016.06.004
6. Haddadi F (2016) Microstructure reaction control of dissimilar automotive aluminium to galvanized steel sheets ultrasonic spot welding. Mater Sci Eng, A 678:72–84. doi:10.1016/j.msea.2016.09.093
7. Haddadi F, Abu-Farha F (2015) Microstructural and mechanical performance of aluminium to steel high power ultrasonic spot welding. J Mater Process Technol 225:262–274. doi:10.1016/j.jmatprotec.2015.06.019
8. Qiu R, Yu X, Zhang H, Zhang K (2011) Joining Phenomena of Resistance Spot Welded joint between titanium and aluminium alloy. Adv Mater Res 230–232:982–986. doi:10.4028/www.scientific.net/AMR.230-232.982
9. Abe Y, Kato T, Mori K (2009) Self-piercing riveting of high tensile strength steel and aluminium alloy sheets using conventional rivet and die. J Mater Process Technol 209(8):3914–3922
10. Hsieh M-J, Lee R-T, Chiou Y-C (2017) Friction stir spot fusion welding of low-carbon steel to aluminum alloy. J Mater Process Technol 240:118–125. doi:10.1016/j.jmatprotec.2016.08.034
11. Zhang CQ, Robson JD, Prangnell PB (2016) Dissimilar ultrasonic spot welding of aerospace aluminum alloy AA2139 to titanium alloy TiAl6V4. J Mater Process Technol 231:382–388. doi:10.1016/j.jmatprotec.2016.01.008
12. Chen YC, Bakavos D, Gholinia A, Prangnell PB (2012) HAZ development and accelerated post-weld natural ageing in ultrasonic spot welding aluminium 6111-T4 automotive sheet. Acta Mater 60(6–7):2816–2828. doi:10.1016/j.actamat.2012.01.047
13. Chen Y, Haddadi F, Prangnell P (2010) Feasibility study of short cycle time friction stir spot welding thin sheet Al to ungalvanised and galvanized steel. Paper presented at the 8th international friction stir welding symposium, Germany, 18–20 May 2010
14. Bakavos D, Prangnell PB (2009) Effect of reduced or zero pin length and anvil insulation on friction stir spot welding thin gauge 6111 automotive sheet. Sci Technol Weld Join 14(5):443–456. doi:10.1179/136217109x427494
15. Harthoon JL (1978) Ultrasonice metal welding, vol 1
16. Jahn R, Cooper R, Wilkosz D (2007) The effect of anvil geometry and welding energy on microstructures in ultrasonic spot welds of AA6111-T4. Metallurg Mater Trans A 38(3):570–583. doi:10.1007/s11661-006-9087-0
17. Kenik E, Jahn R (2003) Microstructure of ultrasonic welded aluminum by orientation imaging microscopy. Microscopy Microanal 9 (Suppl S02):720–721
18. Bakavos D, Prangnell PB (2010) Mechanisms of joint and microstructure formation in high power ultrasonic spot welding 6111 aluminium automotive sheet. Mater Sci Eng, A 527(23):6320–6334. doi:10.1016/j.msea.2010.06.038
19. Panteli A, Chen Y-C, Strong D, Zhang X, Prangnell PB (2011) Optimisation of aluminium to magnesium ultrasonic spot welding. JOM 64(3):2012. doi:10.1007/s11837-012-0268-6
20. Robson JD, Panteli A, Iqbal N, Prangnell PB (2012) modeling intermetallic phase formation in dissimilar metal solid state welding of aluminium and magnesium alloy. Mater Sci Eng: A
21. Panteli A, Robson JD, Chen Y-C, Prangnell PB (2013) The effectiveness of surface coatings on preventing interfacial reaction during ultrasonic welding of aluminum to magnesium. Metallurg Mater Trans A 44(13):5773–5781. doi:10.1007/s11661-013-1928-z
22. Haddadi F, Abu-Farha F (2016) The effect of interface reaction on vibration evolution and performance of aluminium to steel high power ultrasonic spot joints. Mater Des 89:50–57. doi:10.1016/j.matdes.2015.09.121
23. Daniels HPC (1965) Ultrasonic welding. Ultrasonics 3(4):190–196. doi:10.1016/0041-624x(65)90169-1
24. Matheny MP, Graff KF (2015) 11—Ultrasonic welding of metals. In: Power ultrasonics. Woodhead Publishing, Oxford, pp 259–293. doi:http://dx.doi.org/10.1016/B978-1-78242-028-6.00011-9

25. Staff PDL (1997) Chapter 5—Ultrasonic welding. Handbook of plastics joining. William Andrew Publishing, Norwich, NY, pp 35–43b
26. Hetrick T, Baer JR, Zhu W, Reatherford LV, Grima AJ, SCHOll DJ, Wilkosz DE, Fatima S, Ward SM (2009) Ultrasonic metal welding process robustness in aluminum automotive body construction applications. Welding J 88 (7):149–158
27. Tsujino J, Hongoh M, Yoshikuni M, Miura H, Ueoka T (2005) Welding characteristics and temperature rises of various frequency ultrasonic plastic welding. In: 2005 IEEE ultrasonics symposium, 18–21 Sept 2005, pp 707–712
28. Watanabe T, Sakuyama H, Yanagisawa A (2009) Ultrasonic welding between mild steel sheet and Al-Mg alloy sheet. J Mater Process Technol 209(15–16):5475–5480. doi:10.1016/j.jmatprotec.2009.05.006
29. Zhu Z, Lee K, Wang X (2011) Ultrasonic welding of dissimilar metals, AA6061 and Ti6Al4 V. Int J Adv Manuf Technol:1–6. doi:10.1007/s00170-011-3534-9
30. Balle F, Wagner G, Eifler D (2009) Ultrasonic metal welding of aluminium sheets to carbon fibre reinforced thermoplastic composites. Adv Eng Mater 11(1–2):35–39. doi:10.1002/adem.200800271
31. Wagner G, Balle F, Eifler D (2012) Ultrasonic welding of hybrid joints. JOM 64(3):401–406. doi:10.1007/s11837-012-0269-5
32. Vries ED (2004) Mechanics and mechanisms of ultrasonic metal welding. Dissertation, The Ohio State University
33. Cheng X, Li X (2007) Investigation of heat generation in ultrasonic metal welding using micro sensor arrays. J Micromech Microeng 17:273–282. doi:10.1088/0960-1317/17/2/013
34. Prangnell PB, Bakavos D (2010) Novel approaches to friction spot welding thin aluminium automotive sheet. Mater Sci Forum 638–642:6. doi:10.4028/www.scientific.net/MSF.638-642.1237
35. Shawn Lee S, Shao C, Hyung Kim T, Jack Hu S, Kannatey-Asibu E, Cai WW, Patrick Spicer J, Abell JA (2014) Characterization of ultrasonic metal welding by correlating online sensor signals with weld attributes. J Manuf Sci Eng 136 (5):051019–051019. doi:10.1115/1.4028059
36. Harman G, Albers J (1977) The ultrasonic welding mechanism as applied to aluminum-and gold-wire bonding in microelectronics. IEEE Trans Parts Hybrids Packag 13(4):406–412
37. Siddiq A, Ghassemieh E (2009) Theoretical and FE analysis of ultrasonic welding of aluminum alloy 3003. J Manuf Sci Eng 131(4):041007–041011
38. Mariani E, Ghassemieh E (2010) Microstructure evolution of 6061 O Al alloy during ultrasonic consolidation: An insight from electron backscatter diffraction. Acta Mater 58 (7):2492–2503. doi:10.1016/j.actamat.2009.12.035
39. Allameh SM, Mercer C, Popoola D, Soboyejo WO (2005) Microstructural characterization of ultrasonically welded aluminum. J Eng Mater Technol 127(1):65–74
40. Borrisutthekul R, Yachi T, Miyashita Y, Mutoh Y (2007) Suppression of intermetallic reaction layer formation by controlling heat flow in dissimilar joining of steel and aluminum alloy. Mater Sci Eng, A 467(1–2):108–113. doi:10.1016/j.msea.2007.03.049
41. Sun Z, Karppi R (1996) The application of electron beam welding for the joining of dissimilar metals: an overview. J Mater Process Technol 59(3):257–267. doi:10.1016/0924-0136(95)02150-7
42. Barnes TA, Pashby IR (2000) Joining techniques for aluminium spaceframes used in automobiles: Part I—solid and liquid phase welding. J Mater Process Technol 99(1–3):62–71. doi:10.1016/s0924-0136(99)00367-2
43. Haddadi F (2015) Rapid intermetallic growth under high strain rate deformation during high power ultrasonic spot welding of aluminium to steel. Mater Des 66, Part B:459–472. doi: http://dx.doi.org/10.1016/j.matdes.2014.07.001
44. Xu L, Wang L, Chen Y-C, Robson JD, Prangnell PB (2016) Effect of interfacial reaction on the mechanical performance of steel to aluminum dissimilar ultrasonic spot welds. Metallurg Mater Trans A 47(1):334–346. doi:10.1007/s11661-015-3179-7

45. Haddadi F, Strong D, Prangnell PB (2012) Effect of zinc coatings on joint properties and interfacial reactions in aluminum to steel ultrasonic spot welding. JOM 64(3):407–413. doi:10.1007/s11837-012-0265-9
46. Siddiq A, Ghassemieh E (2008) Thermomechanical analyses of ultrasonic welding process using thermal and acoustic softening effects. Mech Mater 40(12):982–1000. doi:10.1016/j.mechmat.2008.06.004
47. Jedrasiak P, Shercliff HR, Chen YC, Wang L, Prangnell P, Robson J (2015) Modeling of the thermal field in dissimilar alloy ultrasonic welding. J Mater Eng Perform 24(2):799–807. doi:10.1007/s11665-014-1342-8
48. Kim W, Argento A, Grima A, Scholl D, Ward S (2011) Thermo-mechanical analysis of frictional heating in ultrasonic spot welding of aluminium plates. Proc Inst Mech Eng, Part B: J Eng Manuf 225(7):1093–1103. doi:10.1177/2041297510393664
49. Zhang C, Li L (2010) Effect of substrate dimensions on dynamics of ultrasonic consolidation. Ultrasonics 50(8):811–823. doi:10.1016/j.ultras.2010.04.005
50. Zhang C, Li L (2009) A coupled thermal-mechanical analysis of ultrasonic bonding mechanism. Metallurg Mater Trans B 40(2):196–207. doi:10.1007/s11663-008-9224-9

Part III
Sustainable Manufacturing

Chapter 9
Perspectives on Green Manufacturing

Varinder Kumar Mittal

Abstract This chapter introduces various perspectives on green manufacturing and its implementation issues. It has been known by different names such as sustainable manufacturing, environmentally conscious manufacturing, environmentally responsible manufacturing and cleaner production. The degradation of air, water and land in addition to unsustainable consumption of natural resources, global warming and landfills are posing critical challenges for the sustainable development of the planet. So the adoption of green manufacturing by the industry became vital at global level. However, there are some issues which influence the adoption of green initiatives by manufacturing industry. Hence, the motivations for, hindrances to, and stakeholders of green manufacturing are importantly discussed. This chapter also presents the current international status and future research directions as regards to the implementation of green manufacturing strategies.

Keywords Economy · Energy · Environment · Green · Global warming · Manufacturing · Production · Sustainable

9.1 Introduction

The 1980s have perceived an essential change in the approach governments and development agencies think about environment and sustainable development. These two issues were no longer regarded as mutually exclusive. It has been acknowledged that a healthy environment is indispensable for strong economy and sustainable development. The broad concept of sustainable development was widely discussed in the early 1980s, but was placed firmly on the international agenda with the publication of Brundtland report titled *"Our Common Future"* in 1987. The attempt to bring environmental and socio-economic issues together was

---

V.K. Mittal (✉)
Department of Mechanical Engineering, Amity University,
Uttar Pradesh, Sector 125, Noida 201313, Uttar Pradesh, India
e-mail: varindermittal@gmail.com; vkmittal@amity.edu

voiced in the Brundtland report's definition of sustainable development as 'meeting the needs of the present without compromising the ability of future generations to meet their own needs' [1]. The concept of sustainable development was the consequence of the growing awareness of the global links among mounting environmental challenges, socio-economic issues (e.g. inequality and poverty) and concerns about a future needs for humanity [2]. The *'Brundtland Report'* has pointed out the planet-wide interconnections of environmental problems. The report offered a valuable documentation on problems related to environment, energy, resources, industry and development [3].

Later, Kyoto Protocol in 1997 and Copenhagen Protocol in 2009 followed by Paris Meet in 2015 brought the international community on a common platform to discuss about the sustainable development. It has been stated that sustainable development has gained importance because of the fact that the humanity since 1985 started consuming resources more than that can be regenerated [4]. If everybody on the earth lives the lifestyle of the people from the technologically developed countries, which is not even one-fifth of the current population, the earth population would consume around 3-6 globes per year [5]. Moreover, the global population is growing at a fast rate and will reach to 9 billion by 2050 [6]. More population means more demand for material and energy which will further lead to the challenges like global warming, climate change, landfill problems, depleting natural resources, unhealthy living conditions because of excessive environmental exploitation and pollutions.

Manufacturing is one of the important elements of sustainable development as it produces goods which are required to cater to the needs of the society. Manufacturing is an input–output system in which the resources are transformed into products or semi-products [7]. Materials and energy are the two basic inputs to the manufacturing which are attained by exploiting the natural resources like material ores, fossil fuels, etc. The emerging, developing and underdeveloped countries are busy uplifting the living style of their rising population. At the same time developed countries do not want to sacrifice their current living standard [8]. Therefore, the average global consumption pattern for goods keeps increasing as the living standards improve, although these consumption patterns may slightly vary from region to region, driven by local cultural, societal and economic factors. This means that the growth of manufacturing is inevitable. This has led to a highly unsustainable situation as consumption of energy by the manufacturing sector has more than doubled in the last 50 years. The manufacturing sector is the main consumer of resources as it currently consumes about half of the world's total energy [9]. The consumption of critical raw materials like steel, aluminium, zinc, nickel, copper, wood, rubber, etc. for industrial use has increased worldwide. The competition for natural resources is accelerating in the BRICS nations (*viz.* Brazil, Russia, India, China and South Africa) and many other developing and underdeveloped economies.

The western world is no longer expected to drive world GDP growth in the decades ahead; India and China are expected to take over as shown in Table 9.1. By the turn of the millennium, China's consistent 10% annual growth rate puts it on the top of the list of countries in the world.

9 Perspectives on Green Manufacturing

Table 9.1 Share of global growth [10]

Period	Share of global growth (in %)		
	Leading nations	Emerging nations	Advanced nations
1982–1987	US (29.8), China (9.9), Japan (10.3)	31	69
1992–1997	US (24.2), China (18.9), Japan (3.8)	46	54
2002–2007	US (12.6), China (23.6), India (7.7)	67	33
2012–2017	US (13.9), China (33.6), India (9.4)	74	26

All activities pertaining to the global growth consequently create the threats to the sustainable development of the planet. The problems associated with the sustainable development can be addressed by the green manufacturing strategies and corresponding techniques. Hence, the adoption of green manufacturing in emerging nations should be the priority to attain the goal of sustainable development of the planet.

Green or sustainable manufacturing is the ability to wisely use natural resources in manufacturing by creating products and services that appreciates to new technology, regulatory measures and consistent social behaviours, which are able to satisfy economic, environmental and social requirements, thus conserving the environment while continuing to advance the quality of human life [11]. This is to further ensure the betterment of people, planet and prosperity [12] as shown in Fig. 9.1.

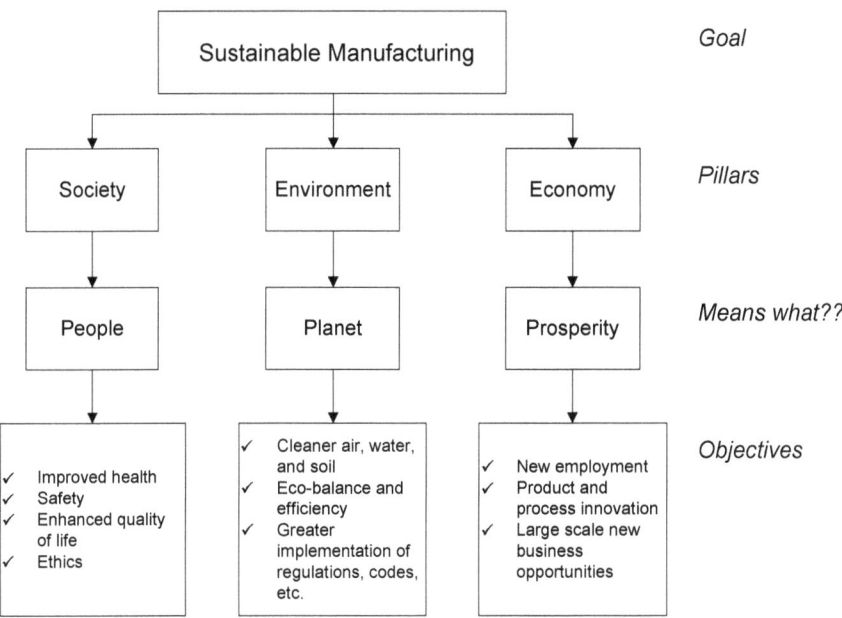

Fig. 9.1 Sustainable manufacturing goals, pillars and objectives [12]

9.2 Taxonomy Across Researchers

The compilation of evolution, connotation and scope of eight terms synonymous to each other and pertaining to the sustainable development are used by various researchers in the past. Intention is to compile the scholarly articles on these terms showing how environmental and societal concerns have been integrated in manufacturing over a period of time.

9.2.1 Sustainable Production (SP)

Holdgate [13] used the term *sustainable production* in the article "The reality of environmental policy", published in the *Journal of the Royal Society of Arts*. This term was first appeared in any title in the masters' thesis of Kowey [14] which was entitled as "An example of planning for sustainable production: the dry-cell battery problem", School of Community and Regional Planning, The University of British Columbia, September, 1990. Sustainable production was also highlighted in United Nations Conference on Environment and Development (UNCED) Conference at Rio de Janeiro, Brazil in 1992 to motivate companies and governments to work for the sustainable development [15].

Basically, sustainable production is stated as 'the creation of goods and services using processes and systems that are non-polluting; conserving natural resources and energy; economically viable; safe and healthy for employees, consumers and communities; and socially and creatively pleasing for all working individuals [16].

9.2.2 Clean Manufacturing (CM)

The term *clean manufacturing* was first used by Nagaraj et al. [17] in 1989. The term clean manufacturing includes incessant incremental improvement of environmental characteristics of products, processes and operations [18]. Mohanty and Deshmukh [19] discussed how green productivity can be increased through clean manufacturing by evolving a mind-set for total waste minimization, creating a sense of urgency for clean manufacturing and directing the efforts in multiple dimensions. As opined by Karp [20], it could be an expanded strategy of lean manufacturing with the inclusion of environmental aspects.

9.2.3 Cleaner Production (CP)

The term *cleaner production* was coined by United Nations Environment Programme Industry and Environment (UNEPIE) in 1989 to create awareness about cleaner production through information dissemination. UNEP IE launched many cleaner production implementation programmes in many countries with the active support of United Nations Industrial Development Organization, Universities, World Bank and other lending organizations. It is a continuous application of an integrated preventive environmental plan to products, processes, operations and services to escalate efficiency and reduce risks to environment and society [21]. It is also a preventive approach to handle pollution and seeks to avoid generation of waste at source rather than treating the symptoms of the waste already generated [22].

9.2.4 Environmentally Conscious Manufacturing (ECM)

The term *environmentally conscious manufacturing* was first used by Granoff [23] in 1991. Environmentally conscious manufacturing refers to the processes that have minimal harmful environmental impacts of manufacturing, including minimization of energy consumption and hazardous waste, higher material utilization efficiency and enhancement of operational safety.

9.2.5 Green Manufacturing (GM)

The term *green manufacturing* in any article was first used by Lewis [24]. The article educates the children about the environmental issues in manufacturing while playing games. The first article with "Green Manufacturing" in the title of the article is found in year 1995 by Dickinson et al. entitled "Green product manufacturing" [25]. Green manufacturing is the application of sustainable science to the manufacturing industry [26]. The term green manufacturing was created to propose the new manufacturing paradigm that uses various green strategies and techniques to become more eco-efficient. This strategy includes creating products/processes that consume minimal energy and material, replacing input materials, reducing waste outputs and converting outputs to inputs through recycling [27].

9.2.6 Environmentally Responsible Manufacturing (ERM)

Larrabee [28] used the term *environmentally responsible manufacturing* at the International Symposium on Semiconductor Manufacturing at Texas for the manufacturing of IC (Integrated Circuits) in an environmentally responsible manner. However, the first article with environmentally responsible manufacturing was found in year 1999 and entitled as "Environmentally Responsible Manufacturing: Past Research, Current Results, and Future Directions for Research" by Curkovic et al. [29]. Environmentally responsible manufacturing is an economically driven, system-wide and integrated approach to the reduction and removal of all waste streams associated with the manufacture, design, usage and disposal of products and materials [30].

9.2.7 Environmentally Benign Manufacturing (EBM)

Allen and Arvizu [31] used the term *environmentally benign manufacturing* in the article titled "Technology Transfer at Sandia National Laboratories" in Proceedings of the 27th Annual Hawaii IEEE International Conference on System Sciences held at Wailea-HI, USA during Jan 4-7, 1994 to discuss the technology transfer from government to the private sector that has assumed important new dimensions with the declining competitiveness of key U.S. industries in world markets. In the same year, Schmitt also used the term environmentally benign manufacturing in the article "Do manufacturing technologies need federal policies" published in the *journal of vacuum science & technology B: microelectronics and nanometer structures*. Environmentally benign manufacturing involves the operational practices, technologies, analytical methods and strategies for sustainable production within the industrial ecology framework.

9.2.8 Sustainable Manufacturing (SM)

Sustainable manufacturing has evolved from the concept of sustainable development to address concerns about economic development, environmental impact, globalization, inequities, etc. [15]. A very first book on sustainable manufacturing was written by Stephen et al. [32] and entitled as "Investing in sustainable manufacturing: a study of the credit needs of Chicago's metal finishing industry".

9.3 Background of Research on Green Manufacturing

Many researchers have conducted studies on various aspects influencing the adoption of environmental initiatives under the name of green manufacturing, cleaner production, eco-design, sustainable manufacturing, etc. These include drivers, motivations, incentives, barriers, hurdles, etc. Few of the important studies are listed here as under:

A study conducted in Indian industry identified 14 drivers and 12 barriers which may influence the adoption of green manufacturing practices [33]. Law and Gunasekaran [34] identified key motivating factors such as strategy/policy, mind-set, system, measures, needs to advance, performance, laws, regulations, social pressure, market trend, competition, and company's willingness and readiness to adopt sustainable development strategies in high-tech manufacturing firms in Hong Kong. Many researchers developed models for the drivers influencing the implementation of green supply chain management using various techniques [35–37]. Later, the drivers for extended supply chain practices in terms of energy saving and emission reduction among Chinese manufacturers were analysed from heavy polluters and high energy consuming industries [38].

The implementation of environmental management system, i.e. ISO 14001, was investigated in Lebanese food industry which would facilitate the policy changes and emerging strategies [39]. The new sustainable business models in China were explored to examine the reality of the level of sustainable development and its drivers [40].

A selective review of research articles from 1997–2007 provided the factors affecting adoption as a primary condition to diffusion and exploitation of cleaner technologies [41]. Yu et al. [42] listed seven drivers and six barriers for adopting eco-design and extended producer responsibility (ERP) in electrical and electronics companies operating in China. The drivers and barriers influencing the implementation of environmentally sound technology were provided through an industrial survey among 9 emerging nations [43]. There are various studies which identified the factors influencing the diffusion of cleaner production practices in manufacturing sector [44, 45].

The drivers and barriers to engage Hong Kong business organizations with voluntary environmental initiatives were analysed and their relevance was compared for companies of different stature [46]. The position of eco-design and manufacturing in SMEs of Midlands, UK is analysed by the results of a questionnaire designed to identify the drivers and barriers faced in their move towards greater sustainability of product design and manufacture in automotive sector [47]. Lawrence et al. [48] examined the motivations other than economic ones which drive sustainability practices in New Zealand SMEs. Dummett [49] identified drivers for corporate environmental responsibility through in-person interviews of 25 senior corporate leaders from key Australian and international businesses using a set of open ended questions. Gutowski et al. [50] identified from the study of Japan, Europe and USA industry, motivating factors for EBM. The motivating factors are regulatory mandates, competitive economic advantages and proactive green behaviour.

Perez-Sanchez et al. [51] developed a strategy for implementing an environmental management system after analyzing drivers of environmental performance in SMEs. Murphy [52] found the drivers for EBM as take-back legislation, material bans, landfill bans, life cycle assessment tool and database development, economic incentives, recycling infrastructure, cooperative and joint efforts with industry, ISO 14000 certification, financial and legal liability, supply chain involvement and EBM as a corporate strategy.

The lack of standardized metrics/performance benchmarks, lack of customers/consumers demand and lack of specific ideas were viewed as the major barriers to sustainability in Spanish companies. However, few critical barriers *viz.* technology risk, top management obligation, trade-offs and low enforcement was

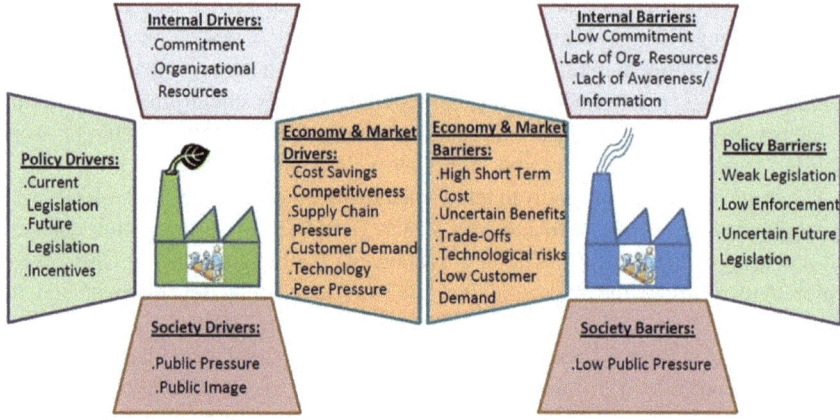

Fig. 9.2 Framework of factors influencing the adoption of ECM [61]

missing [53]. The extra financial liability, lack of time, lack of environmentally sustainable practices information, lack of motivation, stakeholders, inadequate communication, company culture and legal regulations may be the barriers faced by SMEs in adopting environmental practices and strategies in their organizations in USA [54]. Similarly, the studies identified the barriers in SMEs to move towards environmentally benign manufacturing [55]. The issues in the adoption of cleaner production have been widely investigated in various countries through the application of different approaches and methodologies [22, 56–60].

There are various models and frameworks of factors influencing the adoption of green manufacturing and similar approaches. Joint Indo-German researchers developed an environmentally conscious manufacturing framework as shown in Fig. 9.2.

The active involvement of the stakeholders for the implementation of green manufacturing and other similar systems is very important. Many researchers investigated the importance of the role of these stakeholders. Some important studies are listed below:

The specific drives and resources of SMEs were examined from the data of 254 ISO 9000 and ISO 14000 certified SMEs in Canada [62]. Gunasekaran and Spalanzani [63] specified that the pressure promoting sustainable business practices can be categorized as external and internal pressures to the organization. Dey and Cheffi [64] developed and applied an analytical framework for assessing the environmental performance of manufacturing supply chains in three major areas namely supply chain management, environmental management and performance measurement in three manufacturing organizations in the UK. Chang and Chen [65] developed an integral conceptual model of green intellectual capital to explore its managerial implications and determinants by integrating the theories of CSR and green management. Wolf [66] analysed the three competing models of the relationship among sustainable supply chain management, stakeholder pressure and corporate sustainability performance using a dataset of 1,621 organizations for the statistical comparison of these three models of the potential stakeholders.

An analytical framework was suggested that identified various stakeholders that potentially influenced the green efforts of a firm through an integration of the management, marketing and operations literature [67]. A sociological perspective of the insights of corporate environmentalism is held by stakeholder groups relative to each other and the effect that specific firm-level characteristics have on these insights [68]. A study using international sample of 656 large companies investigated the engagement with different stakeholders that promote sustainable innovation [69].

Voinov and Bousquet [70] reviewed the different types of models of stakeholders and compared participatory modelling to other stakeholder participation frameworks. A size moderated stakeholder model was derived and applied for the adoption of proactive environmental practices in firms [71]. The effect of six relevant variables on stakeholder and the environmental pressure is perceived by industrial companies. These effects were theoretically determined by comparing perception capacity and pressure intensity which was empirically tested with a study in 186 Spanish manufacturing organizations [72]. Sarkis et al. [73] established the influence of stakeholder pressure on the adoption of environmental practices in Spanish automotive industry. The study also focused on supporting the relationship between stakeholder and resource-based theory on various stakeholder pressures such as clients, governments, shareholders, employees, NGOs, community and supply chain partners.

The exploration of the extant literature on drivers, hindrances and stakeholders to identify various potential drivers, hindrances and stakeholders which are relevant in the manufacturing sector are presented in the forthcoming sections.

9.4 Motivations for Green Manufacturing

There are several motivational factors also called drivers that in-force the adaption of green manufacturing in the industries. The present section provides a brief report on some of the motivational factors for green manufacturing.

9.4.1 Current Legislation

The European Union (EU) has formulated a number of prescriptive directives surrounding the design, production and treatment of a range of consumer and industrial products [74]. Contrary to the traditional thought of considering market as a main driving force, the command-and-control regulation has over many years promoted diffusion of environmental technologies, such as waste-water treatment plants, chimney emission filters, environmental control technologies, etc. [75]. An analysis of Australian survey data revealed that threat of government legislation is the most important driver to enhance the corporate social responsibility.

9.4.2 Future Legislation

Besides the current legislation, anticipated future legislation is also an important driver to drive industries towards enhancing their environmental performance [76]. In some countries, the stringency of laws is still lacking and future improvements can easily drive companies to improve their environmental performance [43]. The companies in India are not fully complying with the regulatory requirements possibly because of gaps in the laws and ineffective regulations [76].

9.4.3 Incentives

The improvement of the greenness of firms is possible with the provision of financial incentives like tax exemptions for capital investments, loans, grants, etc. Empirical evidences support that financial incentives, like tax breaks or duty free imports, influence the company's investment strategy for environmental technologies [43]. It is worth mentioning that the federal ministry of Germany provides loans and grants to firms for invest in environment friendly production technologies from the state-owned banks [77].

9.4.4 General Awareness

Awareness of the people (citizens) against the environmental exploitation and the pressure created by various stakeholders, i.e. local communities; consumer groups; non-governmental organizations (NGOs); media or green parties; etc. also drive industrial sector to adopt green manufacturing initiatives largely [41].

9.4.5 Economical Concerns

It has become clear to many industries that investments in cleaner technologies can reduce costs, protect the environment and reduce the production cost. Industries can save huge capital by not paying for the carbon credits and all after implementing green manufacturing strategies.

9.4.6 Competitiveness

The accelerated competitiveness due to the implementing of green manufacturing technologies is also one of the important drivers [78]. Green manufacturing provides benefits in the cost structures through a higher efficiency which facilitates them to achieve more independence in the marketplace. A study in the European Union revealed that expected increase in energy costs in future acts as great motivator for eco-innovation and development [79].

9.4.7 Customer Demand

Literature survey highlights that customer demand is not an effective driver, but the trend of demand for green products is continuously increasing, and it would become more important in future. For example, everyone wants to buy a car which consumes less fuel to operate [49]. Consumers are certainly an increasingly important stakeholder that shapes the social responsibility of companies [80]. The consumption behaviour by the consumers is largely shaped by ethical criterion which means that customers choose to purchase more environment friendly manufactured products [78].

9.4.8 Supply Chain Pressure

Companies are part of supply chain network and they have interaction among various different suppliers and distributors. These supply chain network partners can motivate the firms to adopt green manufacturing practices and technologies [43]. Companies attempt to improve their environmental performance, for example, all suppliers are required to alter their processes to enable an enhancement in the overall performance of the entire supply chain network.

9.4.9 Top Management Commitment

Top management commitment has a crucial influence over the organizational culture of any company. One of the important characteristics in forceful environmental management is the active support and participation of top management in environmental protection issues [81]. The involvement and support of top management has an essential impact on the major company initiatives [82], and green manufacturing is no exception.

9.4.10 Company Reputation

The reputation of the company in public is very important to survive in the market. The need to protect the image and to enhance the same is essential part of any future business. Some companies see environmental initiatives not as a responsibility, but their future strategy and survival [49]. The companies in developing countries have already started to pursue a green image, so that they can distinguish themselves from competitors.

9.4.11 Technology

The adoption and usage of newer green technologies is the key part of green manufacturing. The accessibility of verified environment friendly technology to the industry can motivate the companies to adopt green manufacturing. The green technologies have explicit features that foster energy efficiency, resource efficiency and functional effectiveness in manufacturing [41]. For example, usage of energy-saving lights, renewable resource-based metalworking fluids, energy-efficient electric motors, etc. Despite the importance of environmental impact, the performance of the process/technique/machine is more important for manufacturers.

9.4.12 Organizational Resources

The organizational resources are important for any company to adopt and execute green manufacturing technologies and practices. The availability of professionally skilled and self-motivated staff in a company is one among the vital resources [83]. The implementation of green metrics and goals exclusively in the corporate strategy can make the achievements visible and people accountable. Adequate organizational resources have potential to continuously motivate green improvements in any company.

9.5 Hindrances to Green Manufacturing

Manufacturing firms face multiple barriers hindering green manufacturing implementation. The mitigation of these barriers would help industry to effectively implement green manufacturing. This section develops brief descriptions of the 12 identified hindrances based on the literature and the discussion held with experts from industry and academia.

9.5.1 Weak Legislation

The weak environmental legislation is a key barrier to the adoption of green manufacturing. The weak environmental legislation includes absence of environmental legislation, complexity of legislation, ineffective legislation, etc. Green initiatives require stringent, strong, effective and easy to understand and implement environmental legislation which can force the companies to accept them in totality. It has been observed in some developing countries that the environmental legislation is so frail that the cost of complying with the environmental legislation is greater than paying fine for not complying with environmental laws [84].

9.5.2 Low Enforcement

Besides weak legislation, the low enforcement of the environmental regulation is yet another barrier to the adoption of green manufacturing. The enforcement of environmental regulations is a big challenge for concerned authorities in some countries because of different reasons such as high cost of monitoring, lack of organizational infrastructure, lack of skilled human power and dishonest inspectors/officials. In the Netherlands, federal government provides funding to local authorities for monitoring the adoption of environmental regulations and laws; however, this is not evident in countries such as United Kingdom [85].

9.5.3 Uncertain Future Legislation

In addition to existing weak environmental legislation, the uncertainty of new legislations in the future may act as a barrier to the adoption of green manufacturing. The possibility of upcoming new legislations with unanticipated influences can be a threat to them whenever companies think of investing into green technologies and practices [55]. Thus, investments are withheld or delayed for possible prediction of future regulations. The investments and efforts into projects of good environmental standards are withheld due to the fear that the current environmental standards may not be enough in the future for compliance [86].

9.5.4 Lack of Awareness

Lack of awareness is a major barrier to the adoption of green manufacturing. The lack of pressure by key stakeholders like local communities, NGOs, media, banks, insurance companies or politicians may not pose the required thrust for companies to adopt green initiatives. In recent past, the protests and agitations by local communities had forced many governments to act hard on industries polluting at a large scale.

9.5.5 High Short-Term Costs

The high short-term cost in terms of capital investment is a high impact barrier to the adoption of green manufacturing [38]. The investments for new and efficient technology and practices require capital which is generally very high. Also, the changes required for installation of newer technologies in the shop floor lead to higher costs of implementation. Therefore, the manufacturers show reluctance to such changes as it can affect their profits and/or market share [78]. Although green technologies can save costs in the long run, higher initial capital cost of green technologies inhibits companies from adoption of green manufacturing [56].

9.5.6 Uncertain Benefits

Besides, high short-term costs as a big hurdle, the uncertainty in the economic benefits of green manufacturing implementation also acts as barrier [86]. Due to immature, unproven and new technology, the environmentally conscious business costs are still evolving and not yet fully understood. It becomes difficult for the top-level management to economically defend the change from conventional old technologies to newer green technologies.

9.5.7 Low Customer Demand

The lack of customer demand for greener products can be a barrier to adoption of green manufacturing [78]. The customers, particularly in emerging and underdeveloped nations, are price sensitive and willing to buy inexpensive products. The customers are not aware of the production conditions under which a product was manufactured. This is also true for many attributes of green complex products [87]. Studies in European Union nations have revealed that the unclear demand of products from the market is a significant barrier to development and eco-innovation uptake [79].

9.5.8 Trade-Offs

The companies outsourcing their non-environmental friendly manufacturing work to other countries where environmental regulations are less stringent. For example, a lot of manufacturing work by developed countries is being done in India and China, which reduces their share of emissions. In other words, companies have traded their emissions to other nations by outsourcing the non-green manufacturing work. This has resulted into merely trading of the emission rather than reduction of emissions in manufacturing [78]. A growing share of goods is manufactured in BRICS nations for the developed world.

9.5.9 Low Top Management Commitment

Top-level management has substantial ability to influence, support and promote the environmental initiatives in manufacturing. Green manufacturing adoption requires voluntarily commitment of the management. The reluctance to the top-level management hinders the adoption of green manufacturing. There are evidences that companies' particularly small and medium-sized enterprises snub their own environmental impact due to the absence of management will [55]. This may be because of the impression of the top management that green initiatives do not have the potential to yield benefit for them.

9.5.10 Lack of Organizational Resources

The inadequate human, technical and economic resources may discourage the firms to adopt green manufacturing technologies and practices [88]. The financial resources are very vital issue in taking environmental concerns into manufacturing system [89]. The lack of resources, particularly in SMEs, prevents them from investing in designing the products as per the environmental standards and guidelines.

9.5.11 Technological Risk

Since the green manufacturing is relatively new approach, its theories and technologies are still evolving. It is yet to be realized that the adoption of green manufacturing strategies can be used to create extended economic benefits for industry, society and ecology. Consequently, there are various technological risks, including its maintenance, reliability and applicability [26]. The expertise and information on newer technologies, materials, industrial processes and operations is often not available to the companies, predominantly in SMEs [90].

9.5.12 Lack of Awareness

Lack of awareness or information in a company about latest environment friendly manufacturing systems is another major barrier. In emerging countries, plant managers often have insufficient information about the available technology choices [40] and have limited access to green literature or the information diffusion. Even in western economies information seems to be lacking. The British Engineering Employers Federations found that most of their members do not know the sustainability meaning [83]. Lack of exposure is a common problem being faced by the SMEs.

9.6 Stakeholders of Green Manufacturing

This section provides the description and evolution of stakeholders:

9.6.1 Government

Government is an important stakeholder for effective adoption of environmentally conscious manufacturing in industry [91]. Government regulation is a force of law setup by a competent authority, concerning to the actions of those under that control of authority.

9.6.2 Employees

An employee contributes labour, skill and expertise for an employer and is typically hired to perform precise duties. Employees are directly related to a firm and have the ability to impact its core strategies directly. The major foundation for company's success and effective environmental policy planning is provided by the active participation from its employees [81].

9.6.3 Consumers

A consumer, generally known as a customer, purchaser or buyer, is the receiver of a product, good, idea or service obtained from a vendor, seller or supplier for a monetary or other valuable consideration. Several studies are available which justifies the importance of consumer towards the implementation of green manufacturing strategies [92, 93].

9.6.4 Market

A market is one of many varieties of institutions, systems, procedures, infrastructures and social relations by which parties/people engage in exchange. The market facilitates trade and enables the allocation and distribution of resources in the society. Competition among the firms is a major part of market pressure. Competitiveness is identified as one of the major motivations for environmental/ecological responsiveness [94].

9.6.5 Media

The media plays an important role in increasing awareness among public and formation of their views and attitudes toward certain issues. The media plays a vital role to protect natural environment by pressurizing the firms. When environmental crisis occurs, the media can influence society's perception about a company [81].

9.6.6 Local Politicians

Local politicians, i.e. member of local governing bodies or member of legislative assembly or Member of Parliament which are elected by the people, can influence the environmental performance of the nation's company as they are elected from and among the people.

9.6.7 Suppliers

A supplier is an individual/institution/system that supplies goods, products or services. A supplier can utilize its influence on organization by discontinuing delivery or it can pressurize the organization to employ a more environmentally acceptable alternatives. Suppliers influenced the decision to follow certification and standard to certify like ISO 14001 [93]. They contribute to the overall environmental performance of a supply chain [63].

9.6.8 Trade Organizations

A trade association, industry trade group or business association is generally an organization founded and funded by organizations that operate in a specific industry sector. An industry trade association contributes in public relation activities like publishing, advertising, education, political donations and lobbying, but its main focus is collaboration between companies and standardization. These associations may also offer many other services, like organizing conferences, charitable events or networking or offering classes or educational materials. In the context of proactive environmental strategy, industry and trade associations are important stakeholders [71].

9.6.9 Environmental Advocacy Groups

They include all the environmental groups and NGOs. At the plant level the institutional actors like environmental interest groups are most likely to directly influence environmental practices [88]. To improve corporate environmentalism, they can apply strong normative institutional pressure on industrial firms and corporates even though they are not directly linked in the organizations' economic transactions [68, 95].

9.6.10 Investors/Shareholders

A shareholder is an institution, individual or corporation that legally possesses a share of stock in a public or private corporation. Shareholders are the stakeholders which are directly linked to an organization and have the ability to impact its bottom line. A firm is said to be serious about environmental plans if it communicates its plans to employees and shareholders [81]. A company establishes publicly that it is committed to the environment by communicating with shareholders and shows how that commitment could be interpreted as improved environmental performance in the firm [96].

Along with the aforementioned stakeholders, local communities, investors, partners, owners, top officials like CEOs, etc. also play a major role for the encouragement and enforcement of the green manufacturing initiatives at industry level.

9.7 Current International Status

Green manufacturing is relevant to the whole world. However, the level of green manufacturing adoption varies considerably across the countries. The industry in the western world has adopted green manufacturing strategies more comparative to developing countries like of BRICS. We all share the common environment and the pollution does not stop at the borders of any country, so it is required to take international perspective to the implementation of green manufacturing in industry. There is a dire need of diffusion of green technology and resource optimization to developing and underdeveloped countries, so that the entire global industry can adopt it easily. However, economic aspect of the technology keeps the newer manufacturing technology within specific country. The import of green technology suffers from heavy taxation and high cost owing to huge currency differences across continents. The world market is open to all which leads to competition at international level. In order to survive in the international market, it is mandatory to adopt such environmental and social concerns in addition to economy of manufacturing. The change is happening quite fast in developed nations, while developing nations are trying to keep pace with developed nations, and the underdeveloped nations going very slow. Application of international perspective to optimize the use of material and energy resources while protecting the environmental pollution is the need of the coming decades.

The developed countries have shifted their hazardous manufacturing operations to the developing countries like India, China, etc. to gain from the less stringent environmental regulations. So it is more challenging for developing countries to work towards environmental approach. The identification of motivations, hindrances and stakeholders of green manufacturing for every specific country is required owing to different culture, society, economic condition, literacy, etc. Further, it is required to work on process level in order to provide workable solutions to the industry which is more environmental friendly. The incorporation of green concerns in the manufacturing is possible through voluntary adoption of green methods and technologies which should be innovated by research and development in the country.

9.8 Conclusions and Future Research Directions

Nowadays, almost every function within manufacturing organizations is being influenced by internal and external pressures to become green. Issues such as green consumerism, green products, green processes, environmental footprints, etc. have influenced the image of the company in the public. The outdated reactive approaches to these pressures are now being augmented and replaced by more proactive, strategic and competitive approaches. Many businesses have begun to realize that there are economic benefits of green manufacturing in addition to environmental and social benefits. To facilitate the easy and faster adoption and diffusion of green manufacturing in the industry, there is a need to understand and analyse the motivations, hindrances and stakeholders of green manufacturing.

The ambiguity on taxonomy of the naming of incorporating environmental, social and economic perspective is investigated that finds the answers of some questions such as when it appeared in the scientific literature, who introduced it for the first time, how it is different from others and how it is similar to other, etc.

Various researchers investigated these influencing factors and stakeholders for different geographical locations, different industry sectors, etc. which enable to identify the ways to foster the green manufacturing implementation. Due to the different social, economic and cultural status of every nation, the diffusion of these factors largely depends on the implementation models. Plenty of such models are suggested by researchers by taking inputs from different sources. There is need of collaboration of research universities/institutes with the industry.

The analysis of these factors influencing the adoption of green manufacturing in industry which will help the policy makers in the academia, industry, consultancy and government to frame policies and promotions more effective for this green change. The voluntary involvement of the stakeholders because of win–win situation for all would enhance the green manufacturing implementation more easy and smooth.

References

1. World Commission on Environment and Development (WCED) (1987) Our common future. Oxford University Press, Oxford
2. Hopwood B, Mellor M, O'Brien G (2005) Sustainable development: mapping different approaches. Sustain Dev 13(1):38–52
3. Trainer T (1990) A rejection of the brundtland report, IFDA Dossier. No. 77:71–84
4. Neffke PJ, Pant M, Khandelwal G (2008) Sustainable manufacturing: global challenge scenario for developing countries, competitive manufacturing. In: Proceedings of the 2nd international and 23th AIMTDR conference, IIT Madras, Chennai, India
5. Seliger G (2007) Sustainability in manufacturing: recovery of resources in product and material cycles. Springer, Berlin
6. Lutz W, Sanderson W, Scherbov S (2008) The coming acceleration of global population ageing. Nature 451:716–719
7. Liu F, Zhang H, Wu P, Cao HJ (2002) A model for analyzing the consumption situation of product material resources in manufacturing systems. J Mater Process Technol 122(2/3):201–207
8. O'Brien C (2002) Global manufacturing and the sustainable economy. Int J Prod Res 40 (15):3867–3877
9. Ross M (1992) Efficient energy use in manufacturing. Proc Natl Acad Sci USA 89(3):827–831
10. Times of India (2013) Go figure. Bennett Coleman and Co. Ltd. Jaipur dated 09/06/2013, page 13 of 26
11. Garetti M, Taisch M (2012) Sustainable manufacturing: trends and research challenges. Prod Plann Control 23(2/3):83–104

12. Jawahir IS (2008) Beyond the 3R's: 6R concepts for next generation manufacturing: recent trends and case studies. In: Symposium on sustainability and product development, IIT, Chicago, 7–8 Aug 2008
13. Holdgate M (1987) The reality of environmental policy. J R Soc Arts 135(5368):310–327
14. Kowey BN (1990) An example of planning for sustainable production: the dry-cell battery problem. Master' thesis, School of Community and Regional Planning, The University of British Columbia, September 1990
15. Rosen MA, Kishawy HA (2012) Sustainable manufacturing and design: concepts, practices and needs. Sustainability 4(2):154–174
16. Lowell Centre for Sustainable Production (1998) Sustainable production: A working definition, Informal meeting of the committee members
17. Nagaraj HS, Owens BL, Miller RJ (1989) Particulate generation in devices used in clean manufacturing, particles in gases and liquids, vol 1 (edited by Mittal KL), pp 283–293
18. Richards DJ (1994) Environmentally conscious manufacturing. World Class Des Manuf 1(3):15–22
19. Mohanty RP, Deshmukh SD (1998) Managing green productivity, some strategic directions. Prod Plann Control 9(7):624–633
20. Karp HR (2005) Green suppliers network: strengthening and greening the manufacturing supply base. Environ Qual Manage 15(2):37–46
21. United Nations Environmental Programme (UNEP) (1994) Montreal protocol on substances that deplete the Ozone Layer: 1994 Report of the Technology and Economics Assessment Panel. US EPA Identification—EPA/430/K94/023. United Nations Environmental Programme, Nairobi, Kenya
22. Siaminwe L, Chinsembu KC, Syakalima M (2005) Policy and operational constraints for the implementation of cleaner production in Zambia. J Clean Prod 13(10/11):1037–1047
23. Granoff B (1991) Environmentally conscious manufacturing at Sandia National Laboratory, environmentally conscious manufacturing/technology workshop, Albuquerque, NM (USA) during 20–21 Feb 1991
24. Lewis J (1991) The games children play—even Verminous Skumm is made of recycled material. 17 EPA J (September/October):25
25. Dickinson DA, Draper CW, Saminathan M, Sohn JE, Williams G (1995) Green product manufacturing. AT&T Tech J 74(6):26–35
26. Hua L, Weiping C, Zhixin K, Tungwai N, Yuanyuan L (2005) Fuzzy multiple attribute decision making for evaluating aggregate risk in green manufacturing. Tsinghua Sci Technol 10(5):627–632
27. Deif AM (2011) A system model for green manufacturing. J Clean Prod 19(14):1553–1559
28. Larrabee GB (1993) Environmentally responsible manufacturing, international symposium on semiconductor manufacturing, Austin, TX, USA, pp 215–240
29. Curkovic S, Melnyk SA, Calantone R, Handfield R (1999) Environmentally responsible manufacturing: past research, current results, and future directions for research, CiteSeer
30. Curkovic S, Landeros R (2000) An environmental Baldrige? Mid-Am J Bus 15(2):63–76
31. Allen MS, Arvizu DE (1994) Technology transfer at Sandia National Laboratories. In: Proceedings of the twenty-seventh Annual Hawaii IEEE international conference on system sciences, vol 4, Wailea, HI, USA, Jan 4–7, 1994, pp 465–472
32. Stephen B, Maureen H, Kathryn T (1990) Investing in sustainable manufacturing: a study of the credit needs of Chicago's metal finishing industry. jointly published by the Center for Neighborhood Technology and the Woodstock Institute (312/427-8070). Chicago, IL (June 1990)
33. Singh A, Singh B, Dhingra AK (2012) Drivers and barriers of green manufacturing practices: a survey of indian industries. IJ Eng Sci 1(1):5–19
34. Law KMY, Gunasekaran A (2012) Sustainability development in high-tech manufacturing firms in Hong Kong: Motivators and readiness. Int J ProdEcon 137(1):116–125
35. Diabat A, Govindan K (2011) An analysis of the drivers affecting the implementation of green supply chain management. Resour Conserv Recycl 55(6):659–667

36. Walker H, Sisto LD, McBain D (2008) Drivers and barriers to environmental supply chain management practices: Lessons from the public and private sectors. J Purchasing Supply Manag 14(1):69–85
37. Zhu Q, Sarkis J, Geng Y (2005) Green supply chain management in China: pressures, practices and performance IJ Oper Prod Manag 25(5): 449–468
38. Zhu Q, Geng Y (2013) Drivers and barriers of extended supply chain practices for energy saving and emission reduction among Chinese manufacturers. J Clean Prod 40:6–12
39. Massoud MA, Fayad R, Kamleh R, El-Fadel M (2010) Environmental management system (ISO 14001) certification in developing countries: challenges and implementation strategies. Environ Sci Technol 44(6):1884–1887
40. Birkin F, Cashman A, Koh SCL, Liu Z (2009) New sustainable business models in China. Bus Strategy Environ 18(1):64–77
41. Montalvo C (2008) General wisdom concerning the factors affecting the adoption of cleaner technologies—a survey 1990-2007. J Clean Prod 16(1-S1): S7–S13
42. Yu J, Hills P, Welford R (2008) Extended producer responsibility and eco-design changes: perspectives from China. Corp Soc Responsib Environ Manag 15(2):111–124
43. Luken R, Rompaey FV (2008) Drivers for and barriers to environmentally sound technology adoption by manufacturing plants in nine developing countries. J Clean Prod 16(1-Suppl 1): S67–S77
44. Yuksel H (2008) An empirical evaluation of cleaner production practices in Turkey. J Clean Prod 16(1-Suppl 1): S50–S57
45. Gunningham N, Sinclair D (1997) ACEL final report: barriers and motivators to the adoption of cleaner production practices. Australian Centre for Environmental Law, Canberra, p 115
46. Studer S, Welford R, Hills P (2006) Engaging Hong Kong businesses in environmental change: drivers and barriers. Bus Strategy Environ 15(6):416–431
47. Veshagh A, Li W (2006) Survey of eco design and manufacturing in automotive SMEs, In: Proceedings of 13th CIRP international conference on life cycle engineering, Leuven, Belgium, May 31–June 02 2006, pp. 305–310
48. Lawrence SR, Collins E, Pavlovich K, Arunachalam M (2006) Sustainability practices of SMEs: the case of NZ. Bus Strategy Environ 15(4):242–257
49. Dummett K (2006) Drivers for corporate environmental responsibility (CER). Environ Dev Sustain 8(3):375–389
50. Gutowski T, Murphy C, Allen D, Bauer D, Bras B, Piwonka T, Sheng P, Sutherland J, Thurston D, Wolff E (2005) Environmentally benign manufacturing: observations from Japan, Europe and the United States. J Clean Prod 13(1):1–17
51. Perez-Sanchez D, Barton JR, Bower D (2003) Implementing environmental management in SMEs. Corp Soc Responsib Environ Manag 10(2):67–77
52. Murphy CF (2001) Geographic trends, Chapter 2 of WTEC panel report on environmentally benign manufacturing. International Technology Research Institute, World Technology (WTEC) Division, pp 9–22
53. Koho M, Torvinen S, Romiguer AT (2011) Objectives, enablers and challenges of sustainable development and sustainable manufacturing: views and opinions of Spanish companies. In: 2011 IEEE international symposium on assembly and manufacturing (ISAM), 25–27th May 2011. doi:10.1109/ISAM.2011.5942343
54. Herren A, Hadley J (2010) Barriers to environmental sustainability facing small businesses in Durham, NC, Masters project, Duke University
55. Seidel M, Seidel R, Des T, Cross R, Wait L, Hämmerle E (2009) Overcoming barriers to implementing environmentally benign manufacturing practices—strategic tools for SMEs. Environ Qual Manage 18(3):37–55
56. Shi H, Peng SZ, Liu Y, Zhong P (2008) Barriers to the implementation of cleaner production in Chinese SMEs – government, industry and expert stakeholders' perspectives. J Clean Prod 16(7):842–852
57. Mitchell CL (2006) Beyond barriers: examining root causes behind commonly cited Cleaner Production barriers in Vietnam. J Clean Prod 14(18):1576–1585

58. Moors EHM, Mulder KF, Vergragt PJ (2005) Towards cleaner production: barriers and strategies in the base metals producing industry. J Clean Prod 13(7):657–668
59. Zhang TZ (2000) Policy mechanisms to promote cleaner production in China. J Environ Sci Health, Part A: Toxic/Hazard Subst Environ Eng 35(10):1989–1994
60. Cooray N (1999) Cleaner production assessment in small and medium industries of Sri Lanka. In: Ashley Scott J, Pagan RJ (eds) Global competitiveness through cleaner production: proceedings of the 2nd Asia Pacific Cleaner Production Roundtable, 21–23 April 1999, Brisbane, Australia/, pp 108–114
61. Mittal VK, Sangwan KS, Herrmann C, Egede P, Wulbusch C (2012) Drivers and barriers to environmentally conscious manufacturing: a comparative study of Indian and German organizations. Leveraging Technology for a Sustainable World, (eds: Dornfeld and Linke) Springer, CA, USA, pp97–102
62. Roy MJ, Boiral O, Paille P (2013) Pursuing quality and environmental performance: Initiatives and supporting processes. Bus Process Manag J 19(1):30–53
63. Gunasekaran A, Spalanzani A (2012) Sustainability of manufacturing and services: investigations for research and applications. Int J Prod Econ 140(1):35–47
64. Dey PK, Cheffi W (2012) Green supply chain performance measurement using the analytic hierarchy process: a comparative analysis of manufacturing organizations. Prod Plan Control Manag Oper 24(8/9):702–720
65. Chang CH, Chan YS (2012) The determinants of green intellectual capital. Manag Decis 50 (1):74–94
66. Wolf J (2012) The relationship between sustainable supply chain management, stakeholder pressure and corporate social responsibility. J Bus Ethics. doi:10.1007/s10551-012-1603-0. Accessed 24 Feb 2013
67. Cronin JR, Smith JS, Gleim MR, Ramirez E, Martinez JD (2011) Green marketing strategies: an examination of Stakeholders and the opportunities they present. J Acad Mark Sci 39 (1):158–174
68. Shah KU (2011) Corporate environmentalism in a small emerging economy: stakeholder perceptions and the influence of firm characteristics. Corp Soc Responsib Environ Manag 8 (2):80–90
69. Ayuso S, Rodriguez MA, Castro RG, Arino MA (2011) Does stakeholder engagement promote sustainable innovation orientation? Ind Manag Data Syst 111(9):1399–1417
70. Voinov A, Bousquet F (2010) Modelling with Stakeholders. Environ Model Softw 25 (11):1268–1281
71. Darnall N, Henriques I, Sadorsky P (2010) Adopting proactive environmental strategy: the influence of stakeholders and firm size. J Manag Stud 47(6):1072–1094
72. Gonzalez-Benito GJ, Gonzalez-Benito GO (2010) A study of determinant factors of stakeholder environmental pressure perceived by industrial companies. Bus Strategy Environ 19(3):164–181
73. Sarkis J, Torre PG, Diaz BA (2010) Stakeholder pressure and the adoption of environmental practices: the mediating effect of training. J Oper Manag 28(2):163–176
74. Rahimifard S, Coates G, Staikos T, Edwards C, Abu-Bakar M (2009) Barriers, drivers and challenges for sustainable product recovery and recycling. Int J Prod Sustain Eng 2(2):80–90
75. Remmen A (2001) Greening the Danish Industry—changes in concepts and policies. Technol Anal Strateg Manag 13(1):53–69
76. Mejia R (2009) The challenge of environmental regulation in India. Environ Sci Technol 43 (23):8714–8715
77. Federal Ministry for the Environment, Nature Conservation and Nuclear Safety (2011) Federal Cabinet approves draft new law for emissions trading—EU Emissions trading directive into national law, Press Release No. 028/11, Berlin
78. Dwyer J (2007) Unsustainable measures. IET Manuf 86(6):14–19
79. European Commission (2011) Flash Eurobarometer 315. The Gallup Organization—Attitudes of European entrepreneurs towards eco-innovation. http://ec.europa.eu/public_opinion/flash/fl_315_en.pdf. Accessed 01 July 2011

80. Mont O, Leire C (2009) Socially responsible purchasing in supply chains: drivers and barriers in Sweden. Soc Responsib J 5(3):388–407
81. Henriques I, Sadorsky P (1999) The relationship between environmental commitment and managerial perceptions of stakeholder importance. Acad Manag J 42(1):87–99
82. Maidique MA, Zirger BJ (1984) A study of success and failure in product innovation: the case of US electronics industry. IEEE Trans Eng Manag EM-31(4):192–203
83. Schroeder DM, Robinson AG (2010) Green is free–Creating sustainable competitive advantage through green excellence. Org Dyn 39(2):345–352
84. Zhang XH (2005) The economic view and policy choice for government's green procurement. China Gov Procure 12:15–16
85. Rutherfoord R, Blackburn RA, Spence LJ (2000) Environmental management and the small firm: an international comparison. Int J Entrepr Behav Res 6(6):310–326
86. Del Río González P (2005) Analysing the factors influencing clean technology adoption—a study of the Spanish pulp and paper industry. Bus Strategy Environ 14(1):20–37
87. Yim HJ (2007) Consumer oriented development of ecodesign products. Vulkan-Verlag, Essen
88. Hadjimanolis A, Dickson K (2000) Innovation strategies of SMEs in Cyprus, a small developing country. Int Small Bus J 18(4):62–79
89. Min H, Galle WP (2001) Green purchasing (GP) practices of US firms. Int J Prod Oper Manag 21(9):1222–1238
90. Wooi GC, Zailani S (2010) Green supply chain initiatives: investigation on the barriers in the context of SMEs in Malaysia. Int Bus Manag 4(1):20–27
91. Regens JL, Seldon BJ, Elliott E (1997) Modelling compliance to environmental regulation: evidence from manufacturing industries. J Policy Modell 19(6):683–696
92. Gadenne DL, Kennedy J, McKeiver C (2009) An empirical study of environmental awareness and practices in SME's. J Bus Ethics 84(1):45–63
93. Delmas M, Toffel MW (2004) Stakeholders and environmental management practices: an institutional framework. Bus Strategy Environ 13(4):209–222
94. Bansal P, Roth K (2000) Why companies go green: a model of ecological responsiveness. Acad Manag J 43(4):717–736
95. Waddock SA, Graves SB (1997) The corporate social performance—Financial Performance link. Strateg Manag J 18(4):303–319
96. Vos PD, Wittebolle L, Marzorati M, Clement L, Heylen K, Verstraete W, Boon N (2009) Initial community evenness favours functionality under selective stress. Nature 458 (7238):623–626

Chapter 10
Experimental Investigation and Optimization on MQL-Assisted Turning of Inconel-718 Super Alloy

Munish K. Gupta, P.K. Sood, Gurraj Singh and Vishal S. Sharma

Abstract Inconel-718 super alloy primarily used in aerospace, nuclear, marine and energy sectors. Due to the perceived difficulties in machining of this alloy, some advanced machining strategies such as minimum quantity lubrication-assisted machining are used with an objective to improve their machinability along with maintaining the environmental friendliness simultaneously. In minimum quantity lubrication-assisted machining of Inconel-718 alloy, the selection of optimum machining parameters remains a critical concern in order to ensure the quality of the product, reduce the machining cost, increasing the productivity and conserve resources for sustainability. This chapter describes the experimental investigation conducted in turning Inconel-718 alloy with an overall idea of optimizing the process to achieve higher metal removal rate, lower cutting forces and lower power consumption under MQL conditions. The experimental work is based on the analysis and optimization of various process parameters such as cutting speed, feed rate, side cutting edge angle (approach angle) while keeping depth of cut constant using CBN insert tool. The results indicate that the cutting force increases with increase in feed rate, whereas there is no significant change in cutting force as cutting speed and side cutting edge angle increases. The MRR increases with the increase in cutting speed and feed rate, whereas there is no effect of side cutting edge angle on MRR. The power consumption increases with the increase in cutting speed and feed rate, whereas it decreases with the increase in side cutting edge angle. Process parameter optimization enhanced the machinability of Inconel-718 to a great extent.

Keywords MQL · Machinability · Optimization · Super alloy · Sustainability

---

M.K. Gupta (✉) · P.K. Sood
Deptartment of Mechanical Engineering, National Institute of Technology,
Hamirpur 177005, H.P., India
e-mail: munishguptanit@gmail.com

G. Singh · V.S. Sharma
Department of I & P Engineering, National Institute of Technology, Jalandhar, Punjab, India

© Springer International Publishing AG 2017
K. Gupta (ed.), *Advanced Manufacturing Technologies*, Materials Forming,
Machining and Tribology, DOI 10.1007/978-3-319-56099-1_10

10.1 Introduction

Inconel-718 is a nickel-based super alloy, having vast industrial application in the field of aerospace, marine and energy sectors. It is also used particularly in the high temperature area of engine, due to their high temperature and corrosion resistance [1]. It is one of the most difficult-to-machine materials, due to the rapid work hardening, tendency to weld and form a built-up edge with the tool material at high temperature generated during machining and the presence of hard carbides, such as titanium carbide and niobium carbide, in their microstructure. These several properties cause the shorter tool life, increase forces and power consumption, lower production efficiency and various surface damages [2, 3]. Therefore, in order to resolve the above-mentioned issues associated with the machining of Inconel-718 alloy, some innovative sustainable machining strategies including green lubrication systems with environmentally friendly lubricants (as cutting fluids) are used. It is a well-known fact that the cutting fluid represents a considerable proportion of the total costs of production [4, 5]. It plays a vital role in enhancing the surface integrity and tool life, yet they have adverse effects on the environment and contain potentially damaging chemical constituents. Prolonged exposure to coolants during machining may also lead to many health problems. With these concerns, the approach of researchers have been diverted towards an environmentally friendly and sustainable method, i.e. minimum quantity lubrication (MQL) with vegetable oils and synthetic esters which further reduces the use of cutting fluids. It shows very significant results in terms of reduction in machining cost and quantity of cutting fluid consumed. It is also known as near to dry lubrication or micro-lubrication in which the cutting fluids are environmentally friendly and their quantity used for machining is very low as compared to the conventional flood cooling method [4–10].

Many researchers have suggested the beneficial effect of MQL techniques in turning processes. The significant improvement in machinability has been observed using MQL conditions as compared to dry and wet machining [10–19]. Varadarajan et al. examined the influence of machining parameters on hard turning of AISI 4340 steel using dry, wet and MQL conditions in the terms of cutting forces, surface finish, tool life and tool chip contact length. They found that MQL performs superior results as compared to dry and wet machining [10]. Kamata and Obikawa discussed the beneficial effect of MQL over dry and wet conditions during turning of Inconel 718 alloy on tool life and the surface finish [11]. Moreover, the results from experimentation performed by Davim et al. [12] on turning of brass shows that MQL lubrication gives better performance when compared to flood lubrication and MQL lubrication can be an effective replacement for flood lubrication, which results in reduced manufacturing cost and environmental hazards. Settineri et al. carry out the experiments on different coated carbide inserts while turning nickel-based super alloy under dry as well as MQL environments. In their work, coated carbide tools outperformed uncoated carbide tools under both conditions [13]. Obikawa et al. experimentally investigated the performance of micro litter lubrication in machining of Inconel 718 based on the tool life and surface finish. The results proved that micro litter lubrication effectively increasing the tool life as compared to wet machining [14]. Aggrawal et al. conducted the turning experiment on Inconel 718 under MQL conditions and proved that MQL could be used to perform machining of these super alloys. They also presented correlations and confirmation tests for flank wear [15]. Thakur et al. investigate and optimize the lubrication parameters during high speed turning of Inconel 718 with respect to cutting temperature, flank wear and cutting forces. Their experimental results indicate that the use of optimized minimum quantity lubrication parameters leads to minimize the cutting forces, cutting temperature and flank wear [16, 17]. Yazid et al. studied the effect of machining parameters and cutting conditions on surface quality of Inconel 718. They investigated that MQL produces good surface quality by applying various flow rates (50–100 ml/h) as compared to dry machining [18]. Ji et al. scrutinized the effect of MQL on cutting forces, cutting temperature and residual stresses while turning AISI 4130 alloy steel. They also compared the performance of MQL with dry and flood cooling machining. The result shows that MQL significantly reduced the cutting forces and temperature in machining [19].

It is manifested from the published works that an application of minimal quantity cutting fluid in turning process brings prolific outcomes in terms of reducing tool wear, cutting forces, cutting temperature and surface roughness of the machined part. It is seen that the process parameters greatly influence the machining performance. Thus, selecting the optimum combination of the process parameters is essential for the success of the machining process. The literature review also indicates the scarcity of research on use of MQL or other environmentally friendly lubrication strategy for improvement in machinability of Inconel-718 alloy.

The research work discussed in this chapter is a part of detailed experimental work conducted while turning Inconel-718 alloy with an overall idea of optimizing the process to increase metal removal rate, decrease cutting forces and power consumption under MQL conditions.

10.2 Experimental

A highly precise CNC turning centre Sprint 16 Tc equipped with a Siemens control system was used for MQL-assisted turning of Inconel-718. The NOGA make MQL set up is used in the current investigations. The system consists of two inlet pipes and two discharge tubes which are joined at mixing chamber. There are nozzles at the end of both discharge tubes. One inlet pipe is connected to the compressed air pipe while the other is attached to the container containing coolant. Depending on the pressure of the air, the oil from the container is sucked and delivered through the discharge tubes in the form of aerosol. There is a pull-push button which allows the aerosol supply when pulled out and stops the supply when pushed inwards. The arrangement of experimental setup is shown in Fig. 10.1. A three component

Fig. 10.1 Experimental Setup

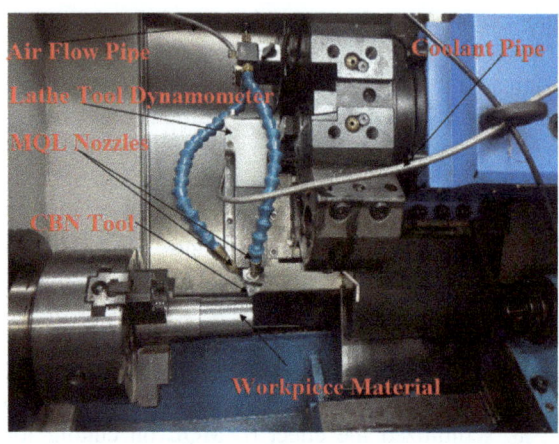

turning dynamometer (TeLC with DKM 2010 software) was used for evaluation of cutting forces and power consumption.

The material removal rate (MRR) is calculated by the following equation:

$$\text{MRR} = 1000 * V_c * f * a_p \tag{10.1}$$

The round bars with 60 mm diameter and 150 mm length of the workpiece material Inconel-718 super alloy) were used in the experimental machining. The chemical composition of Inconel-718 is C 0.034%, Si 0.07%, Mn 0.08%, P 0.01%, S 0.0017%, Cr 18.43%, Mo 2.98%, Ni 52.67%, Al 0.56%, Ti 1.01%, Co 0.11%, Cu 0.04%, B 0.0041% and Fe-Bal. The important mechanical properties of Inconel-718 are: hardness-415 HB, yield strength-1248 MPa, tensile stress-1419 MPa, reduction of area-49% and percentage elongation-20%.

The CCGW 09T304-2 Tips CBN insert having nose radius of 0.4 mm was used for turning Inconel. To avoid the effect of tool wear on given responses, a fresh tool was used for each experiment.

The general purpose "Balmerol make Protosole MQ" soluble cutting oil was used, manufactured from specially selected base stocks with superior performance additive packs. It offers very stable emulsion (20:1) with hard water. The base of the cutting fluid was a commercially available mineral oil. The supply pressure and flow rate for the compressed air were 4 bars and 60 l/min respectively. While the cutting fluid flow rate was 300 ml/hr. The aerosol formed due to the mixing of compressed air and fluid was delivered to the cutting zone through the nozzle which was positioned at a distance of 35–40 mm from the tool tip.

10.3 Results and Discussions

The experiments were designed and conducted based on Box–Behnken design (BBD) method of response surface methodology (RSM) approach to achieve the objectives of this research work. Three important machining parameters, i.e. cutting speed, feed rate and approach angle were varied at three levels each as shown in Table 10.1. Based on BBD technique, all seventeen experimental combinations are given in Table 10.2. Analysis of variance (ANOVA) was used to identify the significant parameters affecting the machinability or responses.

Table 10.1 Process parameters and their levels

Parameters	Coded value	Units	Low level (−1)	Middle level (0)	High level (+1)
Cutting speed (V_c)	A	m/min	200	250	300
Feed rate (f)	B	mm/rev	0.1	0.15	0.20
Approach angle (ϕ)	C	Deg. (°)	60	75	90
Depth of cut (a_p) = 1 mm (fixed)					

Table 10.2 Experimental combinations and corresponding results

Machining parameters				Responses		
Run order	V_c m/min	f mm/rev	ϕ Deg. (°)	Cutting force F_c (N)	MRR (cm³/min)	Power consumption (Watt)
1	200	0.2	75	310.19	39.33	920.38
2	300	0.2	75	280.23	59	1040.69
3	250	0.15	75	252.2	36.88	880.5
4	200	0.1	75	181	20	662
5	250	0.1	90	170	24.58	675
6	250	0.1	60	195	25.58	737.5
7	200	0.15	90	247	29.5	794
8	250	0.15	75	252	35.88	880
9	250	0.15	75	234	37.88	835
10	250	0.2	60	332	49.17	1080
11	300	0.15	90	235	44.52	905
12	250	0.15	75	260	34.88	900
13	300	0.1	75	181	29.5	743
14	300	0.15	60	271	44.25	1013
15	250	0.2	90	303	49.17	1007.5
16	250	0.15	75	252	38.88	880
17	200	0.15	60	271	29.5	842

10.3.1 Regression Modelling

Regression equations for cutting force, MRR and power consumption were established on the basis of experimental data. Established regression equations for all responses are presented in Eqs. (10.2–10.4):

$$F_c = 159.19221 - 0.10490 * V_c + 1246.05000 * f - 0.95000 * \phi \quad (10.2)$$

$$\text{MRR} = 7.78559 - 0.011950 * V_c - 36.72500 * f - 0.078583 * \phi + 1.01700 * V_c * f + 9.00000E - 005 * V_c * \phi + 0.33333 * f * \phi \quad (10.3)$$

$$\text{Power Consumption} = 288.48265 + 1.20827 * V_c + 3077.67500 * f - 2.42500 * \phi. \quad (10.4)$$

Further, to ensure the accuracy and adequacy of developed models, the validation of regression models has been performed by diagnostic tests. Figure 10.2 shows a normal plot of distribution in case of *cutting force*. The normal plot illustrates that for the residuals follow a normal distribution, as residual follows a straight line except few scatter at the upper and lower end which is always present in normal experimental data. Figure 10.3 illustrates the graph of residuals versus predicted values. This demonstrates that the process has the same trend about the residual line and is stable, which proves the acceptability of the model. The same graphs are plotted for the rest of the responses which confirm the adequacy of the developed model.

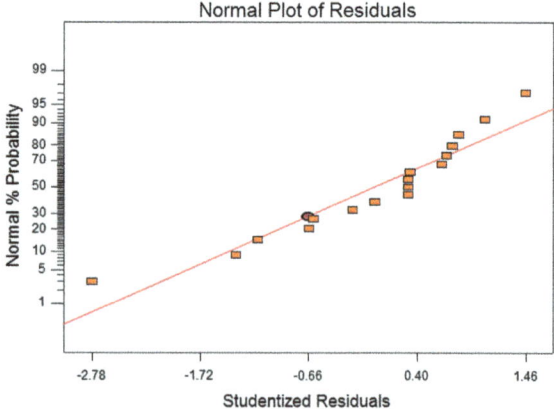

Fig. 10.2 Normal plot of residuals for F_c

Fig. 10.3 Plot of residuals vs. predicted for F_c

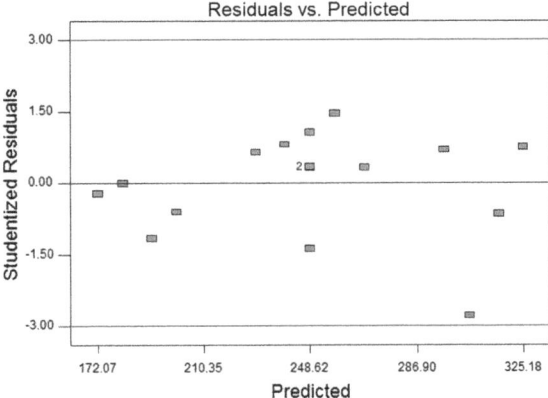

Table 10.3 ANOVA results for cutting force, MRR and power consumption

Factors	Responses		
	F_c (N)	MRR (cm³/min)	Power (watt)
R-Squared	0.954395	0.993662	0.953829
Adj. R-Squared	0.943871	0.98986	0.943175
Pred. R-Squared	0.915136	0.991875	0.91306
Adeq. Precision	28.70329	59.46446	30.24365
Model F-Value	90.68509	261.311	89.52139

10.3.2 Analysis of Variance

The ANOVA has been performed to test the statistical significance of proposed predictive models as shown in Table 10.3. In the current work, the R-squared values for all models are near to unity and the values of Adj-R^2 and Pred-R^2 are in sensible concurrence with each other. Also, the values of adequate precision greater than 4 imply that the established equations are found to be considerable. Further, the F-value for all models indicates that the established models generated through regression methods are statistically significant.

10.3.3 Effect of Process Parameters

This section discusses the effect of machining parameters on the cutting force, *MRR* and power consumption with the help of perturbation and interaction plots (see Figs. 10.4–10.6).

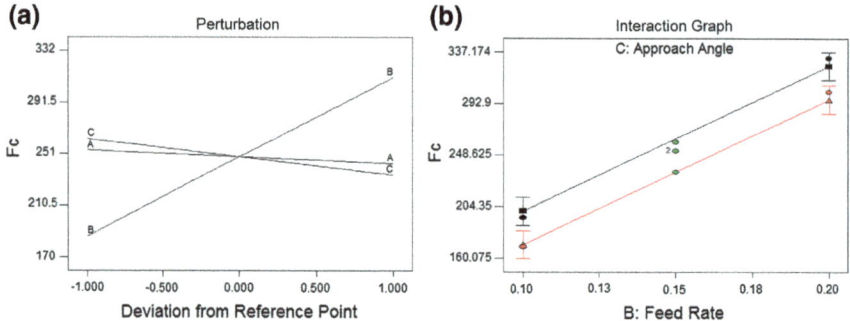

Fig. 10.4 **a** Effect of machining parameters on F_c **b** Interaction graph showing effect of f and ϕ on F_c at $V_c = 250$ m/min

On cutting force: As shown in Fig. 10.4a, the cutting force increases with increase in feed rate from 0.10 to 0.20 mm/rev, wherever there is no significance change in cutting force with increase in cutting speed and side cutting edge angle. These trends are due to the fact that the increase in feed rate (from 0.10 to 0.20 mm/rev) increased the generation of heat on the workpiece surface, damaged the cutting tool and possibly formed a cold weld between chips and tool, which resulted in the formation of build-up-edge. Formation of the build-up-edge on the tool has increased the cutting forces and tool temperature. The interaction plot at cutting speed 250 m/min are shown in Fig. 10.4b indicates that when f = 0.10 mm/rev and $\phi = 90°$, the F_c is 172.072 N, whereas f = 0.10 mm/rev and $\phi = 60°$, the F_c is 200.572 N and when f = 0.20 mm/rev and $\phi = 90°$, the F_c is 296.677 N, whereas f = 0.20 mm/rev and $\phi = 60°$, the F_c is 325.177 N.

On material removal rate: Figure 10.5a depicts that the material removal rate increases with increase in feed rate from 0.10 to 0.20 mm/rev and cutting speed from 200 to 300 m/min. The approach angle has not much effect on MRR. The interaction plot at approach angle 75° are shown in Fig. 10.5b indicates that when $V_c = 200$ m/min and f = 0.10 mm/rev, the MRR is 20.0193 cm^3/min, whereas $V_c = 300$ m/min and f = 0.10 mm/rev, the MRR is 29.663 cm^3/min and when $V_c = 200$ m/min and f = 0.20 mm/rev, the MRR is 39.1868 cm^3/min, whereas $V_c = 300$ m/min and f = 0.20 mm/rev, the MRR is 59.0068 cm^3/min.

On power consumption: It is shown by Fig. 10.6a that power consumption increases with increase in feed rate from 0.10 to 0.20 mm/rev and cutting speed (200 to 300 m/min). The approach angle has mild effect on power consumption. Because it is well-known fact that the power consumption is a function of cutting force and cutting speed (based on the equation $p = F_c * V_c/60,000$). More power or energy is required to plough the material from the resisting surface in case of increased feed and speed.

The interaction plot at approach angle 75° are illustrated in Fig. 10.6b indicates that when $V_c = 200$ m/min and f = 0.10 mm/rev, the power consumption is

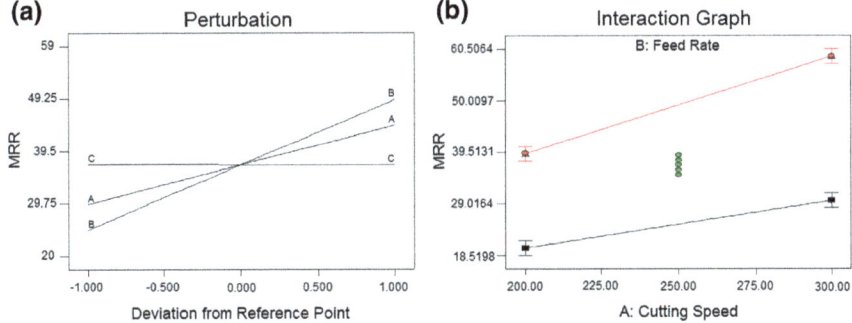

Fig. 10.5 **a** Effect of machining parameters on MRR. **b** Interaction graph showing effect of V_c and f on MRR at $\phi = 75°$

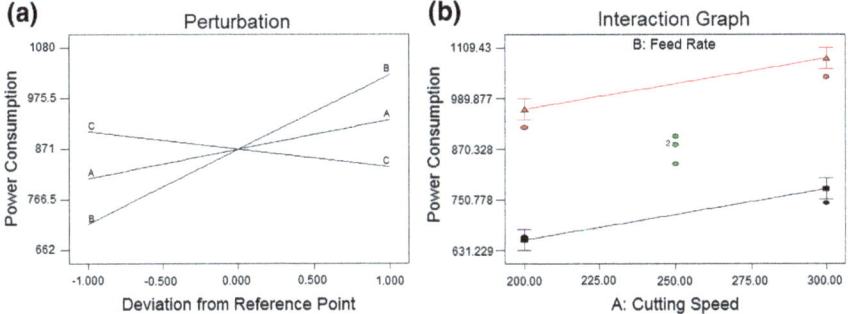

Fig. 10.6 **a** Effect of machining parameters on power consumption. **b** Interaction graph showing effect of V_c and f on power consumption at $\phi = 75°$

656.03 W, whereas V_c = 300 m/min and f = 0.10 mm/rev, the power consumption is 776.858 W, and when V_c = 200 m/min and f = 0.20 mm/rev, the power consumption is 963.798 W, whereas V_c = 300 m/min and f = 0.20 mm/rev, the power consumption is 1084.63 W.

10.3.4 Optimization

A major task of any experimental investigation in the field of machining is to achieve the desired output of the optimal parameters at minimum expenses so as to increase the overall productivity of the concerned process. In recent years, the multi response optimization has been of keen interest to scientists and engineers. Desirability analysis based response surface optimization technique is a good technique to chalk out the best combination of input parameters in any operation for the output with highest desirability.

Table 10.4 Confirmation test for cutting force, MRR and power consumption

Machining parameters			Cutting force (N)			MRR (cm³/min)			Power consumption (Watt)		
V_c m/min	f mm/rev	ϕ Deg. (°)	Actual	Predicted	Error (%)	Actual	Predicted	Error (%)	Actual	Predicted	Error (%)
300	0.15	90	240.23	233.75	2.77	46.987	45.422	3.44	901.2	905.8	0.13
300	0.16	90	245.73	240.29	2.26	48.256	46.986	2.70	912.3	921.934	0.977

The optimization was performed to decide the optimum factors by minimizing the cutting force and power consumption, also maximize the material removal rate. The first step is to choose the target for each process parameter, i.e., maximize the speed, maximize the feed rate and approach angle is selected "within range". This optimization technique combines the individual desirability factors into a single number, and then attempts to minimize this function. The results of optimum working conditions are presented in Table 10.4. The highest value of cutting speed (300 m/min), intermediate value of feed rate (0.15 mm/rev) and highest value of approach angle (90°) represents the optimal parameters.

The final experiment was then performed to confirm the optimum results as shown in Table 10.4. The predicted value and actual experimental value were very close to each other. All the experimental values of confirmation test are within the 95% prediction interval.

10.4 Conclusion

The research work discussed in this chapter contributes towards improvements in the machinability of Inconel-718 super alloys using environmentally friendly lubrication techniques. The chapter discusses the effect of important machining parameters on machinability aspects (cutting forces, MRR and power consumption) during MQL-assisted turning of Inconel-718 superalloys. Empirical models were also established for further prediction and optimization is done to improve the machinability. The following conclusions can be drawn from this research work:

1. The result clearly shows that, cutting force Fc increases with increase in feed rate from 0.10 to 0.20 mm/rev and there is no significance change in cutting force as cutting speed and side cutting edge angle increases.
2. The material removal rate increases with increase in feed rate from 0.10 to 0.20 mm/rev and cutting speed (200 to 300 m/min). The approach angle has mild effect on MRR.
3. The power consumption increases with increase in feed rate from 0.10 to 0.20 mm/rev and cutting speed from 200 to 300 m/min. The approach angle has mild effect on power consumption.
4. The models developed using RSM are reasonably accurate.

The outputs of the research discussed in this chapter will help to sustainable machining of Inconel-718 and other superalloys. Moreover, the future work can possibly be extended to other metal cutting operations such as milling, drilling and grinding, etc.

References

1. Zhou J, Bushlya V, Avdovic P, Ståhl JE (2012) Study of surface quality in high speed turning of Inconel 718 with uncoated and coated CBN tools. Int J Adv Manuf Technol 58:141–151
2. Arunachalam R, Mannan MA (2000) Machinability of nickel based high temperature alloys. Mach Sci Technol 4(1):127–168
3. Ezugwu EO, Bonney J, Yamane Y (2003) An overview of the machinability of aeroengine alloys. J Mater Process Technol 134(2):233–253
4. Gupta K, Laubscher RF (2016) Sustainable machining of titanium alloys: a critical review. Proc Inst Mech Eng Part B J Eng Manuf 1–18. doi:10.1177/0954405416634278
5. Kapil Gupta, RF Laubscher, JP Davim, NK Jain (2016) Recent developments in sustainable manufacturing of gears: a review. J Cleaner Prod 112(4):3320–3330 (Elsevier)
6. Gupta MK, Sood PK, Sharma VS (2015) Machining parameters optimization of titanium alloy using response surface methodology and particle swarm optimization under minimum quantity lubrication environment. Mater Manuf Processes 31:1671–1682
7. Sharma VS, Singh G, Sorby K (2015)A review on minimum quantity lubrication for machining processes. Mater Manuf Processes 30(8): 935–953
8. Sharma VS, Dogra M, Suri NM (2009) Cooling techniques for improved productivity in turning. Int J Mach Tools Manuf 49(6):435–453
9. Gupta MK, Sood PK, Sharma VS (2016) Investigations on surface roughness measurement in minimum quantity lubrication turning of titanium alloys using response surface methodology and box-cox transformation. J Manuf Sci Prod. doi:10.1515/jmsp-2015-0015
10. Varadarajan AS, Philip PK, Ramamoorthy B (2002) Investigations on hard turning with minimal cutting fluid application (HTMF) and its comparison with dry and wet turning. Int J Mach Tools Manuf 42:193–200
11. Kamata Y, Obikawa T (2007) High speed MQL finish-turning of Inconel 718 with different coated tools. J Mater Process Technol 192:281–286
12. Davim JP, Sreejith PS, Silva J (2007) Turning of brasses using minimum quantity of lubricant (MQL) and flooded lubricant conditions. J Mater Process Technol 22:45–50
13. Settineri L, Faga MG, Lerga B (2008) Properties and performances of innovative coated tools for turning Inconel. Int J Mach Tools Manuf 48:815–823
14. Obikawa T, Kamata Y, Asano Y, Nakayama K, Otieno AW (2008) Micro-liter lubrication machining of Inconel 718. Int J Mach Tools Manuf 48:1605–1612
15. Aggarwal A, Singh H, Kumar P, Singh M (2008) Optimizing power consumption for CNC turned parts using response surface methodology and Taguchi's technique—A comparative analysis. J Mater Process Technol 200:373–384
16. Thakur DG, Ramamoorthy B, Vijayaraghavan L (2009) Optimization of minimum quantity lubrication parameters in high speed turning of superalloy Inconel 718 for sustainable development. World Acad Sci Eng Technol 54:224–226
17. Thakur DG, Ramamoorthy B (2010) Investigation and optimization of lubrication parameters in high speed turning of superalloy Inconel 718. Int J Adv Manuf Technol 50:471–478

18. Yazid MZA, Cheharon CH, Ghani JA (2012) Surface integrity of Inconel 718 when finish turning with PVD coated carbide tool under MQL. Procedia Eng 19:396–401
19. Ji X, Li B, Zhang X, Liang SY (2014) The effects of minimum quantity lubrication on machining force, temperature and residual stress. Int J Precision Eng Manuf 15(11):2443–2451

Chapter 11
Dry and Near-Dry Electric Discharge Machining Processes

Krishnakant Dhakar and Akshay Dvivedi

Abstract The electric discharge machining (EDM) is an un-conventional machining process. It is extensively used to generate complex profiles on electrically conductive materials having high temperature resistance and high strength. Due to its wide applicability in manufacturing industries, EDM has become the most popular machining process after conventional machining processes such as turning, milling and drilling, etc. Despite several advantages, EDM process suffers from some limitations such as low material removal rate (MRR), high tool wear rate (TWR) and poor surface integrity in some cases. In past, several attempts have been made to overcome these limitations by augmenting EDM with techniques such as electrode rotation, ultrasonic vibrations and suspensions of powders into the dielectric fluid. Although these techniques are excellent from research perspective, but in practice, are applicable only for fewer applications. Another limitation in EDM process is its possible environmental pollution causing characteristic. During EDM process, material removal is a consequent of thermal energy produced by series of discrete electrical sparks occurring between tool and work electrodes which are immersed in a dielectric medium mostly hydrocarbon oils. These oils produce serious toxic fumes causing health hazard to the machine operator and environmental pollution. To overcome these limitations of EDM process, dry and near-dry variants of EDM were introduced. Dry EDM process utilizes a pressurized gaseous medium as a dielectric whereas, near-dry EDM process uses a combination of liquid and gas (two phase) as a dielectric medium and are environmentally friendly. This chapter provides an insight into both these environmentally friendly variants (i.e. dry and near-dry EDM) of EDM process along with the research progress in this area.

K. Dhakar (✉) · A. Dvivedi
Mechanical and Industrial Engineering Department,
Indian Institute of Technology Roorkee, Roorkee, India
e-mail: kk_jec@yahoo.co.in

A. Dvivedi
e-mail: akshaydvivedi@gmail.com

Keywords Dry-EDM · Near-dry EDM · Green EDM · Surface integrity · Dielectric · Environment · Sustainability

11.1 Introduction

11.1.1 Un-conventional Machining Processes

Un-conventional machining processes are those processes which utilize different forms of energy, such as mechanical, electrical, thermal, chemical and electro-chemical or combination of two energy forms for material removal or machining of different geometries and shapes. The greatest advantage of un-conventional machining processes is that there is generally no direct contact takes place between the tool and workpiece. These processes are also known as advanced machining processes. The un-conventional machining processes are preferred, when conventional machining is not feasible mainly under the following conditions:

- Enormously hard and brittle materials.
- Delicate and difficult to hold workpieces.
- Slender and flexible work materials.
- Intricate profiles to be machined on work materials.
- Requirement of high dimensional accuracy and precision.
- Temperature rise or residual stresses are unacceptable.

Different un-conventional machining processes are being used according to the required machining conditions. These processes increase the efficiency and produce precise and accurate products and therefore play a vital role in the aircraft and automobile components, medical equipment, tool and die making industries, etc. [1].

The following forms of energies are used to produce machining action on advanced engineering materials in un-conventional machining processes:

- Mechanical energy
- Thermal energy
- Chemical energy
- Electrochemical energy

The mechanical action is used in Ultrasonic Machining (USM), Water-Jet Machining (WJM), Abrasive Jet Machining and Abrasive Water-Jet Machining (AWJM) processes. The Electric Discharge Machining (EDM), Plasma Arc Machining (PAM), Laser Beam Machining (LBM), Electron Beam Machining (EBM) and Ion Beam Machining (IBM) use thermal energy for machining. In Chemical Machining process (CHM) the chemical dissolution takes place under the influence of chemical energy while, electrochemical energy causes anodic dissolution in Electrochemical Machining (ECM) processes.

11.2 Electric Discharge Machining (EDM)

Amongst the un-conventional machining processes, EDM is the most popular un-conventional machining process because it is capable to machine any electrically conductive material irrespective of its hardness and toughness; and can cut complex geometries, shapes and features efficiently. It is also called as spark erosion machining. It has a wide area of application in different fields like mold and die manufacturing, aerospace and automotive industries, electronics and medical instruments, etc. With increase in demand of products made from hard metals and alloys especially difficult-to-machine materials, more interest has gravitated towards the EDM process.

11.2.1 Principle of EDM

The principle of EDM is based on thermoelectric erosion where the electrical energy is converted into thermal energy to remove the material from the work surface by a series of discrete electric discharges between the anode (workpiece) and cathode (tool) [2]. In this process, tool and workpiece electrodes are submerged in dielectric fluid as shown in Fig. 11.1.

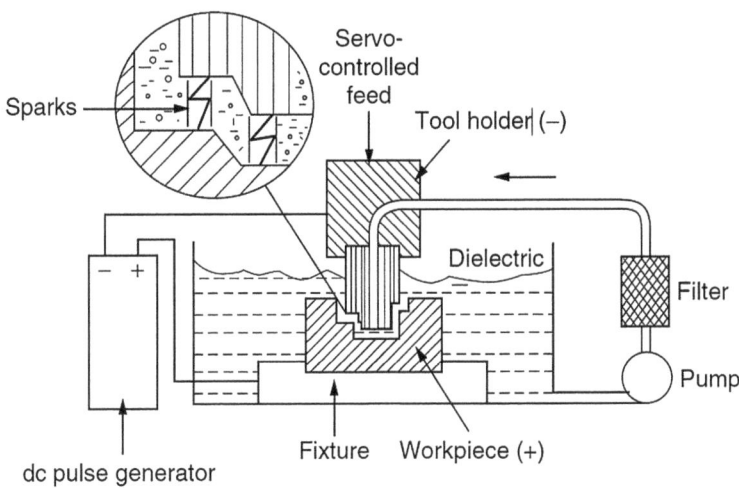

Fig. 11.1 Electric discharge machining

Generally, potential difference of 20–120 V is applied across the tool and workpiece [1]. A typical gap of few microns (15–150 μm) between tool and workpiece is maintained that causes the breakdown of dielectric. The discharge location is generally the closest point of the electrodes gap, however it also depends on the impurities present in the dielectric medium between inter electrode gap (IEG) [3]. When voltage reaches the discharge voltage value, electrons are emitted from the tool electrode. The emitted electrons accelerate towards the anode under the effect of electrical field force. While accelerating towards the anode, the electrons strike with the neutral molecules of dielectric, and break them into electrons and positive ions. Subsequently, these ions produce a narrow conductive channel called "plasma channel". Due to low electrical resistance of plasma, large numbers of electrons start flowing from the cathode to anode. In this phenomenon, electrons and ions crashing among them and therefore creating high temperatures as well as a formation of gas envelope around the plasma channel. This gas envelope grows till discharge occurs [4]. Figure 11.2 shows the different phases of electric discharge.

In EDM, the localized heat generation due to the extremely high temperature (8000–12000 °C) causes the melting and evaporation of electrode material [5]. The amount of material removed from the anode is comparatively more than that of the cathode because the momentum with which the negative ions strike the anode surface is much more than the momentums of positive ions strike the cathode surface [6]. During pulse off time, the potential difference is withdrawn from the electrodes and consequently the plasma channel no longer exists. It causes the gas envelope around plasma channel to collapse and produce shock waves. These shock waves eject the melted material from the discharge location and generate a crater on the work material [4]. Due to impulsive decrease in inside pressure of gas envelope, the dielectric liquid around it causes the envelope to implode. It ejects out the molten material from the gap/envelope and subsequently the cratered EDM surface is formed on the workpiece [4].

11.2.2 Applications of EDM

The electric discharge machining has found the following applications:

- in die and tool making industries;
- to drill and fabricate tiny holes, orifices and fragile features;
- to produce complex shapes which are extremely difficult to make by another processes;
- in removal of broken taps, drills studs, reamers and pins;
- to manufacture dies for extrusion, stamping and wire drawing that require machining of through holes, shapes and cavities.

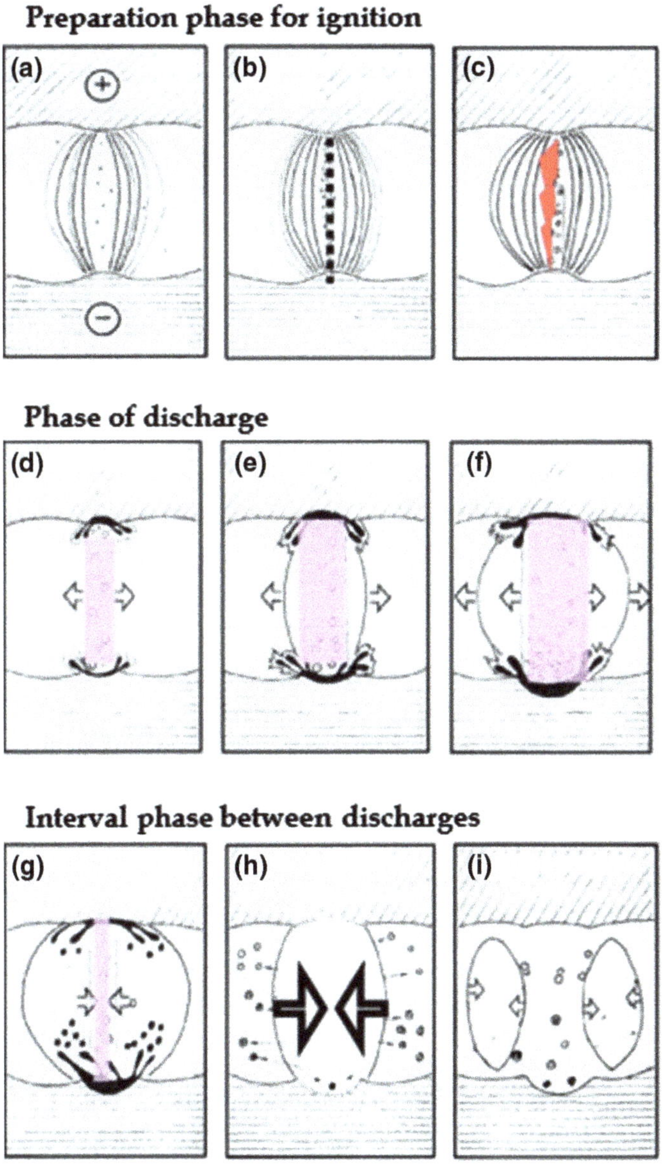

Preparation phase for ignition

Phase of discharge

Interval phase between discharges

Fig. 11.2 Phases of electrical discharges [3], with kind permission from Elsevier

11.3 Classification of EDM

Despite being a choice of manufacturers due to its special characteristics, it suffers from some inherent limitations such as low MRR, high tool wear, poor surface integrity in some instances, and environmental pollution. These EDM limitations

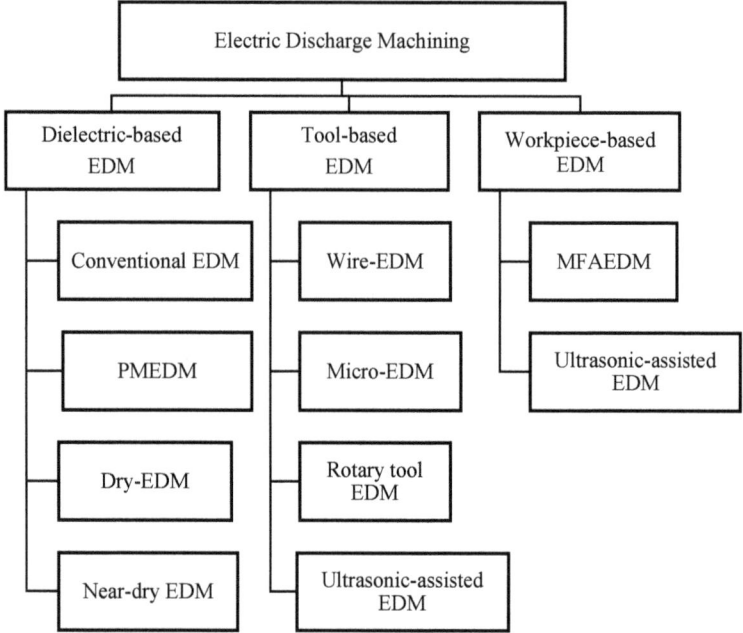

Fig. 11.3 Classification of electric discharge machining processes

bound its areas of application. Researchers have tried to overcome these problems by modifying the EDM process. These modifications, in turn, evolved newer EDM process variants. These process variants can be classified into three categories such as dielectric based, tool based and workpiece based. Figure 11.3 shows flow diagram of classification of electric discharge machining processes. There are several categories of EDM processes according to the type of dielectric medium used. These variants are conventional EDM, powder mixed electric discharge machine (PMEDM), dry EDM and near-dry EDM. Wire EDM, micro EDM and rotary EDM are tool based EDM process variants. Whereas, magnetic-field assisted EDM (MFAEDM) is workpiece based EDM process. Furthermore, the ultrasonic assisted EDM comes under both categories (tool or workpiece).

11.3.1 Conventional EDM

The commercially available die sinking EDM is in general referred as conventional EDM. Generally, conventional EDM uses hydrocarbon oil as a dielectric medium. It plays an essential role in the EDM process, such as insulating medium between two electrodes to induce discharge, influences the plasma channel expansion and material erosion during the discharge, flushing the debris from discharge gap and reconditioning after the discharge. In conventional EDM, different types of dielectric medium such as hydrocarbon oils, synthetic oils, deionized water etc. have been used. These dielectric medium are a great cause of environmental pollution. These dielectric fluids generate toxic fumes, aerosols and odors and can initiate fire hazards during machining. Thus, a series of investigations have been done on use of green dielectric medium as sustainable alternatives of hydrocarbon oils to increase the efficiency of conventional EDM and keep the environment clean and green. These previous investigation were performed with an objective of possible increase in the efficiency of conventional EDM and are discussed in the subsequent sections.

11.3.2 Powder Mixed EDM (PMEDM)

Powder mixed electric discharge machining (PMEDM) is one of the novel innovations to enhance the capabilities of EDM process. In PMEDM, the electrically conductive abrasive powder is mixed with the dielectric fluid, with the objectives to reduce the dielectric insulating strength and increase the spark gap between the tool and workpiece electrodes. As a result, the process becomes more stable, thereby, improving the MRR and surface finish. Figure 11.4 depicts schematic illustration of PMEDM process. The advantages offered by PMEDM are; higher MRR and

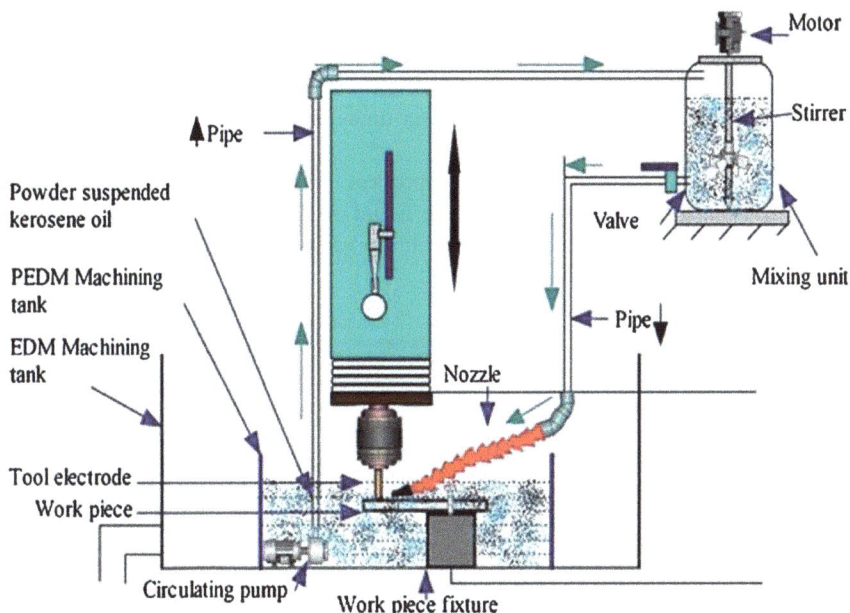

Fig. 11.4 Schematic illustration of PMEDM

improved surface finish, possible to achieve near-mirror finish (with reverse polarity); stable machining and comparatively shorter machining time. On the other hand, PMEDM is expensive and environmentally unfriendly as the disposal of used dielectric is a serious environmental concern.

Conventional and PMEDM both are environmentally unfriendly and are great cause of pollution. To overcome the limitations of EDM, Dry and near-dry EDM were developed as their sustainable substitutes. These EDM variants are discussed in the subsequent sections.

11.3.3 Dry and Near-Dry EDM

Dry and near-dry EDM processes are sustainable process variants of EDM. These processes are classified according to the use of dielectric mediums. The EDM conducted with gas as a dielectric medium is generally called as dry EDM as shown in Fig. 11.5. A high velocity gas flow is supplied through tubular electrode, causing molten workpiece material to be removed and flushed out from the IEG. During the pulse off time, high velocity gas blows off the plasma formed by the preceding spark and decreases the temperature of the sparking zone [7]. Firstly, National Aeronautics and Space Administration (NASA) investigated on feasibility of use of inert gas, as dielectric medium [8]. Subsequently, Kunieda et al. [9] investigated on possibility of use of combination of different gases as dielectric medium in EDM. Successively, this process variant of EDM was termed as dry EDM. It applies high flow rate gaseous dielectric fluid such as oxygen, nitrogen, hydrogen and compressed air through tubular electrode between the inter electrode gap. It tends to eliminate environmental problem and resulted enhance machining performance. The MRR of dry EDM was measured as six times higher than conventional EDM, at the same parametric conditions [10].

Another parallel approach on EDM process performance improvement was initiated by Tanimura et al. in 1989 where liquid and gaseous mixture (two phase) were utilized as dielectric medium [11]. It was initially called as EDM-in-mist.

Fig. 11.5 Dry EDM process

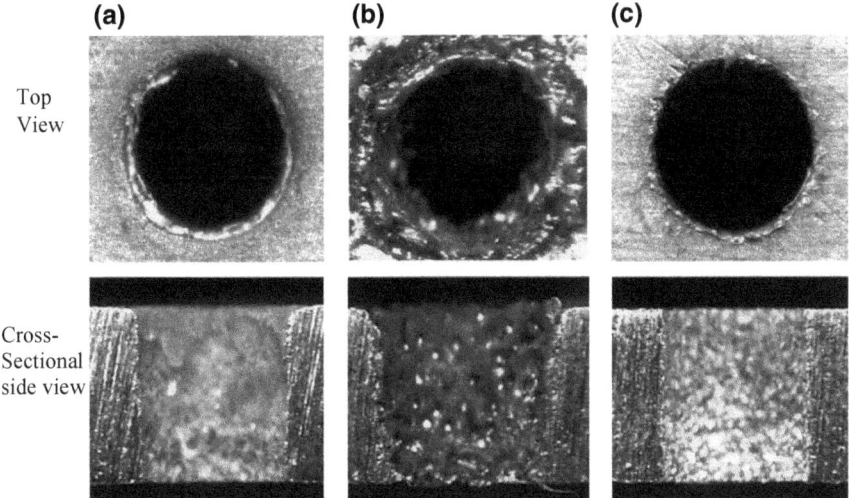

Fig. 11.6 Optical micrographs on holes drilled on 1.27 mm Al6061: **a** wet, **b** dry and **c** near-dry EDM conditions [13] with kind permission from Elsevier

Later, this process evolved as near-dry EDM. In this process, mist (liquid and air) or combination of gases like nitrogen, helium, argon, etc. are utilized as dielectric medium. The two-phase dielectric medium provided stable machining as compared to dry EDM. Further, it was investigated by Kao et al. that near-dry EDM successfully drills straight holes with sharp edges while the dry EDM had a serious debris reattachment issue, subsequently turn into the creation of tapered hole [12]. The taper in conventional EDM was also found to exist but not as substantial as in dry EDM (refer Fig. 11.6). Furthermore, near-dry EDM results in better surface finish (up to 0.09 μm) and machining consistency [13].

The near-dry and dry EDM may render low cost of dielectric medium with fewer environmental concerns. Another, advantage of near-dry and dry EDM would be the elimination of requirement of huge quantity of dielectric medium and its handling system (circulation and filtration). Thus, the floor-space requirements of commercially available EDM can also be reduced. The schematic of near-dry EDM is illustrated in Fig. 11.7.

11.3.3.1 Process Parameters

The process parameters of dry and near-dry EDM are broadly classified in four categories which are as given below:

- Electrical parameters
- Non-electrical parameters

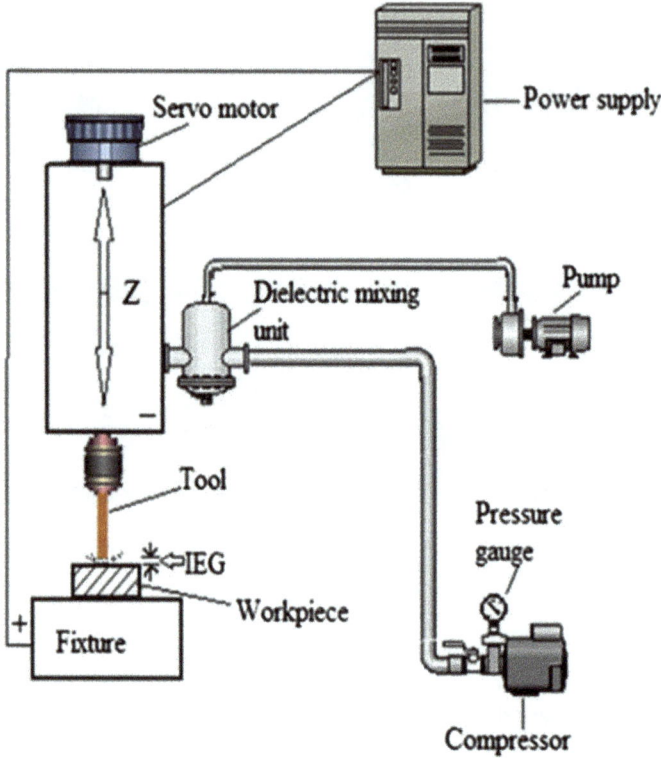

Fig. 11.7 Near-dry EDM process [14], with kind permission from Taylor and Francis

- Tool electrode based parameters
- Near-dry based (dielectric) parameters

Each category involves various sub categories of process parameters. The possible effects of these process parameters on the response characteristics are illustrated in the Ishikawa cause-and-effect diagram as shown in Fig. 11.8.

Electrical Parameters

The current, gap voltage, duty factor, pulse on time, pulse off time and polarity are the electrical parameters which dominantly affect the output characteristics in near-dry and dry EDM processes. The current is a flow of electric charge which is carried by ions and electrons in plasma. Gap voltage is the potential difference generated between two electrodes (tool and workpiece) for breaking down the dielectric strength of fluid at inter electrode gap (IEG). Finally, this results in spark generation. Pulse on time is defined as the time during which the discharge occurs and machining is done. Pulse off time is the time period of no discharge and reionization of the dielectric takes place. Duty factor is the ratio of on time and the addition of on and off time.

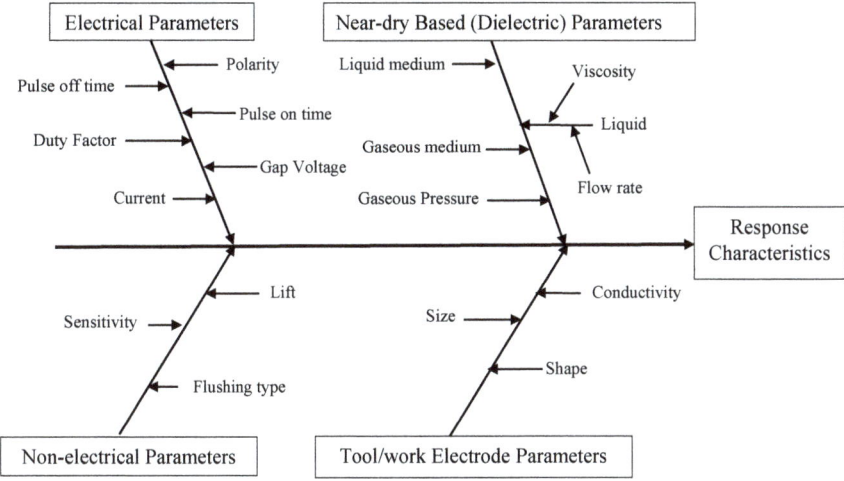

Fig. 11.8 Cause and effect diagram (Ishikawa diagram)

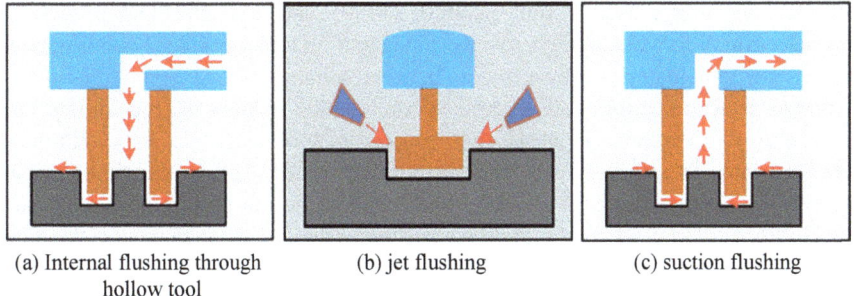

(a) Internal flushing through hollow tool   (b) jet flushing   (c) suction flushing

Fig. 11.9 Type of flushing techniques in EDM

Non-electrical Parameters

Non-electrical parameters such as flushing, lift and sensitivity also affect the response characteristics to a great extent. Flushing plays an important role in EDM process. It is the method of removing waste particles (debris) from the spark gap. The flushing when ineffective, results in lower MRR and poor surface finish. An effective flushing can greatly increase the MRR. Literature study revealed several techniques for flushing in EDM namely internal flushing through tool electrodes, jet flushing and suction flushing [15]. Figure 11.9 shows different flushing techniques. Internal flushing through tubular tool electrode is used in near-dry EDM.

The lift parameter is the degree of "Z" axis (Z axis is the moving direction of tool electrode) retraction after erosion time. The servo power on the cutting axis is controlled by the sensitivity of the servo. This power resists the pressure increase

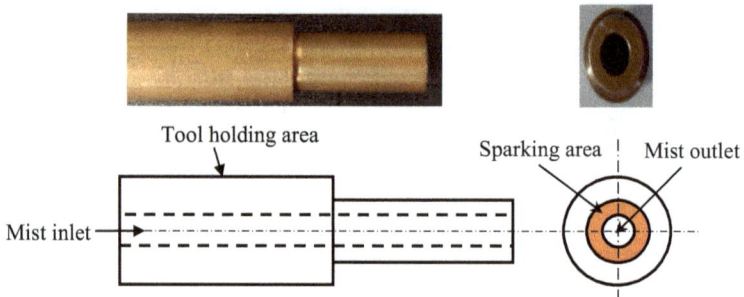

Fig. 11.10 Tubular tool electrode used in dry and near-dry EDM

within the spark gap caused during machining. Higher sensitivity means faster advancement of tool towards workpiece.

Tool Electrode Based Parameters

According to the working principle of EDM, any electrically conductive material can be used as tool electrode. Literature study reveals that copper, brass, graphite and steel are mostly used as tool electrodes in dry and near-dry EDM processes. The selection of tool electrode is mainly dependent on tool-workpiece combination, where one tool material may provide better results in comparison with others. A proper selection of tool and workpiece combination reduces the machining cost considerably. Literature suggests that copper as tool electrode may eliminate problems such as ignition delay and short circuiting etc. [1]. In addition, copper receives much attention as a tool electrode material due to its properties to attain stock removal, low wear and stable sparking. The size and shape of the tool electrode greatly affects the responses in dry and near-dry EDM. Generally tubular tool electrodes (as shown in Fig. 11.10) are used in these processes so that internal diameter of tool can alter the velocity of dielectric fluid.

Near-Dry Based (Dielectric) Parameters

The near-dry parameters that influence the response characteristics are gaseous pressure, gaseous medium, liquid medium, viscosity and flow rate. Gaseous pressure is an important parameter in near-dry EDM, high gaseous pressure helps to easily atomize liquid dielectric into mist. Further, it improves the flushing efficiency of the process and results in the increased MRR. Dielectric fluid in EDM plays role of an insulator between the tool and workpiece electrodes; however, debris flushing is also one of its important work. Thus, viscosity of liquid is an important factor for selecting the dielectric fluid in EDM. In EDM process debris accumulation is a key issue; thus, dielectric liquids with low viscosity are preferable. However, high gaseous pressure assists in debris flushing consequently eliminates this problem in near-dry EDM. Thus, higher viscous liquid can also be used in this process. Liquid flow rate is the flow of liquid constituent which mixes with air in a controllable manner.

11.3.3.2 Mechanism of Material Removal

Material removal mechanism of near-dry EDM is discussed with the help of three different liquid constituent (glycerin, EDM oil and water,) utilized in dielectric medium in a systematic experimental research. EDM oil and water are commonly used dielectric medium in EDM [16]. Water has good cooling property while EDM oil has good corrosion resistance property. Both, water and EDM oil possess low viscosity. On contrary, the glycerin possesses a very high viscosity coupled with low thermal conductivity in comparison with other dielectric fluids.

The fluid droplets in near-dry EDM get quasi-homogeneously distributed (meshing) in IEG as shown in Figs. 11.11a and 11.12. This meshing between the fluid droplets occurs in the direction of flow of current. It facilitates bridging between IEG. Afterward, these droplets attain charge (up to a threshold limit) due to potential difference between tool and work electrodes (see Fig. 11.11b). These charged droplets while moving towards opposite charged workpiece accelerate and transfer their charge over the workpiece surface. Subsequently, they explode on the work surface as shown in Fig. 11.11c. This causes distributed explosions in the IEG with series of discharges.

These combined actions promote flushing of IEG with high rate of material removal. Thus, faster and distributed sparking within IEG causes uniform erosion from the workpiece surface. Material removal rate increases while providing better surface integrity (in terms of surface roughness and HAZ at lower value of pulse current). In case of water and EDM oil as dielectric medium, both of them have low viscosity. Thus, they atomize in smaller droplets (range of droplet diameter 50–120 μm as shown in Fig. 11.12a) whereas, glycerin has high viscosity therefore it atomizes in bigger droplets (range of droplet diameter 150–250 μm as shown in Fig. 11.12b) [17, 18]. Further, glycerin has bigger thermal boundary layer than velocity layer. It may be due to absorption of heat from spark by the larger droplets which explodes on the work surface. It results in large craters at the work surface with glycerin-air dielectric.

Figure 11.13a, b shows spark gap during machining with different combinations of dielectric medium. In near-dry EDM, the MRR obtained by glycerin-air

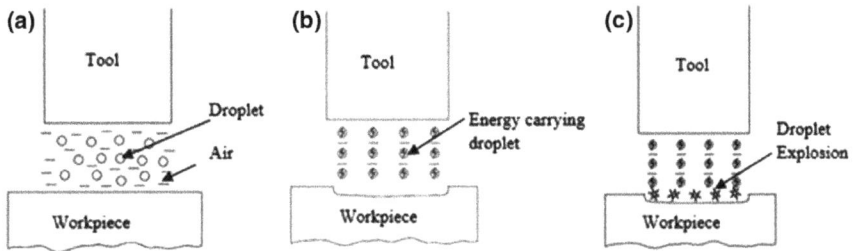

Fig. 11.11 Schematic illustrations of material removal mechanism (**a, b, c**) [17]

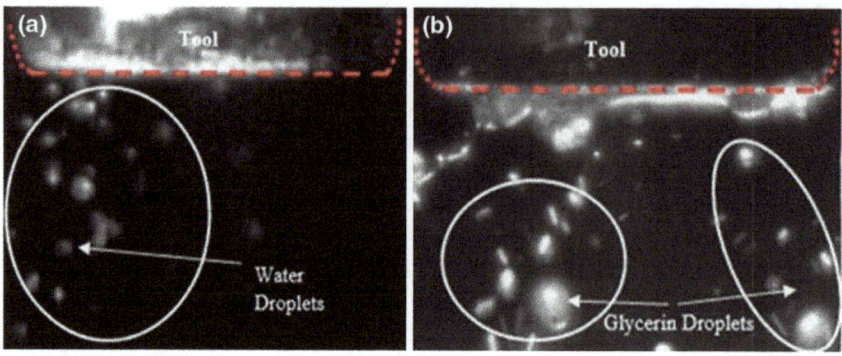

Fig. 11.12 High speed imaging (7100 fps, resolution 320 × 240) of **a** water-air mist and **b** glycerin-air mist [17]

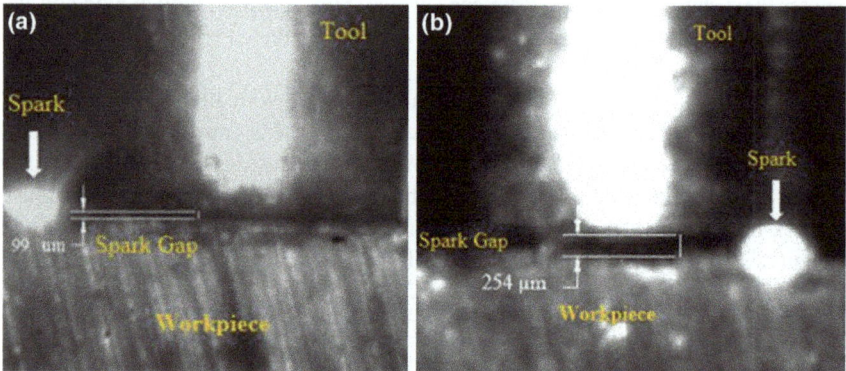

Fig. 11.13 High speed imaging (8113 fps, resolution 256 × 256) of spark gap with **a** water-air and **b** glycerine-air combination of dielectric mediums [17]

dielectric medium is approximately three times higher than the other two investigated dielectric mediums at the best parameters setting.

The compressed air in dry EDM has very low dielectric constant (3 MV/m) and to break this dielectric strength (to initiate spark), the IEG required (at constant voltage) would be very small. A similar observation was revealed from imaging of dry EDM, where measured IEG was 64 μm. This small IEG is beneficial for generation enormous thermal energy and consequently high MRR. However, results in poor surface quality, coupled with deposition of debris on tool tip and edges of workpiece as shown in Fig. 11.14.

Fig. 11.14 Tool and workpiece after dry-EDM process

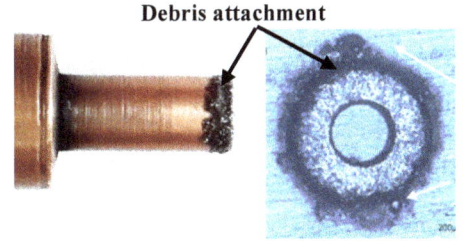

Fig. 11.15 MRR with different dielectrics at high discharge energy (current 15 A, duty factor 0.80, flushing pressure 80 psi) [17]

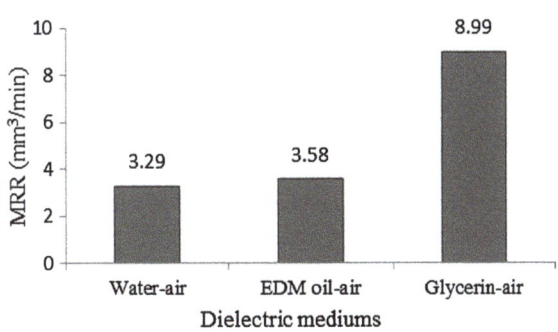

11.3.3.3 Dielectric Mediums in Dry and Near-Dry EDM

Generally, dry-EDM utilizes compressed air as a dielectric medium. However, some researchers have used oxygen gas as dielectric medium in dry-EDM process [19–21]. It was found that the MRR is increased due to the enlarged volume of discharged crater and more frequent occurrence of discharges when using oxygen. The NASA reported usage of argon and helium as a dielectric medium in dry EDM.

In near-dry EDM, mixture of liquid and gas used as dielectric medium. Tao et al. [22] mixed different gases such as helium, oxygen and nitrogen with water. It was found that mixture of oxygen-water provided highest MRR and surface roughness as well. Further, Dhakar and Dvivedi [17] used different combinations of dielectric mediums viz. glycerin-air, water-air and EDM oil-air. The MRR produced by glycerin-air dielectric medium was approximately three times higher (refer Fig. 11.15) than the EDM oil-air and water-air combinations at best parametric settings (current 15 A, duty factor 0.80, flushing pressure 80 psi).

Glycerin generates high thermal energy and gaseous pressure in IEG when it decomposes with discharge. It is apparent that glycerin generates concentrated explosion at IEG, thus increasing material removal. However, glycerin-air dielectric medium produced slightly higher TWR than other dielectric mediums but it was approximately negligible because wear ratio (ratio of TWR and MRR in percentage) of the process was less than one percent.

Furthermore, it was observed that recast layer produced by EDM oil-air and water-air were 0.52 µm and 0.40 µm, respectively, while combination of glycerin-air did not produce any measurable recast layer [17].

11.4 Comparison of Dry, Near-Dry and Conventional EDM

This section highlights the advantages of dry and near-dry EDM over conventional EDM in terms of various machinability aspects as given below.

Productivity: The MRR of near-dry EDM was nearly 50–60% higher than conventional EDM. In conventional EDM, carbon particles and other debris particles are generated during spark erosion [19]. Further, due to inefficient flushing they do not flush away from the IEG effectively. Subsequently, they disturb the erosion process resulting in ineffective sparks which results in low MRR. In near-dry EDM high pressure dielectric medium provides better flushing than conventional EDM. This reduces debris accumulation problem at IEG resulting in higher MRR. Several investigations have been conducted where it was found that near-dry EDM achieves higher MRR as compared to conventional EDM [23–25].

The experimental investigation conducted by Dhakar [24] reveals that the MRR of near-dry EDM was marginally higher than dry EDM (see Fig. 11.16). It can be attributed to the fact that two phase (liquid and air) dielectric medium eliminates debris reattachment problem. Consequently, lower down chances of possible short circuiting. This phenomenon improves MRR in near-dry EDM. It was also interestingly observed that, conventional EDM produced very high TWR at higher values of current than near-dry and dry EDM process.

Cost concerns: The near-dry and dry EDM are economically efficient processes as compared to conventional EDM. Because, the cost of conventional EDM dielectric is $17.68 per gallon (SHELL MACRON EDM 135) while the cost of tap water and air/gas (used in dry and near-dry EDM) is almost negligible [25].

Environmental aspects: These processes are green because there is no generation of hazardous gases/fumes; and no waste is produced from the dielectric liquid. There is no risk of fire hazards because no flammable dielectric is used.

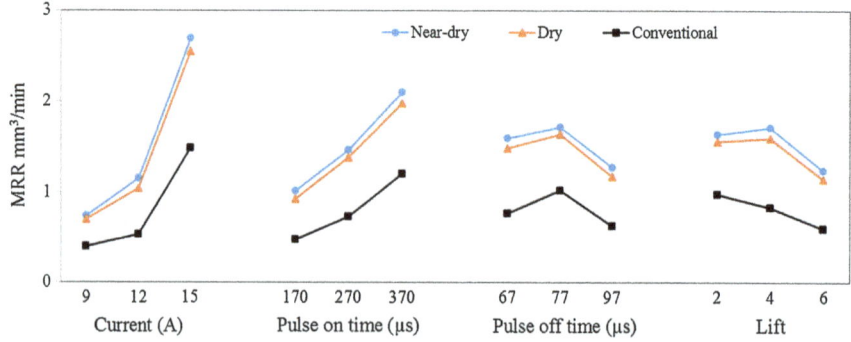

Fig. 11.16 Effect of process parameters on MRR in near-dry, dry and conventional EDM processes [24]

Efficiency: Highly efficient due to high MRR and fine surface finish. Further, TWR in dry and near-dry EDM processes is negligible.

Space requirements: These processes do not require large floor space because they do not need a huge dielectric circulation unit.

11.5 Summary

In essence, dry and near-dry EDM are the environmentally friendly variants of EDM. These processes are capable to generate high surface integrity and to achieve high productivity along with minimum hazard, wastages and pollution, and thus attain overall sustainability which makes them the sustainable substitutes to the conventional EDM.

References

1. Jain VK (2002) Advanced machining processes. Allied Publishers Pvt. Ltd., New Delhi, pp 126–159
2. Davim JP (2013) Nontraditional machining processes. Research Advances Springer. doi:10.1007/978-1-4471-5179-1
3. Schumacher BM (2004) After 60 years of EDM the discharge process remains still disputed. J Mater Process Technol 149(1–3):376–381
4. Puertas I, Luis CJ (2003) A study on the machining parameters optimization of electrical discharge machining. J Mater Process Technol 143–144:521–526
5. El-Hofy H, Youssef H (2009) Environmental hazards of nontraditional machining. In: Proceedings of the 4th IASME/WSEAS international conference on energy & environment (EE'09), pp 140–145
6. Pandey PC, Shan HS (1980) Modern machining processes. Tata McGraw-Hill Education, ISBN:13: 978-0070965539
7. Kunieda M, Lauwers B, Rajurkar KP, Schumacher BM (2005) Advancing EDM through fundamental insight into the process. CIRP Ann Manuf Technol 54(2):64–87
8. NASA (1985) Inert-gas electrical-discharge machining. NASA Technical Brief No. NPO-15660
9. Kunieda M, Yoshida M (1997) Electrical discharge machining in gas. Ann CIRP 46(1):143–146
10. Yu ZB, Jun T, Masanori K (2004) Dry electrical discharge machining of cemented carbide. J Mater Process Technol 149:353–357
11. Tanimura T, Isuzugawa K, Fujita I, Iwamoto A, Kamitani T (1989) Development of EDM in the mist. In: Proceedings of ninth international symposium of electro machining (ISEM IX). Nagoya, Japan, pp 313–316
12. Tao J, Shih AJ, Ni J (2008) Near-dry EDM milling of mirror-like surface finish. Int J Electr Mach 13:29–33
13. Kao CC, Tao J, Shih AJ (2007) Near dry electrical discharge machining. Int J Mach Tools Manuf 47(15):2273–2281
14. Dhakar K, Dvivedi A (2016) Parametric evaluation on near-dry electric discharge machining. Mater Manuf Process 31(4):413–421

15. Makenzi MM, Ikua BW (2012) A review of flushing techniques used in electrical discharge machining. In: 4th Proceedings of the mechanical engineering conference on sustainable research and innovation
16. Chakraborty S, Dey V, Ghosh SK (2015) A review on the use of dielectric fluids and their effects in electrical discharge machining characteristics. Precis Eng 40:1–6
17. Dhakar K, Dvivedi A, Dhiman A (2016) Experimental investigation on effects of dielectric mediums in near-dry electric discharge machining. J Mech Sci Technol 30(5):2179–2185
18. Jayasinghe SN, Edirisinghe MJ (2002) Effect of viscosity on the size of relics produced by electrostatic atomization. Aerosol Sci 33(10):1379–1388
19. Govindan P, Joshi SS (2010) Experimental characterization of material removal in dry electrical discharge drilling. Int J Mach Tools Manuf 50(5):431–443
20. Kunieda M, Furuoya S (1991) Improvement of EDM efficiency by supplying oxygen gas into gap. Ann CIRP 40(1):215–218
21. Liqing L, Yingjie S (2013) Study of dry EDM with oxygen-mixed and cryogenic cooling approaches. Procedia CIRP 6:344–350
22. Tao J, Shih AJ, Ni J (2008) Experimental study of the dry and near-dry electrical discharge milling processes. J Manuf Sci Eng 130(1):11002–11009
23. Tao J. (2008) Investigation of dry and near-dry electrical discharge milling processes. A Dissertation, The University of Michigan
24. Dhakar K (2016) Investigation on near-dry EDM and analysis of process performance. Ph.D. thesis MIED Indian Institute of Technology Roorkee, INDIA
25. Dhakar K, Pundir H, Dvivedi A, Kumar P (2015) Near-dry electrical discharge machining of stainless steel. Int J Mach Mach Mater 17(2):127–138

Chapter 12
Laser Metal Deposition Process for Product Remanufacturing

Rasheedat M. Mahamood, Esther T. Akinlabi and Moses G. Owolabi

Abstract Remanufacturing is a process of bringing a damaged part back to its perfect working condition. The cost of remanufacturing equipment or parts is cheaper than the cost of buying a new one. The conventional manufacturing method involves processes which are energy inefficient that causes lots of emissions, thereby contributing immensely to the global warming problems. An alternative manufacturing process is inevitably required which is capable of reducing the environmental impact during various phases of product life cycle that helps in cost saving by extending the service life of equipment or parts which will in turn greatly improve the country's economy. The advanced remanufacturing techniques such as laser metal deposition (LMD) process that belongs to a class of additive manufacturing processes can overcome the limitations of conventional manufacturing processes and capable to repair engineered parts having complex features that are difficult to access at the repair site. LMD can be used to fabricate part directly from its three-dimensional (3D) computer-aided design (CAD) model just by adding materials layer by layer. This technology offers design flexibility to engineers by allowing modification of an existing design without having to start from the scratch which is a basic requirement in product remanufacturing. This chapter discusses the capability of LMD process for restoring and remanufacturing high-valued components back to their perfect working conditions. Some of the research works that demonstrate the capability and effectiveness of using LMD for remanufacturing of high-valued products that is sustainable, cheap, and above all energy efficient are briefed. A case study to demonstrate the metallurgical integrity and properties of titanium alloy powder deposited on Ti6Al4V substrate using the LMD process and its sustainable aspects is also discussed in this chapter.

R.M. Mahamood (✉) · E.T. Akinlabi
Department of Mechanical Engineering Science, University of Johannesburg, Johannesburg, South Africa
e-mail: mahamoodmr2009@gmail.com

R.M. Mahamood
Department of Mechanical Engineering, University of Ilorin, Ilorin, Nigeria

M.G. Owolabi
Department of Mechanical Engineering, Howard University, Washington DC, USA

© Springer International Publishing AG 2017
K. Gupta (ed.), *Advanced Manufacturing Technologies*, Materials Forming, Machining and Tribology, DOI 10.1007/978-3-319-56099-1_12

Keywords Additive manufacturing · Laser metal deposition · Remanufacturing · Repair · Sustainability · Titanium

12.1 Introduction

Product repair is the process of bringing a failed or broken down device or equipment parts to an acceptable operational state. The restoration of damaged equipment or part to a state where the whole equipment or part can work properly again is referred to as 'repair'. Repair usually involves the replacement of damaged or worn-out part; or the joining together of broken parts to restore it to the original state. The simplest repair is the one in which the damaged, broken, or worn-out parts are just replaced with another similar parts that has the same specification with the damaged ones following the correct procedure. The more challenging repair is the one in which the broken or the damaged part has to be restored using one or more manufacturing processes. Repair of this nature is labor intensive and the life of the repaired part may not be long as may be expected [1]. The use of traditional joining techniques, for example, the traditional welding such as arc welding, is characterized by high heat energy input which results in a large heat affected zone area. This heat affected zone is found to alter the properties of the part that surrounds the repair site area which will contribute to the weakness of these heat affected zone areas and hence results in another kind of failure arising when the part is in operation, which can further contribute to another downtime. This is one of the reasons why companies preferred to carry out repair by replacing the broken parts instead of going through the tedious process of repair that has resulted in a large amount of waste with high negative environmental impact on the recycling process of these damaged parts. The consciousness of the whole world about the damage to our planet through pollution by human activities, which include the waste generated through the disposal of damaged equipment has made most of the developed world to come up with the regulations on equipment remanufacturing.

Product 'remanufacturing' is the process of restoring an equipment to its perfect working condition by complete disassembly of the equipment and carry out series of processes such as recovery, repair, and replace or redesigning of worn out or obsolete parts both at the module level and at component level. Remanufactured equipment could be made to outperform the original equipment through the replacement of obsolete or underperforming component with a better design and with improved surface property that, for example, reduce the wear rate of parts involved in rubbing action with other part. The use of traditional manufacturing process to achieve this objective is limited to large or bulky product where the influence of the heat affected zone could have less impact on the equipment. Repair of part with small thickness cross section and with intricate parts can be made possible with the use of some advanced manufacturing techniques including laser metal deposition (LMD) process.

Laser metal deposition process (LMD) is an advanced manufacturing method and an ideal repair technique that is used to restore damaged part with less heat affected zone and hence negligible damage to the surrounding material of the repair site [1–4]. LMD belongs to a class of additive manufacturing technology as recently grouped by the international standard organization committee on additive manufacturing (committee F42) [5] that can produce three-dimensional (3D) object as well as has the ability to repair worn-out or broken parts. It is one of the few additive manufacturing technologies that can be used in the quality repair process. The ability to produce a new part on an existing part with sound metallurgical properties is another plus for this technology and its unique place in achieving product remanufacturing with performance that can exceed that of the initial product. This is because old equipment which was discarded because some of its parts are obsolete can be remanufactured with better designed and better performing components using the LMD process. The missing part can easily be fabricated on the remaining part if the 3D image of the missing part can be acquired and sent into the machine. The remanufactured component can hardly be identified because the process is so unique that the transition zone is hardly visible and the microstructure produced is a continuous one which results in a better mechanical property of the component [6].

In this chapter, the capabilities of LMD process for the repair and remanufacturing of high-valued component parts are presented. Some of the research works in this field are also presented. A case study that demonstrates the metallurgical integrity and properties of titanium alloy powder deposited using the LMD process is discussed. The sustainability aspects and the future research direction on the use of LMD for product remanufacturing are also analyzed and the chapter ends with summary.

12.2 Introduction to Laser Metal Deposition Process

This section presents the working principle, process details, process capabilities, and application areas of LMD process. The use of laser metal deposition for repair of high-valued components and the prospect of this technology for product remanufacturing are discussed. The detailed description of the LMD process is presented in the next subsection.

12.2.1 Working Principle of Laser Metal Deposition Process

There are three types of LMD processes, namely powder-based LMD process, wire-based LMD process, and powder-/wire-based process. LMD process is a member of the direct energy deposition class of additive manufacturing technologies according to the grouping performed by the International Organization for

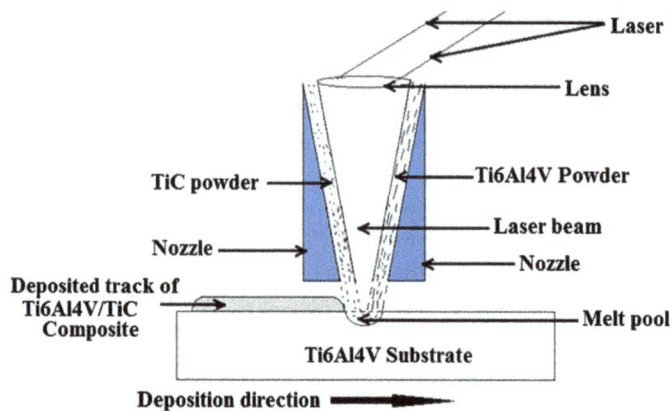

Fig. 12.1 Schematic diagram of powder-based LMD process [8] with kind permission from Elsevier

Standardization (ISO) committee on additive manufacturing (committee F42) [3]. This is the only class of additive manufacturing technology that can be used to carry out repair of parts that were not repairable or that were prohibitive to repair due to intricacies in the part or if the part is so small that repairing it through the traditional techniques could result in more damage because of the large heat affected zone.

In this process, the deposition is achieved by the coherent laser beam that creates a melt pool on the substrate and the powder or wire materials to be deposited are delivered into the melt pool. After solidification, the track of solid material is left on the laser path. Figure 12.1 depicts the schematic representation of LMD process.

The use of LMD for the production of 3D part involves five basic steps given as follows:

- *Loading* The loading of the CAD image of the component to the LMD machined is the first step of this process.
- *Conversion* The conversion of the loaded CAD file into the language understood by the LMD machine is done at this stage. The file format that the CAD file is converted to is known as additive manufacturing file (AMF). The old file format is the standard triangulation language (STL). The superiority of the AMF as regards to define the part structure replaced the old file format STL [3]. The AMF format provides a representation of the 3D surface assembly of planar and curved triangles that are made up of the co-ordinates of the vertices of these triangles.
- *Slicing* The slicing of the AMF into planes of two-dimensional (2D) profile sections is defined according to the geometry of the CAD model as well as the building orientation that was chosen before the slicing operation is performed at this stage. The building orientation is the direction required to follow by the building process. For example, building orientation could be chosen such that the building process starts from the bottom to the top, or from the left-hand side to the right-hand side. Usually, the software chooses the building orientation

Fig. 12.2 Flowchart showing various steps of laser metal deposition process

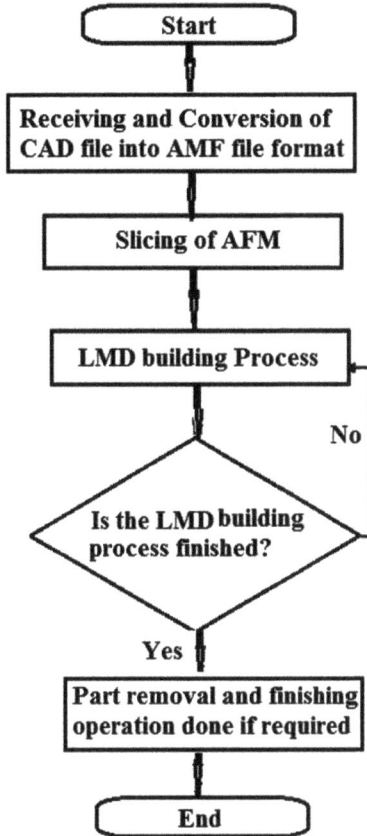

according to the convenience of the building process or the stability of the build object. Moreover, during the slicing process, the software generates the support structures for the build object if the object is of non-stable shape [7]. The flowchart of the stages involved in the LMD process is shown in Fig. 12.2.
- *Building* After slicing, the building of the part begins. The building process of the LMD is achieved by the creation of the melt pool on the surface of the substrate by the laser beam, then the powder or wire materials are fed into the melt pool to create a solid mass that is representing the 2D sliced section of the solid 3D model of the part. The laser continues to follow the path dictated by the 2D sliced sections and deposit the powder or wire materials layer after layer until the part building is completed.
- *Part removal and finishing* The final step is the removal of the built part from the machine and to perform finishing operations according to the service requirement of the part. The support structures are removed if they were generated during the slicing process and any form of heat treatment is also conducted on the part at this stage if required.

The process parameters are of great importance in the LMD process because they affect the physical as well as the mechanical properties of the deposited material [6, 8]. There are four key processing parameters that influence the LMD process, namely laser power, scanning velocity, powder flow rate, and gas flow rate. These processing parameters need to be adequately controlled in order to achieve desired properties. The next subsection discusses the capabilities of the LMD process.

12.2.2 Capabilities of Laser Metal Deposition Process for Repair and Remanufacturing

LMD process is an important additive manufacturing processes that can revolutionize the manufacturing industry ranging from helping to change the way products are designed to the way the waste materials are recovered. Additive manufacturing in general has offered the product designer the high needed flexibility in product design as against the design for manufacturing which used to be the practice when the traditional manufacturing processes are used. That is, apart from the designing of the product to satisfy the functional requirement of the product, also the designer need to consider the ease with which the product can be made. LMD enables the product to be made without specific regard for the ease of manufacturing. Only the functionality of the product is considered in the design of the product. LMD is also a promising manufacturing process that can be used to repair high-damaged or worn-out components/parts that were not repairable and discarded in the past. Some parts that are very small such as telephone parts or part made of high-carbon ferrous materials (e.g., high pressure valve) were not repairable in the past because of the damage caused by the heat affected zone generated during the repair process using the traditional welding processes such as arc welding. The recovery of end-of-life part used to be through recycling process which is energy intensive and environmentally unfriendly.

The end-of-life parts can be recovered by cutting off the damaged part and a new part can be built on the existing part using the LMD process. This is the most important characteristic of the LMD process because of its ability to produce a sound metallurgical bonding of the deposited powder on the substrate material [8–10]. All these characteristics are important in product remanufacturing process and sustainability issues of remanufacturing. The aerospace industry is going to have the lion's share of these benefits because the much needed reduced *buy-to-fly ratio* in the aerospace industry can easily be achieved with the LMD process [11].

Buy-to-fly ratio is the ratio of the stock material to the final material that made it into the flying component. Some aerospace parts are so complex that the percentage of the material that ended as scrap when the component is made through the traditional manufacturing route could be as high as 80%. The LMD process manufactures by the addition of materials, this makes it an economic process with

Table 12.1 Difference between LMD process and conventional welding processes

LMD process	Conventional welding processes
Superior metallurgical bonding	Less superior metallurgical bonding
Low heat input	High heat input
Low dilution is possible	High dilution rate
Low heat affected zone	High heat affected zone
Process cannot be controlled manually	Manual process control possible
Higher thermal diffusivity and conductivity	Lower thermal diffusivity and conductivity
High equipment cost	Low equipment cost
Surface quality can be controlled	Uncontrollable
Good for repair and remanufacturing of 3D part or equipment	May not be applicable for repair of certain parts due to heat damage it will cause to the part
Can be used to produce 3D part from the scratch	Cannot be use to build a 3D part

less material wastages. With proper process parameters control, the material wastage can be reduced to the barest minimum [12–16]. Also, because a new part can be made on an existing part with sound metallurgical integrity, the ease with which a product can be remanufactured is phenomenal and the use of LMD for remanufacturing is also sustainable. LMD provides an excellent alternative manufacturing method for the processing of difficult to machine materials such as titanium [17]. It is also possible to change an obsolete designed component in existing component and adding an improved design on a product through the process of remanufacturing using the LMD process. Table 12.1 highlights the important points of the differences between the LMD process and the conventional welding processes.

Remanufacturing is the bringing back of a damaged product to its perfect working condition and with appropriate warrantee. Not only does product remanufacturing can help to save revenue, but it can also help to limit environmental impacts. Remanufacturing is an important strategy for sustainable manufacturing and also for improving any country's economy. Remanufacturing provides a great opportunity to restore broken or discarded products to its like new condition. Remanufacturing is achieved through a series of steps such as repair, replace, rebuild, and bring back the product to its standard working condition by following the relevant standards. Traditional welding has been used in the past to restore broken engineering components, but such repair does not bring back the product to its standard condition because it is a matter of time before the same part is broken again. This traditional welding process usually creates a nonuniform microstructure between the filler materials and the base material. This mismatch always result in poor bonding between the damaged part and the filler material and generation of high residual stress that causes the part to fail again at the same location. Also traditional welding cannot be used to rebuild a three-dimensional component. An important characteristic of the laser metal deposition process is that it can be used to build a new part on an existing component. This is the key properties of this process in remanufacturing. This makes it possible to remanufacture a three-dimensional component as well as permitting redesigning of an existing component for better performance. Obsolete equipment can be redesigned and remanufactured with the help of laser metal deposition process because it is able to produce sound metallurgical bonding between the deposited materials with continuous microstructure. This is demonstrated from a number of researches in the literature [8–11] and it is also demonstrated in the case study which is presented later in this chapter. Laser metal deposition process capabilities not only enables product remanufacturing, but also helps in achieving improvement in an existing product. It can help to redesign an existing product without having to start from the scratch.

A number of research works have been done on repair of high-valued components such as turbine and various aerospace parts with excellent results. Some of the research studies on repair using LMD process are briefly discussed in the following sections.

12.2.3 Review of the Past Work on Use of LMD Process for Repair and Remanufacturing

Repair is usually cheaper when it is achievable than replacement of the broken part. It has been demonstrated through extensive research that laser metal deposition process can be used to repair high-valued parts and at a higher rate too. Some broken equipments are needed to be repaired as quickly as possible because idle time of some equipment could cost millions of dollars in 24 hours. For example, it could cost about $1.32 million of lost in revenue for a 350 MW combined cycle power plant in one day of outage [18]. It is usually more economical to repair damaged parts instead of replacing them with new ones. The quality of repair is also another factor of concern because it determines the life of the repaired part as well as influence the future downtime. It is therefore important that the repaired parts should have properties close to those of the substrate in order to ensure proper service performance and long service life.

Aerospace industries are in dear need of repair methods that can help to prolong the service life of their critical components. The main interest of this industry is to improve the structural integrity as well as properties of their critical parts through repair [19–21]. Some of the critical aerospace components are very difficult to repair because of their shape or size using the conventional repair technology such as arc welding because of the high heat affected zone generated by this process. Because of this problem, the usual practice was to replace the damaged part. *Rolls-Royce* has been using laser-assisted repair techniques to carry out repair on gas turbine blades in order to improve the wear resistance behavior of the blades since mid-1980s [22]. The repaired blades were found to have low heat affected zone and also of better quality and improved wear resistance property. A number of research works have appeared in the literature on the possibility of using the LMD process for the repair of high-valued parts and some of them are discussed in this section.

Lourenç et al. [23] studied the fatigue and fracture behavior of repair of high-valued complex aerospace component using the LMD process. The study was conducted by depositing AerMet 100 alloy powder on AerMet 100 substrate. The studies conducted on the deposited samples include the microstructure, residual stress, hardness, and fatigue. Moreover, the effects of heat treatment on the samples were also studied. The results obtained in this study show that laser metal deposition process significantly improves the fatigue life of the samples at 184,290 cycles when compared to the baseline sample with a notch of 16,087 cycles. The baseline samples were provided with an artificial notch in order to simulate damaged condition. The compressive residual stress of 300–500 MPa was observed in the clad region and the heat affected zone. The post heat-treated samples were found to be ineffective in improving the fatigue life. The fracture modes of the post heat treatment samples showed mainly decohesive rupture, which was found to reduce the fatigue life. The low fatigue life for the post heat-treated samples was also associated with their brittleness as shown by the increase in hardness in these set of samples.

In another study, Wen et al. [24] attempted to repair an important load-bearing aerospace part made by precipitation-hardening martensitic stainless steel and studied the properties of repaired part. The repair was performed using preheated wire in LMD process. The wire was preheated by resistance heating and multiple layers were deposited on the surface of substrate. They investigated the microstructure and the mechanical properties of the deposited layers and the results were compared with the properties of the substrate. Mixed microstructures were observed in the deposited layer, which consist of martensite with coarse laths grain structure as shown in Fig. 12.3a and a small number of precipitates as shown in Fig. 12.3b.

The tensile strength was found to be very close to those of the substrate, while the ductility and impact toughness were found to be lower than that of the substrate.

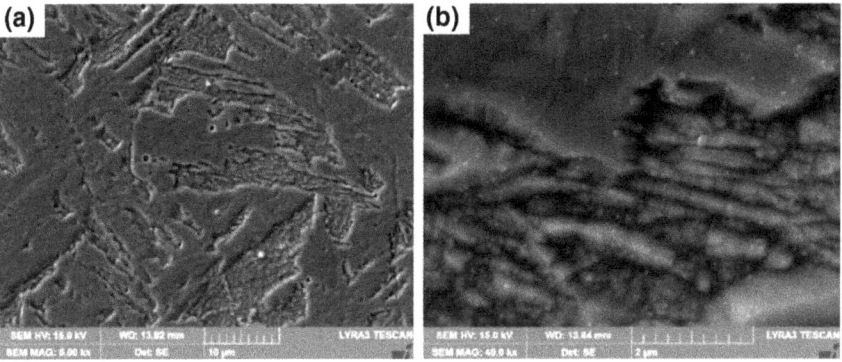

Fig. 12.3 Micrograph showing the **a** martensite with coarse laths **b** precipitate iron particles [24] with kind permission from Elsevier

The effect of laser rescanning of each layer after layer during the deposition process was also investigated. The microstructures of the deposited layers were found to be changed to fine martensite uniformly distributed with more precipitates. An improvement in the ductility and toughness was also observed for the deposited layer and were found to be close to those of the substrate. The tensile strength was found to be a bit lower than the tensile strength of the substrate. TiC nanoparticles were added to the deposits and laser scanning was performed. Addition of TiC nanoparticles causes the precipitation strengthening effect to be improved. The microstructure was found to be refined in the deposited layer, while the strength, ductility, and toughness were also found to be further improved. High-quality repaired layer with even superior mechanical properties to those of the substrate using the preheated wire and LMD process and without post deposition heat treatment were achieved. Finally, it was concluded that high-quality repair can be achieved with equivalent or even superior mechanical properties and that this method can be used as a valuable technique for repair of PH-MSS.

Pinkerton et al. [25] investigate the advantages and potential problems of using LMD for repair of internal cracks and defects in metallic parts. The first step toward the repair of such defect was to machine a groove or slot to the point where the defect is located and then refill the groove or slot as shown in Fig. 12.4. The study investigates the results obtained from two different types of groove geometries of rectangular and triangular cross section. H13 tool steel was used as the substrate and the H13 powder was used in the study. A number of process parameters were studied and each of the samples was analyzed for microstructure, porosity, heat affected zone, and microhardness.

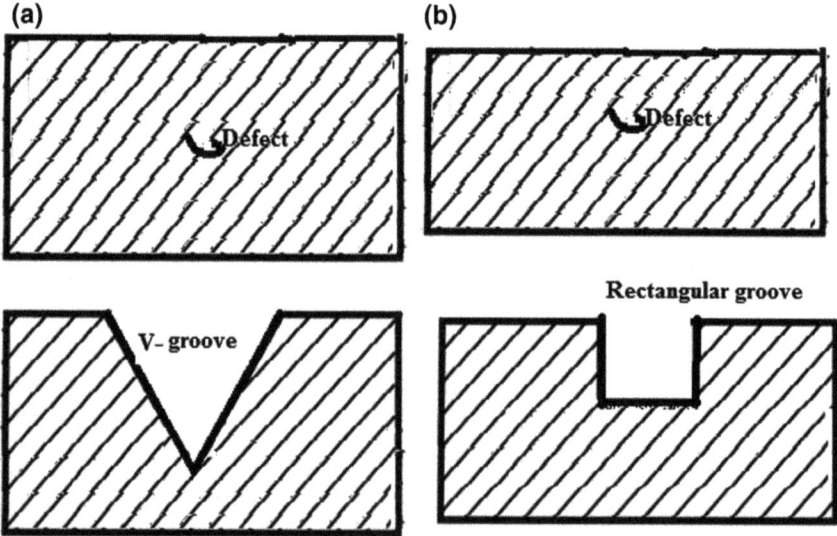

Fig. 12.4 Schematic diagram of defects and cutting of groove to reach the defect showing **a** V-groove **b** rectangular groove

It was observed that all the samples that were studied composed of tempered martensite with high hardness values. The heat affected zones were found to consist of untempered martensite. Porosities were observed in all the samples and that the porosity increases as the laser power and powder flow were increased for the two types of grooves and the porosity has a detrimental effect on the strength of the repaired area. The cause of the porosity was attributed to the vertical sidewalls of a square slot that shielded the powder and prevented direct laser material interaction with the powder. The sharp angles within a slot, such as the lowest point of a V-slot were also found to be responsible for the porosity. They concluded that the method can be used to produce high-quality repairs, but porosity at the boundaries between the original part and the added material was a problem. This problem can be overcome by avoiding the shape of the groves that will prevent proper laser–material interaction that can promote porosity formation.

In a similar study, the feasibility of using the LMD process for the repair of milled grooves was investigated by Graf et al. [26]. The powders of stainless steel and titanium alloy were deposited into different types of groove shapes. The influence of processing parameters on the microstructure and the heat affected zone produced during the repair process were investigated. The study was able to achieve good results without porosity and defects. Good sidewall fusion was also achieved and the hardness was found to be increased. It was concluded that if the groove was made big enough for proper powder deposition, then a porosity free repair can be achieved. A deposition strategy was developed in this study that produces a deposition free of defect. The V-groove and U-groove show good side-wall fusion as shown in Fig. 12.5.

Wang et al. [27] carried out a study on the repair of titanium alloy using the LMD process with the aid of a surface response design of experiment. The influence of processing parameters on the properties of the repair was investigated. The process parameters studied were the laser powers, scanning speeds and powder flow rate. The titanium alloy powder was used for the repair process on the surface circular groove defects. The properties investigated were the microstructure, tensile strength, relative density, microhardness, elemental composition, and internal defects. The results of this study showed that the tensile strength could be as high as

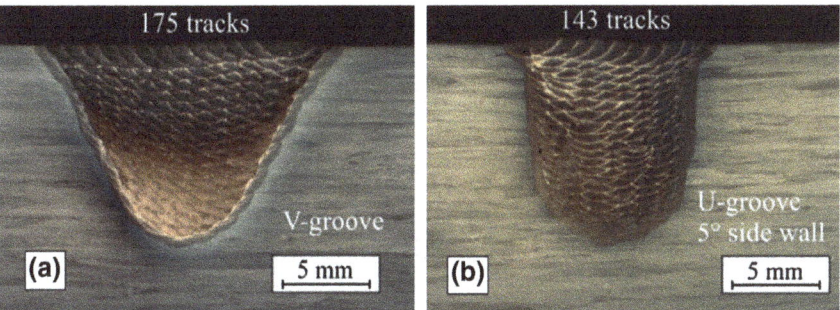

Fig. 12.5 Cross section of repaired samples showing **a** the V-groove and **b** the U-groove [26] with kind permission from Elsevier

1100 MPa and percentage elongation as high as 91–98% than that of the substrate material. The optimal process parameters achieved were 1000 W laser power and 8 mm/s of scanning peed. At these optimum process parameters, the microhardness of the deposited zone, the heat affected zone and base material were found to be between 370 and 390 HV 500. Different chemical compositions were observed in the deposition zone and the base material was found to be less than 70.15% and the relative density was found to be as high as 99%, which was found to increase as the powder feeding rate was reduced. The repaired zone microstructure was investigated to consist of typically columnar and dendritic crystal microstructures, while the microstructure of the heat affected zone was consisting of coarse equiaxial crystalline structure.

The study of repair of titanium alloy part using LMD process was undertaken by Dey [28]. This research investigated the mechanical properties of the repaired titanium alloy samples that include the yield strength, ultimate tensile strength, and percentage elongation. The results revealed that the properties of the repaired material were enhanced by the LMD process. The ductility of the repaired samples was found to be slightly higher than those of the substrate material when no heat treatment was conducted on the repaired samples. The study concluded that the LMD process can be used to repair titanium alloy components in aerospace parts.

Capello et al. [29] investigated the repair of sintered tool using laser metal deposition process with wire. Defects in sintered tool are generally as a result of the sintering process such as porosity, cracks and distortion, and other forms of damage could be as a result of high wear rate during the service life. The feasibility of repairing the sintered tools and dies using direct LMD process were investigated. The results revealed by the study showed that if optimum process parameters are used, the LMD process can be used to achieve multilayered repair of sintered crack and porosity free tools and dies with strongly metallurgical integrity.

Leunda et al. [30] also studied the repair of tool steel using the LMD process. CPM 10 V and Vanadis 4 Extra tool steel powders were used in their study. The volumetric fraction of retained austenite and microhardness profiles were measured. The results showed that a crack-free deposited layer was achieved with microhardness as high as 700 HV. The microstructure of carbides embedded in a martensite plus retained austenite matrix were also observed. This indicates that there was sound metallurgical bonding between the deposited powder and the damaged substrate with improved mechanical properties.

Kattire et al. [31] studied the repair of H13 tool steel using the laser metal deposition process. CPM 9 V steel powder was deposited on H13 tool steel plate in order to repair the damaged die surface using a continuous wave CO_2 laser. The influence of the processing parameters repair quality was investigated. The results showed the presence of vanadium carbide particles embedded in martensite and retained austenite microstructure. The microhardness was found to be higher as a result of this hard vanadium carbide particles and it was found to be four times greater than that of the substrate. It was also observed that compressive residual stresses were generated in deposited layer and this was said to be desirable for repair applications because it would impede the crack propagation and would result in enhanced die life. The process parameters were found to have significant influence on the resulting properties of the deposited layer. The deposited height was found to decrease as the scanning speed was increased and also increases with increase in the powder feed rate. The study also revealed that the residual stresses had increasing trends with laser power and decreasing trend with scanning speed.

Torims [32] described the use of LMD for repair and reconditioning of various mechanical components. The use of LMD process and the basic principle for repair of high-valued parts were discussed. The state of the art of this technology and trends for its industrial application were outlined with the emphasis on the use of the technology for repair. A case study was also presented that demonstrates the capabilities of LMD process to repair crankshaft.

Bendeich et al. [33] proposed the repair of worn-out low-pressure turbine blades used in hydro power plant using the LMD process. The turbine blades were worn as a result of erosion damage from the water impingement. This study was aimed to repair the damaged blades in order to extend the service life of these blades. The repair was carried out by depositing Stellite material on the worn part. The study revealed that the addition of Stellite material was found to generate residual stresses in the parent metal because of contraction during the rapid cooling and also as a result of the differences in the thermal expansion between the two materials. For further reading on other various studies on the use on laser metal deposition for repair, literature from [34–45] can be referred.

The major advantages of using laser metal deposition process for repair and in product remanufacturing are as follows:

- The heat input into the system is controllable.
- High-dimensional accuracy is achievable.
- It offers a lot of flexibility in terms of its ability to handle multiple materials to produce parts with functionally graded materials [8, 42, 46, 47].

Apart from the ability to restore damaged components back to their original physical and mechanical properties, laser metal deposition process is suitable to repair small components which were not repairable using the conventional repair process because of the large heat affected zone that tend to damage the whole component. These components were discarded and replaced in the past but can now be repaired with laser metal deposition process. Laser metal deposition has also been used to improve the surface properties of high-valued parts [48]. The 3D laser remanufacturing based on reverse engineering and laser metal deposition process was investigated by Nan et al. [48]. Laser remanufacturing has been proved in this study as a technology that can be used to extend the life of aging dies, ships, vehicles, and weapon systems. The investigation was carried out by reconstructing a worn part through the production of use of a 3D digital image of the worn part which helps in gaining the surface data for CAD model reconstruction. The CAD data was used to generate the path for the laser metal deposition which was used for the reconstruction process. This process has been proven to reduce lead time and associated costs. The life spans of critical military equipment parts have been restored to their original working condition with the help of the LMD process which has saved millions of dollars of United States military [49]. The low heat input from the process made it possible to restore small and thin components such as turbine blades, vanes, and impellers [49–51]. Remanufacturing of obsolete or failed tooling is made possible with the laser metal deposition technology even for component that were not repairable in the past. Remanufacturing was seen to be cost-effective, increase productivity as well a reduce environmental impact and remanufacturing in the United States alone was about $56 billion per annum [52] even when LMD process was not used.

Wilson et al. [53] studied the remanufacturing of turbine airfoils with LMD process. The defective voids in the turbine airfoils were refilled according to the semi-automated geometric reconstruction algorithm used in the study. The study revealed that the LMD process is effective for the remanufacturing of this high-valued engineering component and it has the potential to adapt to a wide range of part defects. The tensile test results showed that the samples failed at points far away from the repaired region and also showed that the repaired samples were as strong as the original material. This shows that the repair zone is not the weakest part and demonstrated the improved mechanical properties obtainable with the LMD process. These furthered strengthened the capability of laser metal deposition process in component remanufacturing.

A study on the energy consumption and environmental impacts by remanufacturing was also conducted. About 36% saving in terms of revenue was achieved with the remanufactured component when compared to the replacement of the part. A savings of 45% of energy was also made when compared to the energy required to manufacture a new blade. The pictorial diagram of the remanufactured blade is shown in Fig. 12.6.

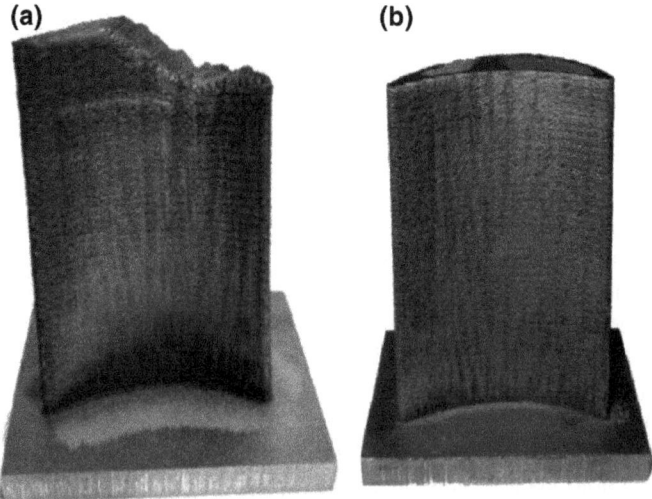

Fig. 12.6 Damaged turbine blade remanufactured using laser direct deposition. **a** damaged blade **b** restored blade [53] with kind permission from Elsevier

Damages in high-valued engineering parts are usually in the form of wear, voids, or cavities which could be as a result of service condition the part is subjected to or the inherent error created through the manufacturing process. These types of damages prevent the part from performing well as per the specified designed standard.

Remanufacturing of components using the LMD process showed that the parts are able to regain their designed performance efficiencies. The main challenges in the remanufacturing process using the LMD process are the unavailability of the CAD data of the part because in most cases the original equipment manufacturers are not interested in remanufacturing and may not make this information available to industries that are interested in this remanufacturing process. A number of researches have been conducted to create path for this remanufacturing process to be possible even if the CAD information about the components or equipments are not available [54–59]. In the event that the 3D model of the part is not available, the method used in such instance include the reverse engineering process and a surface reconstruction algorithm are often used to reconstruct a geometric representation of the missing shape or part which can then be used to define the required missing or worn-out part geometry [58].

Apart from the LMD process, there are also other types of additive manufacturing technologies that can be used in product remanufacturing which include selective laser sintering, selective laser melting, and 3D printing. Some equipments are obsolete and can still function very well because a part is damaged and cannot be replaced as the manufacturers no longer produce the part. Such parts can be manufactured conveniently using any of the additive manufacturing technologies and the equipment can be remanufactured even with better functionalities. Also with additive manufacturing technologies existing parts can be improved to have better surface properties simply by changing the surface material or adding advanced materials to the surface and hence longer service life. A case study to demonstrate the metallurgical integrity of LMD process is presented in the next section.

12.3 A Case Study on Laser Metal Deposited Titanium Alloy Powder

12.3.1 Introduction

Ti6Al4V is the most widely used titanium alloy due to its excellent properties such as the high strength-to-weight ratio and high corrosion resistance properties [60]. These properties make Ti6Al4V to be favored in a number of industries which include aerospace and automobile. In spite of these excellent properties, Ti6Al4V is often termed as difficult to machine material and its repairing through traditional manufacturing route is challenging [61]. The use of laser metal deposition process for fabricating, repair and remanufacture, this material will go a long way for the aerospace and the automobile industries as some of the components in this industries are of high value and are subjected to wear because of the service condition they are exposed to. Considering the high cost and widespread applications of titanium and its alloys, an attempt has been made to perform LMD for Ti6Al4V. The details about the process setup, experimentation, and post-experimental investigations/observations are given in the next subsections.

12.3.2 Experimental Method

The Ti6Al4V powder and the Ti6Al4V substrate used in this study are both 99.6% pure. The Ti6Al4V powder used is of particle size range between 150 and 250 μm. The LMD process was achieved through an experimental setup that is available at the National Laser Center, CSIR, Pretoria- South Africa. The experimental setup consists of a Kuka robot that is carrying Nd–YAG laser in its end effector. Coaxial powder nozzles are also attached to the end effector of the robot. The maximum available power of the Nd–YAG laser is 4.0 kW. The pictorial diagram of the experimental setup is shown in Fig. 12.7.

The schematic representation of the LMD process achieved is given in Fig. 12.1 of Sect. 12.2.1.

The focal length between the substrate and the laser was maintained at a distance of 195 mm throughout the experiment which helps to keep the laser spot size at 2 mm. The powder flow rate and the gas flow rate were also maintained at constant values of 1.44 g/min and 2 l/min respectively. The scanning speed was varied from 0.2 to 0.005 m/s and the laser power was varied between 400 and 800 W. This was done in order to see the processing parameters that could give a porous free deposited sample. The processing parameters are given as in Table 12.2. After the deposition process, the samples were cut across the deposition direction in order to reveal the cross section of the sample and to show if there is porosity or not. The cut sample was mounted in resin, ground, polished, and etched according to standard metallurgical sample preparation of titanium and its alloys. The etched samples were studied under optical microscope to analyze its microstructural characteristics.

12.3.3 Results and Discussion

The micrograph of the Ti6Al4V substrate is shown in Fig. 12.8a, while Fig. 12.8b shows the morphology of the Ti6Al4V powder. The Ti6Al4V powder consists of a spherical gas atomized powder while the microstructure of the substrate consists of a mixture of beta phase (dark colored) grain structures in the matrix of the alpha phase (bright colored) that is typical of any Ti6Al4V. The micrographs of the sample at a laser power of 400 W and a scanning speed of 0.1 m/s is shown in

Fig. 12.7 Pictorial diagram of the LMD experimental setup

Table 12.2 Experimental matrix

Sample No	Laser power (kW)	Scanning speed (m/s)	Powder flow rate (g/min)	Gas flow rate (l/min)
1	400	0.2	1.44	2
2	400	0.1	1.44	2
3	400	0.01	1.44	2
4	400	0.005	1.44	2
5	800	0.2	1.44	2
6	800	0.1	1.44	2
7	800	0.01	1.44	2
8	800	0.005	1.44	2

Fig. 12.9. The sample shows some degree of porosity as a result of improper melting of the powder due to the low laser power and high scanning speed. The deposited powder is not properly bonded to the substrate because of the porosity that was seen at the interface because of the improper melting of the powder due to the low laser power as well as the low laser–material interaction time. This

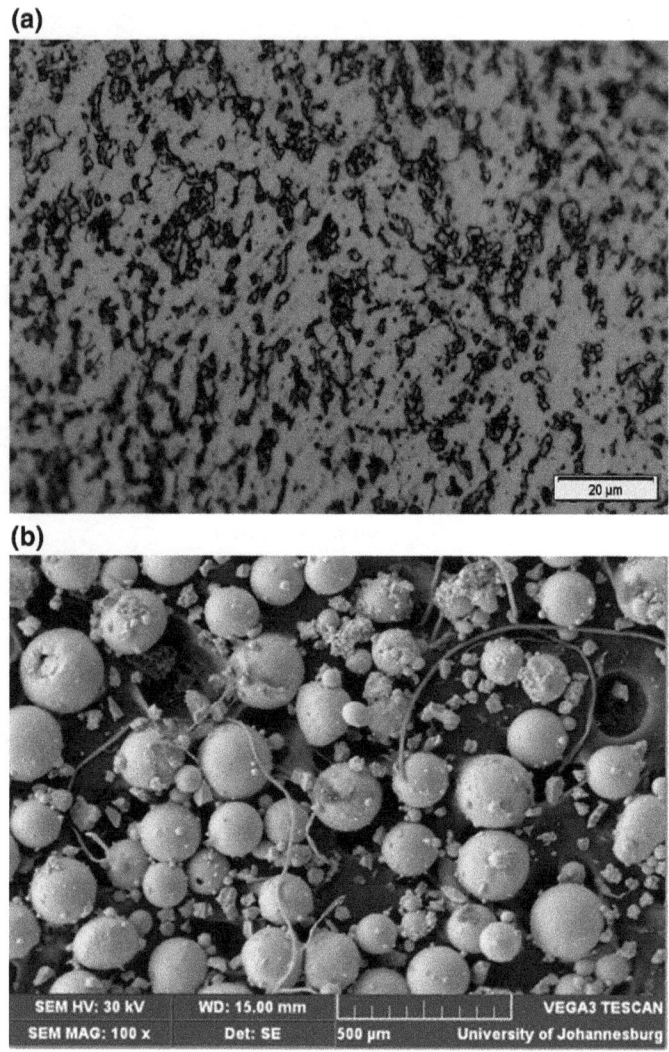

Fig. 12.8 The micrograph of the **a** Ti6Al4V substrate **b** Ti6Al4V powder

combination of processing parameters is not suitable for repair application because of the poor metallurgical integrity obtained.

The degree of porosity was found to decrease with increased laser power and decreased scanning speed. As the scanning speed was decreased and the laser power was increased, a fully dense deposit was achieved. At a laser power of 800 W and scanning speed of 0.005 m/s a fully dense deposit was achieved as shown in Fig. 12.10.

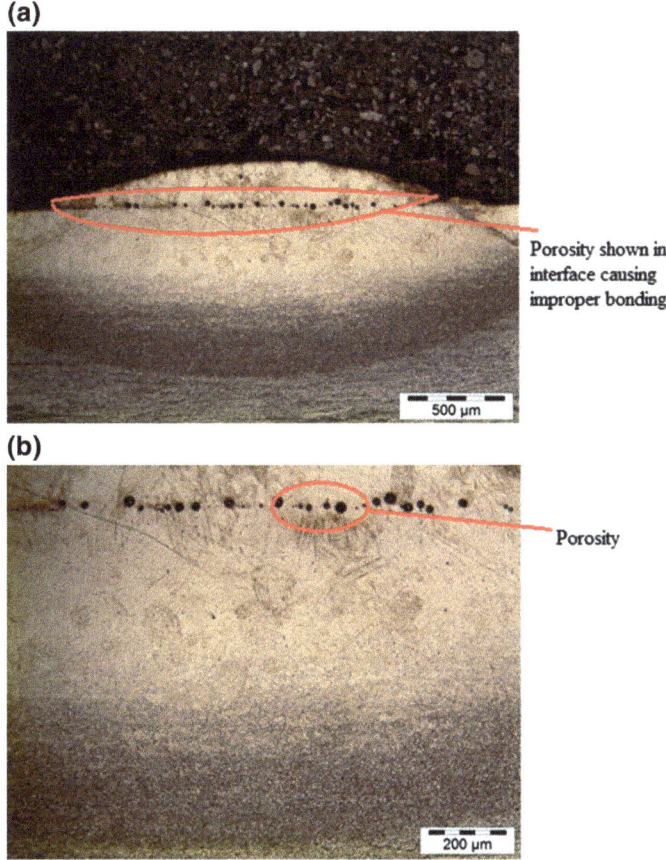

Fig. 12.9 The micrograph of sample with porosity in the interface between the deposit zone and the substrate **a** at low magnification **b** at higher magnification [62]

It can also be seen in Fig. 12.10 that there is continuity in the interface between the deposited powder and the substrate which shows that there is proper bonding between the substrate and the deposit. This is desirable in any successful repair process with the processing parameters at a laser power of 800 W and scanning speed of 0.005 m/s a sound metallurgical bonding of Ti6Al4V powder and Ti6Al4V substrate can be achieved. The reason behind this proper bonding is that the laser power was high enough and the scanning speed was low which makes the laser–material interaction time to be longer that promotes adequate melting of the powder and hence proper bonding of the deposited powder with the substrate. The sound metallurgical integrity achievable in laser metal deposition process makes this technology an important one especially when new part is required to be built on an existing part. The continuous microstructure makes it difficult to distinguish between the new part and the old base material which makes the remanufactured

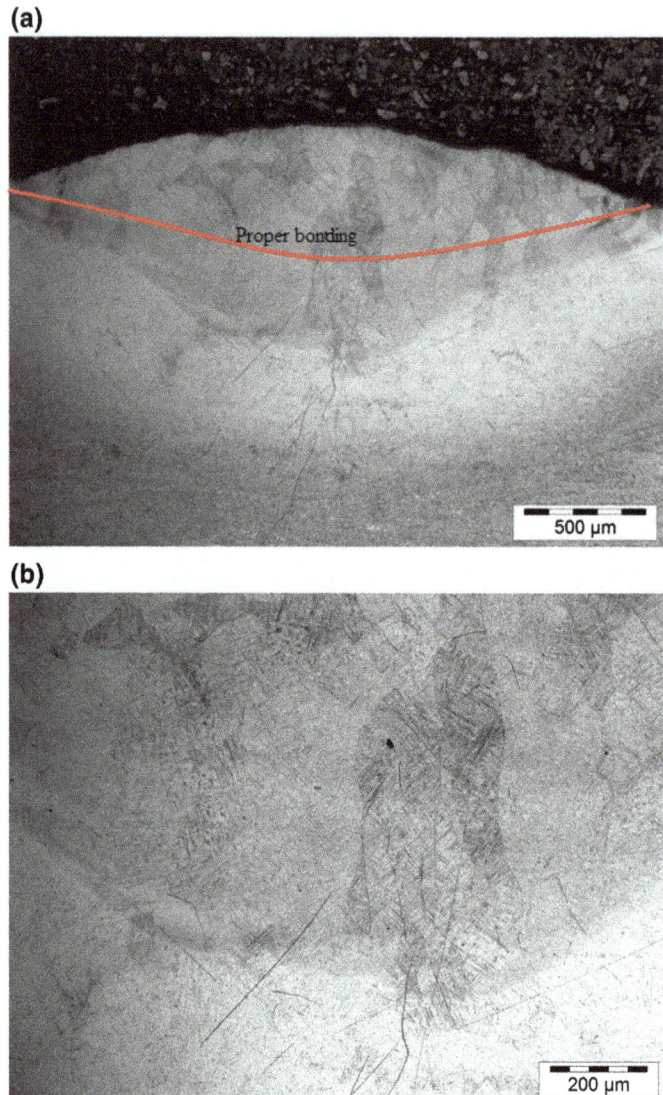

Fig. 12.10 Micrograph of sample without porosity **a** at low magnification **b** at higher magnification [62]

material to function as new without any interface. Based on these results, it can be concluded that laser metal deposition process is capable of repairing as well as remanufacture high-valued engineering components efficiently and effectively.

12.4 Sustainability Aspects of Laser Metal Deposition Process

Discarding or replacing components or equipment as a result of damages presented a rising challenge over the high emission rate those results from waste management and recycling such equipment. This is one of the reasons behind the campaign of product remanufacturing. Remanufacturing and specially the remanufacturing by LMD are seen as a promising method to reduce the environmental impact of products. By extending the product life span through remanufacturing the replacements can be reduced which are not only costly in terms of financial expenditure but also costly in terms of energy consumption and also detrimental to the environment. LMD process has made a number of equipments to be restorable, thereby saving companies millions of dollars in terms of revenue. The equipments worth over $100 million in one of the Army Depots which can be restored through laser-based remanufacturing was reported to be evaluated [63]. LMD process as well as other additive manufacturing technologies have been proved to be energy-efficient manufacturing processes and are termed as green manufacturing which are also sustainable [64, 65]. No doubt that the LMD process has positioned itself to make product remanufacturing an economic and viable option. Product remanufacturing using LMD will certainly help to boost global economy through energy and material saving, and protecting environment that are key to sustainable development and manufacturing. Remanufacturing process using the LMD process is truly a green process from both financial end environmental point of view by increasing the length of time that a product stays out of the recycling stream is a venture that has significantly reduced pollution through energy saving. LMD certainly helps to create opportunities for enhancing product and material life cycles and thereby improving sustainability [66].

12.5 Summary

LMD process has been identified as one of the key driving manufacturing methods in sustainable product remanufacturing and discussed thoroughly in this chapter. Some of the past efforts as regards to repair and remanufacture of engineered parts by LMD are also discussed. The ability of the process to produce the sound metallurgical bonding between the deposited material and the substrate which is an important and desirable characteristic and useful in product remanufacturing has been demonstrated through a case study. The mechanical and metallurgical properties of the repaired components were found to be similar or in some cases exceeding those of the parent materials. LMD process has proved that remanufacturing can be achieved in an effective, efficient, and in a timely manner too. The concentrated heat from the laser beam helps to direct the heat only to the needed area, thereby limiting the heat affected zone that minimizes the microstructural damage to the base material. This helps to ensure that the metallurgical integrity of the repaired part is sound and properly bounded to the base material which helps to prolong the life of the part. In essence, it can be concluded that LMD process is a sustainable alternate to the conventional processes for product remanufacturing and would revolutionize the future of product remanufacturing.

Acknowledgements This work was supported by University of Johannesburg Research Council and the L'OREAL-UNESCO for Women in Science. The authors would also like to acknowledge the financial support from the Department of Defense through the research and educational program HBCU/MSI (contract # W911NF-15-1-0457) under the direct supervision of Dr. Joycelyn S. Harrison (Program Manager, AFOSR Complex Materials and Devices Program).

References

1. Lant T, Robinson DL, Spafford B, Storesund J (2001) Review of weld repair procedures for low alloy steels designed to minimise the risk of future cracking. Int J Press Vessel Piping 78 (11–12):812–818
2. Pinkerton A, Li L (2004) Multiple-layer cladding of stainless steel using a high-powered diode laser: an experimental investigation of the process characteristics and material properties. Thin Solid Films 453–454:471–476
3. Song J, Deng Q, Chen C, Hu D, Li Y (2006) Rebuilding of metal components with laser cladding forming. Appl Surf Sci 252(22):7934–7940
4. Capello E, Colombo D, Previtali B (2005) Repairing of sintered tools using laser cladding by wire. J Mater Process Technol 164–165:990–1000
5. Scott J, Gupta N, Wember C, Newsom S, Wohlers T, Caffrey T (2012) Additive manufacturing: status and opportunities. Science and Technology Policy Institute. https://www.ida.org/stpi/occasionalpapers/papers/AM3D_33012_Final.pdf. Accessed 11 July 2016
6. Mahamood RM (2016) Laser metal deposition process. In: Akinlabi, ET, Mahamood, MR, Akinlabi SA (eds) Advanced manufacturing using laser material processing. IGI Global, pp 46–59
7. Boboulos MA (2010) CAD-CAM & rapid prototyping application evaluation. Ph.D & Ventus publishing Aps. http://www.bookBoom.com. Accessed 1 Aug 2016
8. Mahamood RM, Akinlabi ET, Shukla M, Pityana S (2013) Scanning velocity influence on microstructure, microhardness and wear resistance performance on laser deposited Ti6Al4V/TiC composite. Mater Des 50:656–666
9. Mahamood RM, Akinlabi ET (2016), Microstructure and mechanical behaviour of laser metal deposition of titanium alloy. Lasers Eng. In press
10. Mahamood RM, Akinlabi ET, Shukla M, Pityana S (2013) Laser metal deposition of ti6al4v: a study on the effect of laser power on microstructure and microhardness. In: International multi-conference of engineering and computer science (IMECS 2013) March 2013, pp 994–999
11. Brandl E, Michailov V, Viehweger B, Leyens C (2011) Deposition of Ti–6Al–4 V using laser and wire, part I: microstructural properties of single beads. Surf Coat. Technol. 206:1120–1129
12. Mahamood RM, Akinlabi ET, Shukla M, Pityana S (2013). The role of transverse speed on deposition height and material efficiency in laser deposited titanium alloy. In: 2013 International multi-conference of engineering and computer science (IMECS 2013), March 2013, pp 876–881

13. Mahamood RM, Akinlabi ET, Shukla M, Pityana S (2012). Effect of laser power on material efficiency, layer height and width of laser metal deposited Ti6Al4V. In: World congress of engineering and computer science. San Francisco 2012, 24–26 Oct 2012, pp 1433–1438
14. Mahamood RM, Akinlabi ET, Shukla M, Pityana S (2013). Material efficiency of laser metal deposited ti6al4v: effect of laser power. Eng Lett 21:1, EL_21_1_03. http://www.engineeringletters.com/issues_v21/issue_1/EL_21_1_03.pdf
15. Mahamood RM, Akinlabi ET (2016) Process parameters optimization for material deposition efficiency in laser metal deposited titanium alloy. Lasers Manuf Mater Process 3(1):9–21. doi:10.1007/s40516-015-0020-5
16. Akinlabi ET, Mahamood RM, Shukla M, Pityana S (2012) Effect of scanning speed on material efficiency of laser metal deposited Ti6Al4V. World Acad Sci Technol Paris 2012 (6):58–62
17. Wang ZM, Ezugwu EO (1997) Titanium alloys and their Machinability a review. J Mater Process Technol 68:262–270
18. Kannan P, Amirthagadeswaran KS, Christopher T, Rao BN (2013) Failures of high-temperature critical components in combined cycle power plants. J Fail Anal Prev 13(4):1–11
19. Fallah V, Alimardani M, Corbin SF, Khajepour A (2019) Impact of localized surface preheating on the microstructure and crack formation in laser direct deposition of Stellite 1 on AISI 4340 steel. Appl Surf Sci 257:1716–1723
20. Alimardani M, Fallah V, Khajepour A, Toyserkani E (2010) The effect of localized dynamic surface preheating in laser cladding of Stellite 1. Surf Coat Technol 204:3911–3919
21. Bhattacharya S, Dinda GP, Dasgupta AK, Mazumder J (2011) Microstructural evolution of AISI 4340 steel during direct metal deposition process. Mater Sci Eng A 528:2309–2318
22. Draper CW, Mazzoldi P (1986) Laser surface treatment of metals, 1st edn. Martinus Nijhoff Publishers, Dordrecht
23. Lourenç JM, Sun SD, Sharp K, Luzin V, Klein AN, Wang CH, Brand M (2016) Fatigue and fracture behavior of laser clad repair of AerMet_ 100 ultra-high strength steel. Int J Fatigue 85 (2016):18–30
24. Wen P, Cai Z, Feng Z, Wang G (2015) Microstructure and mechanical properties of hot wire laser clad layers for repairing precipitation hardening martensitic stainless steel. Opt Laser Technol 75:207–213
25. Pinkerton AJ, Wang W, Li L (2008) Component repair using laser direct metal deposition. In: Proceedings IMechE, vol 222 Part B, pp 827–836. (J. Eng Manuf)
26. Graf B, Gumenyuk A, Rethmeier M (2012) Laser metal deposition as repair technology for stainless steel and titanium alloys. Phys Procedia 39:376–381. doi:10.1016/j.phpro.2012.10.051
27. Wang Y, Zheng H, Tang K, Li H, Gong S (2016) TC17 titanium alloy laser melting deposition repair process and properties. Opt Laser Technol 82:1–9
28. Dey ND (2014) Additive manufacturing laser deposition of Ti-6Al-4V for aerospace repair application. Master Thesis, Missouri University of Science and Technology
29. Capello E, Colombo D, Previtali B (2005) Repairing of sintered tools using laser cladding by wire. J Mater Process Technol 164–165:990–1000
30. Leunda J, Soriano C, Sanz C, Navas VG (2011) Laser cladding of vanadium-carbide tool steels for die repair. Phys Procedia 12:345–352
31. Kattire P, Paul S, Singh R, Yan W (2015) Experimental characterization of laser cladding of CPM 9 V on H13 tool steel for die repair applications. J Manuf Process 20(3):492–499
32. Torims T (2013) The application of laser cladding to mechanical component repair, Renovation and Regeneration. In: Katalinic B, Tekic Z (eds) DAAAM international scientific book. DAAAM International, pp 587–608
33. Bendeich P, Alam N, Brandt M, Carr D, Short K, Blevins R, Curfs C, Kirstein O, Atkinson G, Holden T, Rogge R (2006) Residual stress measurements in laser clad repaired low pressure turbine blades for the power industry. Mater Sci Eng A 437(1):70–74
34. Song J, Deng Q, Chen C, Hu D, Li Y (2006) Rebuilding of metal components with laser cladding forming. Appl Surf Sci 252:7934–7940

35. Koehler H, Partes K, Seefeld T, Vollertsen F (2010) Laser reconditioning of crankshafts: from lab to application. Phys Procedia Part A 5: 387–397
36. Nowotny S (2011) Current use of laser technology for build-up welding applications. Surf Eng 27(4):231–233
37. Torims T (2013) Laser cladding device for in-situ repairs of marine crankshafts. Adv Mater Res 712–715:709–714
38. Torims T. (2013) Device and method for the in-situ repair and renovation of crankshaft journal surfaces by means of laser build-up. Patent of the Republic of Latvia no. B24B5/42
39. Vishnevetskaya IA, Denisov VA, Solovyov AV (1996) Tribotechnical efficiency of journal-bearing connection of crankshaft renewed by laser built-up welding. Proc SPIE 2713(301):301–305
40. Weisheit A, Backes G, Stromeyer R, Gasser A, Wissenbach K, Poprawe R. Aachen (2016) Powder injection: the key to reconditioning and generating components using laser cladding. http://ilt.fraunhofer.de/ilt/pdf/eng/paper1232.pdf. Accessed 7 Aug 2016
41. Woodyard D (ed) (2009) Pounder's marine diesel engines and gas turbines, 9th edn. Butterworth Heinemann, Oxford
42. Mahamood RM, Akinlabi ET (2015) Laser metal deposition of functionally graded Ti6Al4V/TiC. Mater Des 84:402–410
43. Jhavar S, Paul CP, Jain NK (2014) Causes of failure and repairing options for dies and molds: a review. Eng Fail Anal 34:519–535
44. Paul CP, Bhargava P, Kumar A, Pathak AK, Kukreja LM (2014) Laser rapid manufacturing: technology, applications, modeling and future prospects. In: Paulo Davim J (ed) Lasers in Manufacturing. Wiley-ISTE, UK
45. Gu D (2015) Laser additive manufacturing (am): classification, processing philosophy, and metallurgical mechanisms. In: Laser additive manufacturing of high-performance materials. Springer, Berlin Heidelberg
46. Mahamood RM, Akinlabi ET, Shukla M, Pityana S (2012) Functionally graded material: an overview. In: Proceedings of the world congress on engineering (2012), WCE 2012, vol. III. London, U.K, 4–6 July 2012, pp 1593–1597
47. Shukla M, Mahamood RM, Akinlabi ET, Pityana S (2012) Effect of laser power and powder flow rate on properties of laser metal deposited Ti6Al4V. World Acad Sci Technol 6:44–48
48. Nan LL, Liu WJ, Zhang K (2010) Laser remanufacturing based on the integration of reverse engineering and laser cladding. Int J Comput Appl Technol 37(2):116–124
49. Grylis R (2003) Laser system saves damaged military parts from the scrap heap. The Fabricator. http://www.thefabricator.com/article/lasercutting/laser-system-saves-damaged-military-parts-from-the-scrap-heap. Accessed 8 Aug 2016
50. Plourde R (2003) Laser-based repair system reclaims high value military components. In: RTO AVT specialists' meeting on the control and reduction of wear in military platforms. Williamsburg, USA, 7–9 June 2003. (published in RTO-MP-AVT-109 1-4)
51. Dutta B, Singh V, Natu H, Choi J, Mazumder J (2009) Direct metal deposition six-axis direct metal deposition technology enables creation/coating of new parts or remanufacturing of damaged parts with near net-shape. Adv Mater Process:29–31
52. Lund B (1996) The remanufacturing industry: hidden giant. Boston University Press, Boston, MA
53. Wilson JM, Piya C, Shin YC, Zhao F, Ramani K (2014) Remanufacturing of turbine blades by laser direct deposition with its energy and environmental impact analysis. J Clean Prod 80 (1):170–178
54. Gao J, Yilmaz O, Noble D, Gindy N (2008) An integrated adaptive repair solution for complex aerospace components through geometry reconstruction. Int J Adv Manuf Technol 36:1170–1179
55. Gao J, Chen X, Zheng D (2010) Remanufacturing oriented adaptive repair system for worn components. In: Proceedings of Responsive Manufacturing Green Manufacturing ICRM, 5th International Conference, pp 13–18

56. Bremer C (2005) Automated repair and overhaul of aero-engine and industrial gas turbine components. In: Proceedings of the ASME turbo expo. Reno-Tahoe, Nevada, USA
57. Hamed A, Tabakoff W, Wenglarz R (2006) Erosion and deposition in turbomachinery. J Propul Power 22:350–360
58. Yilmaz O, Gindy N, Gao J (2010) A repair and overhaul methodology for aeroengine components. Robot Comput Integr Manuf 26(2):190–201
59. Yilmaz O, Noble D, Gao J (2005) A study of turbomachinery components machining and repairing methodologies. Aircr Eng Aerosp Technol Int J 77(6):455–466
60. Gupta K and Laubscher RF (2016) Sustainable machining of titanium alloys- a critical review. In: Proceedings IMechE. doi:10.1177/0954405416634278. (Part B J Eng Manuf)
61. Pramanik A (2014) Problems and solutions in machining of titanium alloys. Int J Adv Manuf Tech. 70:919–928
62. Mahamood RM, Akinlabi ET, Shukla M, Pityana S (2013) Characterizing the effect of processing parameters on the porosity properties of laser deposited titanium alloy. In: International multi-conference of engineering and computer science (IMECS 2014)
63. Hedges M, Calder N (2006) Near net shape rapid manufacture & repair by LENS. In: Proceedings of cost effective manufacture via net-shape processing. Neuilly-sur-Seine, France, pp 13–21
64. Mani M, Lyons KW, Gupta S (2014) Sustainability characterization for additive manufacturing. J Res Nat Inst Stand Technol 119:419–428. doi:10.6028/jres.119.016
65. Mahamood RM, Akinlabi ET, Shukla M, Pityana S (2014) Evolutionary additive manufacturing: an overview. Lasers Eng 27:161–178
66. Ford S, Despeisse M (2016) Additive manufacturing and sustainability: an exploratory study of the advantages and challenges. J Clean Prod. doi:10.1016/j.jclepro.2016.04.150. Accessed 9 Aug 2016

Index

A
Abrasive, 25, 32, 33, 38, 39, 41, 77–79, 81, 83–85, 87–93, 95, 143, 250, 255
Additive manufacturing, 267, 269, 270, 272, 281, 287

B
Biomedical, 4, 23, 24

C
Ceramic, 11, 23, 24, 27, 55, 57, 78, 82, 83, 104, 127, 156, 175, 199
Clinching, 102, 105
Composite, 38, 52, 77–83, 85, 91, 95, 103–105, 107, 119, 125, 127, 156, 161, 171, 175, 177–179, 186, 190

D
Dielectric, 15, 24, 249, 252, 254–258, 260–262, 264
Difficult-to-machine, 49–51, 238, 251, 273, 281
Dry-EDM, 249, 254, 256, 257, 262–264

E
EDM, 4, 8, 14, 15, 17, 82, 249–252, 254, 256, 259, 260, 264
Electron beam welding, 115, 118–120, 128, 129

F
Friction stir welding, 120, 121, 123, 132, 162–164, 167, 170, 171, 173, 181, 182

G
Glass, 8, 23–27, 29, 31, 32, 34, 35, 37–40, 42–44, 87, 88, 127, 128, 199
Green manufacturing, 213, 215, 217, 218, 220–227, 229, 231, 232, 287

Grinding, 5–8, 19, 25, 26, 29, 30, 32, 33, 40, 42, 78, 142, 247

K
Kerf, 39, 86, 87, 91

L
Laser, 4, 11, 12, 19, 25, 42, 43, 49, 51–59, 65, 68, 70, 82, 110, 115–119, 128, 130–132, 139, 146–150, 152, 154–157, 268, 271, 274, 277–280, 282–285, 287

M
MEMS, 4, 23, 42, 128
Metal matrix composite, 55, 58, 78, 82, 83, 170, 178
Minimum quantity lubrication, 237–239

N
Nano-joining, 102, 127, 129, 130
Near-dry EDM, 249, 254, 256–261, 263, 264

P
Plasma, 25, 43, 49, 51, 52, 65–67, 71, 90, 108, 110, 113, 115, 132, 146, 157, 252, 258
Polymer, 71, 77, 78, 82, 83, 91, 95, 104, 105, 116, 121, 127, 128, 185, 188, 189

R
Remanufacturing, 137, 139, 146, 267–269, 272–274, 279–281, 287

S
Superalloys, 57
Surface integrity, 42, 51, 55, 56, 81, 83, 238, 249, 253, 261, 265
Sustainability, 219, 221, 228, 237, 265, 269, 287

U

Ultrasonic, 6, 25, 31, 36, 37, 78, 82, 120, 125–127, 129, 130, 132, 133, 185, 187–192, 194, 195, 197, 199, 202, 204, 206

W

Wire-EDG, 10, 11, 14, 19

Wire-EDM, 10, 19

Lightning Source UK Ltd.
Milton Keynes UK
UKHW02n1242250718
326268UK00002B/139/P